26.25

THE ODONATA OF CANADA AND ALASKA

The O D O N A T A
of Canada and Alaska

By

EDMUND M. WALKER
Professor Emeritus of Zoology, University of Toronto
Honorary Curator of Zoology, Royal Ontario Museum

and

PHILIP S. CORBET
Professor of Biology, University of Waterloo, Waterloo, Ontario

VOLUME THREE

PART III: THE ANISOPTERA—THREE FAMILIES

UNIVERSITY OF TORONTO PRESS
TORONTO AND BUFFALO

Library of Congress Cataloging in Publication Data

Walker, Edmund Murton, 1877–1969.
  The Odonata of Canada and Alaska.

  Includes bibliographies.
  Vol. 3 by E. M. Walker and P. S. Corbet.
  CONTENTS: v. 1. General. The Zygoptera – damselflies.
  – v. 2. The Anisoptera – four families. – v. 3. The
  Anisoptera – three families.
  1. Odonata – Canada. 2. Odonata – Alaska. I. Corbet,
  Philip S. II. Title.
  QL520.2.C2W34   595.7′33′0971   54-4344
  ISBN 0-8020-5321-1

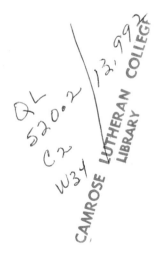

Edmund M. Walker (1877–1969)

FOREWORD

In 1948 Professor Edmund M. Walker set out to assemble results of his fifty years' study of dragonflies into a faunistic monograph, *The Odonata of Canada and Alaska*. The first two volumes were completed and were published by the University of Toronto Press, in co-operation with the Royal Ontario Museum, in 1953 and 1958. Professor Walker was eventually forced by illness to give up his work on the third volume, and he died in 1969 at the age of 91. This book, with Professor Philip Corbet as co-author, is Volume III, the concluding part of the monograph.

It was particularly heartening for Professor Walker, and for all of us concerned about the future of this large undertaking, when Professor Corbet agreed to accept responsibility for completion of Volume III. Through publication of many scientific papers on dragonflies, a book on the British species, and his well-known work *A biology of dragonflies*, Professor Corbet's background for the task was firmly established. His eligibility became complete when his professional career brought him to Canada, where he worked as a research scientist and later a research director with the Canada Department of Agriculture, before joining the staff of the University of Waterloo.

Professor Walker's work on the monograph was supported by the Department of Entomology of the Royal Ontario Museum, where he was Honorary Curator, and where his collections are deposited. Professor Corbet's early work on Volume III was encouraged and facilitated by the Canada Department of Agriculture when he was a member of the staff, and his progress with the manuscript owes much to that support. At a later stage in the project, funds to engage assistants for technical and illustrative work were provided by a grant from the Canadian National Sportsmen's Show to the Department of Entomology, Royal Ontario Museum. This support came at a time when funds were not available elsewhere, and is gratefully acknowledged. The National Research Council of Canada and the Publications Fund of the University of Toronto Press made substantial grants towards the cost of publication.

In bringing the monograph to completion with this volume, Professor Corbet has contributed a great deal of effort for which those concerned in any way with the biology of dragonflies are indebted to him. It is especially gratifying that the large task on which the last twenty years of Edmund Walker's life were concentrated is now complete.

July 1973
GLENN B. WIGGINS
Curator, Department of Entomology,
Royal Ontario Museum

PREFACE

Professor Edmund M. Walker's accomplishments in the science and teaching of biology, which were both manifold and distinguished, have been reviewed in a specially prepared tribute volume (Wiggins, 1966). Of particular interest to students of the Odonata are the list of his publications, an autobiographical sketch (Walker, 1966a) and an account of his work on dragonflies (Corbet, 1966). The last of these items reveals clearly the measure of his contribution to our current knowledge of Canadian Odonata—a contribution covering a period of about seventy years and culminating in the monograph of which this is the third and final volume.

When, in April 1964, I accepted the invitation to complete Volume III of *The Odonata of Canada and Alaska* I did not expect that this task would take nearly as long as it has. Two circumstances have been largely responsible for this: first, the amount of work needed to revise and complete the manuscript was considerable, and required a familiarity with the North American dragonfly fauna that I did not at first possess; and second, the administrative responsibilities that I have assumed since 1967 have made it difficult to maintain the continuity of application needed for the prompt completion of such an undertaking. That this task is now finished is therefore largely due to the assistance that I have received and which it is a pleasure for me to acknowledge here.

The financial resources needed for my work on this volume have come mainly from the Canada Department of Agriculture, as a consequence of my employment in its Research Branch until 1971, and the facilities that were so readily placed at my disposal there. Significant support has come also from the Department of Biology, University of Waterloo, and from a grant from the Canadian National Sportsmen's Show to the Royal Ontario Museum.

I am glad of this opportunity to record my gratitude to the people who have given me valued assistance. My principal debt is to my wife, Dr. Hildegard Corbet, who has helped extensively with most aspects of the work. I thank Miss Barbara M. Rogers and Miss Patti Spry for thorough, accurate and expert assistance with the extraction and systematic presentation of source information, and Miss Rogers also for taking responsibility for the production of wing photographs, many of which were taken by Mr. P. Brian O'Donovan to whom my thanks are due. I am grateful to Mr. Marty Bueno de Mesquita and Miss Sally M. Melville for preparing most of the line drawings. Miss Jane Peterson has helped to prepare the index to this volume.

For the provision of information and the loan of many specimens I am greatly indebted to Dr. Minter J. Westfall, Mr. J. E. H. Martin and Dr. Glenn

B. Wiggins. I thank Dr. R. Jean Musser and Mr. Steve Jensen for the loan of the drawing reproduced in plate 39: 3, Drs. Dennis R. Paulson and Olavi Sotavalta for providing information, and Dr. George H. Bick, Dr. José I. Furtado, and Mr. Kenneth J. Deacon for commenting on parts of the text of this and previous volumes.

As well as thanking persons mentioned in Volumes I and II, Dr. Walker and I wish to acknowledge the specimens or records received from Mr. Kenneth J. Deacon, Mr. William M. M. Edmonds, Dr. J. B. Falls, Dr. José I. Furtado, Mr. Alan Hayton, Miss Jane Peterson, Dr. Jean-Guy Pilon, Mr. Robert A. Restifo, Father J. C. E. Riotte, Dr. F. A. Urquhart and Dr. Harold B. White.

Throughout this project I have greatly appreciated the encouragement that I have received from Miss Marjory A. Ford.

Credits for the photographic illustrations are as follows: plates 2, 7, 20, 22, 26 and 27 are from *A manual of the dragonflies of North America (Anisoptera)* by James G. Needham and Minter J. Westfall (1955), originally published by the University of California Press and reprinted here by permission of The Regents of the University of California; Mr. P. Brian O'Donovan took the photographs for plates 1, 4, 23, 24, 28 and 37.

Credits for the line drawings are as follows, the convention "10: 8" signifying "plate 10: figure 8": Mr. Marty Bueno de Mesquita for 3: 4, 5; 6: 3–7; 8: 1–12; 9: 1–12; 10: 1–7, 9–12; 11: 1–4, 7–13, 16; 12: 1–7; 13: 1–6; 14: 1–3, 5–7, 9; 21: 2, 3; 25: 1, 2, 5–9; 29: 1–4; 30: 4; 31: 14–16; 32: 1–6; 33: 1–3, 4; 34: 1, 3–7, 10; 35: 7; 36: 1–3, 5, 7; Miss Sally M. Melville for 3: 6, 7; 5: 1, 3, 5–8; 6: 1, 2; 19: 1–7; 21: 7, 8; 30: 3, 6, 7, 9–14; 31: 1, 2, 4–8, 11; 36: 8–10; 38: 1–8; 39: 2; 42: 1–6; 43: 1–9; 44: 1–6; 45: 4–8; Dr. Hildegard Corbet for 10: 8; 11: 5, 6, 14, 15; 13: 7; 14: 4, 8; 21: 1, 5, 6; 25: 3, 4; 29: 5; 30: 1, 2, 5, 8; 31: 3, 9, 10, 12, 13; 33: 4, 6; 34: 2, 8, 9, 11, 12; 35: 1–6, 8; 36: 4, 6; 45: 1–3; Dr. E. M. Walker for 3: 1–3; 5: 2, 4, 9, 10; 21: 4; 39: 1; 40: 1–4; 41: 1–6.

In the fifteen years that have elapsed since the publication of Volume II there have been several developments of interest to students of Canadian Odonata. Three years ago White and Raff (1970) discovered the nymph (larva) of *Williamsonia lintneri* (Hagen); thus there is now no *genus* of the Canadian dragonfly fauna that remains unknown in the nymphal stage. Remaining to be found and described is the nymph of the only Canadian *species* of *Williamsonia*: *W. fletcheri* Williamson—a prize for some fortunate collector! Other recent work which we may hope will soon be extended to involve Canadian species and populations is that on adult behaviour (e.g. Bick and Bick, 1961, 1963; Johnson, 1962a, 1962b) and on the responses of nymphs to day-length and temperature that regulate the timing of seasonal emergence (e.g. Lutz and Jenner, 1964; Lutz, 1968). Four species have been added to the Canadian list, and the known distribution of many Canadian species has been extended (such information has been included in the Addenda to Volumes I and II on p. 279).

Another noteworthy development that we have witnessed, and one with fundamental implications, is the rapidly growing human pressure on land in the southern region of Canada. Because of the progressive and usually irreversible habitat destruction it entails, this offers as a prospect in the years to come a significant diminution of the Canadian dragonfly fauna. This process will probably become evident first in southwestern Ontario, which contains more species of dragonfly than any comparable area in Canada and is also subject to the greatest pressures from agricultural, industrial and urban growth. When this growth stops—as eventually it must—it is to be hoped that the remaining freshwater habitats will still sustain a diverse and vigorous dragonfly fauna—a reliable, and delightful, indicator of a healthy environment.

July 1973   P.S.C.

Readers who wish to correspond with the second author should note that in May 1974 his address became: Department of Zoology, University of Canterbury, Christchurch, New Zealand.

ABBREVIATIONS

Defined below are abbreviations appearing in the text. Abbreviations used in any plate are defined in the caption to that plate. Those used for measurements (of adults or nymphs) have a precise meaning that is specified in the appropriate section of the Introduction.

GENERAL

abd.—abdomen, abdominal
alc.—in alcohol
ant.—anterior
anx—antenodal cross-veins
app(s).—appendage(s)
E.—east longitude
e.—east
excl.—excluding
exuv.—exuvia, exuviae
f.w.—fore wing
Ft.—Fort
hd.—head
h.f.—hind femur
h.w.—hind wing, hind wing-sheath
I(s).—Island(s)
in litt.—by letter
incl.—including

inf.—inferior
lat.—lateral
mm.—millimetres
Mt(s).—Mountain(s)
N.—north latitude
n.—north
pnx—postnodal cross-veins
Pt.—Point
pt.—pterostigma, pterostigmata
Rlwy.—Railway
s.—south
seg(s).—segment(s)
spt.—supratriangle
vs.—versus
W.—west longitude
w.—west or width

PROVINCES OF CANADA

Alta.—Alberta
B.C.—British Columbia
Labr.—Newfoundland (Labrador)
Man.—Manitoba
Nfld.—Newfoundland (Island)
N.B.—New Brunswick

N.S.—Nova Scotia
Ont.—Ontario
P.E.I.—Prince Edward Island
Que.—Quebec
Sask.—Saskatchewan

TERRITORIES OF CANADA

N.W.T.—Northwest Territories   Yukon T.—Yukon Territory

STATES OF UNITED STATES

(states not listed here are not abbreviated)

Ala.—Alabama	Nebr.—Nebraska
Ariz.—Arizona	Nev.—Nevada
Ark.—Arkansas	N.H.—New Hampshire
Calif.—California	N.J.—New Jersey
Colo.—Colorado	N.Mex.—New Mexico
Conn.—Connecticut	N.Y.—New York
Del.—Delaware	N.C.—North Carolina
D.C.—District of Columbia	N.Dak.—North Dakota
Fla.—Florida	Okla.—Oklahoma
Ga.—Georgia	Oreg.—Oregon
Ill.—Illinois	Pa.—Pennsylvania
Ind.—Indiana	R.I.—Rhode Island
Kans.—Kansas	S.C.—South Carolina
Ky.—Kentucky	S.Dak.—South Dakota
La.—Louisiana	Tenn.—Tennessee
Me.—Maine	Tex.—Texas
Md.—Maryland	Vt.—Vermont
Mass.—Massachusetts	Va.—Virginia
Mich.—Michigan	Wash.—Washington
Minn.—Minnesota	W.Va.—West Virginia
Miss.—Mississippi	Wis.—Wisconsin
Mo.—Missouri	Wyo.—Wyoming
Mont.—Montana	

CONTENTS

PART III (continued)

THE ANISOPTERA OF CANADA
AND ALASKA

INTRODUCTION

The two preceding volumes of this monograph deal, respectively, with the Zygoptera (Walker, 1953) and with four families of the Anisoptera: the Aeshnidae, Petaluridae, Gomphidae and Cordulegastridae (Walker, 1958). Volume III deals with the superfamily Libelluloidea of the Anisoptera, which encompasses three families: the Macromiidae, Corduliidae and Libellulidae. The known Canadian representatives of these three families comprise 76 species belonging to 20 genera. At the time that this volume went to press the species of Odonata that had been recorded from Canada (including the additions listed on p. 279) amounted to 193, of which 51 are Zygoptera.

The contents of Volume III are arranged and treated almost exactly as in Volume I and II. To comply with the first author's intention regarding this volume and to conform with his practice in preceding volumes, species have been listed in the order that reflects Walker's assessment of their affinity. So that descriptions of given species can be located readily, I have made liberal use of page references in the text.

In the rest of this introduction I describe the procedures followed, and define the terms employed in each part of the text so as to reduce the likelihood of misunderstanding or imprecision when the descriptions are being used. So far as is possible, remarks are placed in the same sequence and under the same headings as will be found in the text.

Abbreviations of terms used generally in the text are defined on page xi; those limited to particular sections are defined or referred to below.

Use of the pronoun "I" in the Preface and Introduction signifies a statement by the second author alone; if used elsewhere in the text its authorship will be evident from the context. Use of the pronoun "we" signifies an observation by both or either of the authors (but usually by the first author alone), thus preserving a usage adopted in Walker's previous publications.

SUPERFAMILY, FAMILIES AND GENERA

The superfamily Libelluloidea and each family and genus are prefaced by a section describing the principal distinguishing features of each, first of adults and then of nymphs. This section is normally short but is most detailed for the three genera that are primarily north-temperate in distribution and that are therefore particularly strongly represented in Canada and Alaska, namely *Somatochlora*, *Sympetrum* and *Leucorrhinia*.

Amplifying statements regarding keys to species, which appear below, apply equally to keys to families and genera.

Species

Adult

Keys and descriptions. The attributes used in the adult keys have been derived from a wide variety of sources, including the authors' own observations. For this reason, and because several sources sometimes contribute to a single couplet, it is impracticable to cite or identify the original authorship of a diagnostic character. However, it is appropriate here to mention the books by Needham and Westfall (1955) and Robert (1963) which were used most frequently, the key by Borror (1945) for Libellulidae, and that by Gloyd (1958) which was adopted without change for *Tramea*. The other publications used are listed in the Bibliography but not necessarily cited elsewhere in the text.

Where restrictions govern the use of a key, these are mentioned in the preamble to that key. General morphological terms have been defined in Volume I (pp. 4–23), or, if convenience requires this, by an illustration in this volume cited in the relevant couplet of a key (see plates 21, 29, 30, 35 on pp. 138, 208, 212, 242). Terms used for venation are *not* those used in Volume I, which follows the Comstock-Needham system. In Volumes II and III the Tillyard-Fraser system has been adopted (see remarks below under *Venation*).

In the keys the abbreviations given for measurements are those defined below under *Measurements*. Use of such an abbreviation signifies that the numerical value given is a dimension. Whenever *segments* are mentioned without further qualification, these are abdominal segments.

The purpose of the species descriptions is to amplify and supplement the diagnostic characters given in the keys and illustrations. The descriptions vary in content, length and detail according to the species and the ease with which it can be identified. For some genera of Libellulidae that are not well represented in Canada and Alaska it has sometimes proved difficult to obtain adequate material for examination; in such cases use has been made, to a varying degree, of published descriptions, particularly by Needham and Westfall (1955) but also by Garman (1927), Byers (1930) and other authors.

Venation. Entries that I have had to provide have been derived, where possible, from five or more specimens, and by preference from Canadian material.

It had been the first author's manifest intention to use the system of Tillyard and Fraser (1957) in this volume, as evidenced by the completed parts of the text and by his remarks in Volume II on pages vii and 5–9. To make it easier for the reader to relate the Tillyard-Fraser system to that used by Comstock, Needham, Westfall and others, the first author illustrated the salient features of each in plate 8 on p. 70 in Volume II.* However, since it is

*"Cu₂" in the upper figure of this plate should be "CuP."

necessary to illustrate certain venational characters of the Libelluloidea in Volume III in any case, the opportunity has been taken to include a complete list, with illustrations and abbreviations, of the terms used in this volume (pl. 1, p. 6), and also a list of the main terms that differ in the two systems.

Readers who require further information on venation should consult Volume I (pp. 10–17) and Volume II (pp. 5–9, 70, 71). Excellent illustrations devoted to special features of venation are given by Needham and Westfall (1955) on pp. 20, 63, 347 and 423–425.

The numbers of antenodal and postnodal cross-veins (anx and pnx) are expressed as ranges (with occasional or less common values given in parentheses) with counts for the two wings being given in the form "fore wing/hind wing." Fractional values (e.g. in *Erythrodiplax* and certain *Sympetrum*) denote the existence of an incomplete cross-vein, at the distal end of the first series. An incomplete antenodal cross-vein is typically present in this position in the fore wing of *Pantala* and *Tramea*; however, the frequent lack of correspondence between the cross-veins of the first and second series in these genera renders fractional values less informative than in other genera. Therefore in *Pantala* and *Tramea* the values given for anx refer only to those in the first series (e.g. $11\frac{1}{2}$ for a genus like *Sympetrum* [expressed here as "11.5"] would be recorded as 12 for *Tramea*). An entry in the form "(NW: ♂♀ hind wing 5–6)" indicates that the range given by Needham and Westfall (1955) extends beyond the smallest or greatest values for the specimens on which the counts for this volume were based.

Measurements. Unless obviously incorrect (e.g. because of a transcription error) the values in the first author's original manuscript have been included without alteration. None of the first author's fractional values have been rounded to integers, since he alone was aware of the precision that accompanied his original measurements. Measurements that I have had to provide have been derived, where possible, from five or more specimens, and by preference from Canadian material; and they have been made to the nearest 0.1 millimetre.

With negligible exceptions the same measurements are provided for all species within a genus, though they may differ for species from different genera. If the range for the length of the hind wing (h.w.) given by Needham and Westfall (1955) extends beyond the smallest or greatest dimensions for specimens treated here by more than 0.5 millimetre, the range is given after our entry in the form "(NW: ♂♀ 10–15)." In the few (seven) other cases where measurements are provided from, or qualified by, values from other sources, these are identified by an asterisk for which an explanation is provided at the end of the paragraph. The abbreviations used for measurements are defined as follows: *Total length*: the distance between the anterior surface of the head (excluding the antennae) and the posterior margin of

PLATE 1

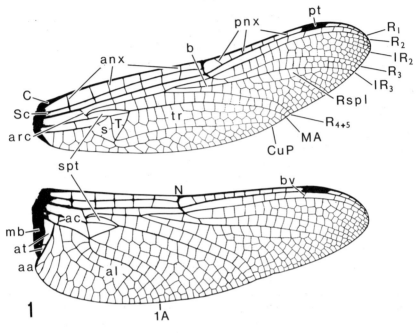

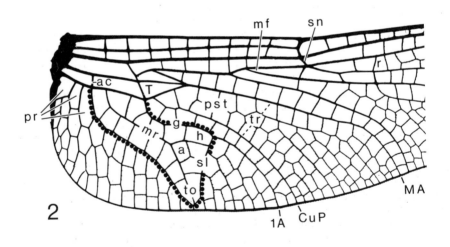

abdominal segment 10 (that is, excluding the anal appendages or ovipositor). *Length of abdomen* (abd.): the distance between the anterior margin of segment 1 and the posterior margin of segment 10 (that is, excluding the anal appendages or ovipositor). *Width of abdomen* (w.abd.): the greatest width. *Length of hind wing* (h.w.): the distance between the sclerotized basal hinge (at its point of articulation) and the apex of the wing. *Length of hind femur* (h.f.): the distance between proximal and distal ends (coxa and trochanter excluded). *Width of head* (w.hd.): the greatest width. *Length of pterostigma* (pt.): the greatest length, measured along the costal margin of the wing (that is, not along the diagonal of the pterostigma); where applicable, an oblique stroke divides the entries for fore and hind wings; the absence of such division implies that values apply to both wings.

When using these measurements it is important to remember that they should be considered only as a guide to the identity of specimens but (unless mentioned as such in the key) not as diagnostic criteria. When comparisons are being made with measurements given by other authors the way the measurements are defined should be carefully noted: unlike Walker (see

---

PLATE 1

The principal venational characters of the Libelluloidea according to the Tillyard-Fraser system, which is used in this volume: (1) *Cordulia shurtleffi*, fore and hind wings; (2) *Erythemis simplicicollis*, base of hind wing showing venation in the expanded anal region characteristic of the Libellulidae; a dotted line indicates the margins of the anal loop.

Abbreviations

(Equivalent notations in the Comstock-Needham system are included in parentheses where the two systems differ.)

1A—anal vein (Cu₂)
a—ankle cell
aa—anal angle
ac—anal crossing
al—anal loop
anx—antenodal cross-veins, first or anterior series
arc—arculus
at—anal triangle
b—bridge
bv—brace vein
C—costal vein
CuP—cubital vein or posterior cubitus (Cu₁)
g—gaff
h—heel cell
IR₂—radial intercalary vein (M_{1A})
IR₃—radial intercalary vein (Rs)
MA—anterior media (M₄)
mb—membranule
mf—middle fork
mr—midrib or bisector

N—nodus
pnx—postnodal cross-veins, first or anterior series
pr—paranal cells
pst—post-trigonal cells
pt—pterostigma
R₁—branch of radius (R)
R₂—branch of radius (M₁)
R₃—branch of radius (M₂)
R₄₊₅—branch of radius (M₃)
r—reverse vein
Rspl—radial supplement or planate
s—subtriangle
Sc—subcostal vein
sl—sole
sn—subnodus
spt—supratriangle or supertriangle
T—triangle
to—toe
tr—trigonal interspace, between MA and CuP

1953, p. 61), several authors *include* the anal appendages in measurements of length of abdomen (e.g. Byers, 1930, p. 19; Needham and Westfall, 1955, p. 52).

Nymph

Keys and descriptions. Attributes used in the nymphal keys have been derived from a wide variety of sources, including the authors' own observations. For this reason, and because several sources sometimes contribute to a single couplet, it is impracticable to cite or identify the original authorship of a diagnostic character. However, it is appropriate here to mention those publications that were used most frequently when the nymphal keys were prepared. In addition to those of Walker (1916b, 1917b and 1925), for *Leucorrhinia, Sympetrum* and *Somatochlora* respectively, these publications were those by Garman (1927), Wright and Peterson (1944), Needham and Westfall (1955) and Musser (1962). These and many other works were consulted when the descriptions of individual species were being constructed. If the description given here is derived predominantly from that of other authors this is indicated in the form "(Description from . . .)." The other publications used, for the keys or descriptions, are listed in the Bibliography but not necessarily cited elsewhere in the text.

It must be emphasized that the characters in the keys and descriptions are intended *only* for the diagnosis of nymphs in the final instar, a stage which (with a little practice) can be recognized readily and reliably by noting the relative lengths of wing-sheaths and abdomen (see pl. 39: 1–3, p. 258). It is possible that certain of the characters listed *may* serve to distinguish a species in earlier instars also, but this cannot be assumed without independent evidence; moreover, it is known that certain of these characters (e.g. the setae on the labium) undergo marked changes during nymphal development.

General morphological terms have been defined in Volume I (p. 24); those specifying the anal appendages and the parts of the labium are defined in this volume in pl. 19: 3 (p. 128) and pl. 38: 5, 7, 8 (p. 254) respectively. As was the first author's intention (see Volume II, pp. 9–11) the terminology for the labium has been changed (from that used in Volume I and by Needham and Westfall (1955)) to conform with current views on the homology of parts of the larval labium in insects (Snodgrass, 1954). This terminology, as it relates to Odonata, is that defined by Corbet (1953).

With few exceptions, characters are described in the same sequence as that followed in Volumes I and II, and the same conventions have been used. A few terms require mention here.

The hairlike setae borne on the dorsal surface of the body, especially in *Somatochlora* and *Libellula*, are for convenience referred to as "hairs."

Whenever setae on the prementum or palpus of the labium are mentioned in keys, and when a single value is given for either character, this refers to the complement of setae *on each side* of the labium. Thus the expression "palpal setae 10" means "palpal setae 10 & 10" in the setal formula defined by Corbet (1953). On the palpus: the "setella," if present, lies at the proximal end of (and usually out of line with) the main series of palpal setae and is much smaller than any of them; and the crenations described are the deepest ones, usually near the middle of the distal margin of the palpus.

Whenever *segments* are mentioned without further qualification, as for instance when the existence of dorsal hooks or lateral spines is being specified, these are abdominal segments. The length of a dorsal hook is usually expressed as a proportion of the mid-dorsal length of a specified segment. The length of a lateral spine on segment 8 or 9 is usually expressed as a proportion of the length of the segment that bears it. The dimension used for this purpose is occasionally the segment's mid-dorsal length or its dorsal length mesial to the base of the spine at its inner edge; but in most cases it is the length of the lateral margin of the segment. The lateral margin may be measured in one of two distinct ways: *excluding* the spine (as, for example, in all species of *Somatochlora* and *Leucorrhinia*); and *including* the spine (for example all species of *Libellula* and *Sympetrum*). To avoid ambiguity the measurement that is used has been specified wherever it is mentioned in a key couplet or description. Here, as in certain other instances, two or more systems of measurement have been used because only in this way can the integrity of original descriptions (often based on more material than we had access to) be preserved.

The procedure followed for making and recording measurements was closely similar to that adopted for adults. The abbreviations used for measurements are defined as follows: *Total length* (length): the distance between the anterior surface of the head (excluding the antennae) and the posterior tip of the anal appendages. *Length of abdomen* (abd.): the distance between the anterior margin of segment 1 and the posterior tip of the anal appendages (best measured along the ventral surface). *Width of abdomen* (w.abd.): the greatest width. *Length of hind wing-sheath* (h.w.): the distance between the anterior limit of the costal margin and the posterior tip of the hind wing-sheath. *Length of hind femur* (h.f.): the distance between proximal and distal ends (coxa and trochanter excluded). *Width of head* (w.hd.): the greatest width.

When measurements are based on an exuvia (exuv.), or nymphal skin, allowance should be made for the state of the exuvia. If it is in alcohol (alc.) the total length and length of the abdomen may be greater than in a nymph; if it is dry these dimensions may be less than in a nymph on account of telescoping of the abdominal segments.

The nymphal descriptions are intended to amplify and supplement the diagnostic characters given in the keys and illustrations; even so, they are not always sufficient to enable all individuals of a species to be identified with confidence, especially in certain of the larger genera. Species of *Sympetrum* and *Leucorrhinia*, for example, are known to show overlapping variation in the very characters that, perforce, are still used to assist a diagnosis. This situation is not improved by the fact that descriptions of nymphs are often based on very few specimens; thus the prospect exists that examination of additional material will show that characters which are now regarded as discrete actually overlap. It should be noted also that the nymphs of eight species* treated in this volume are as yet unknown. For these several reasons, then, must the nymphal keys and descriptions be regarded as provisional.

The descriptions of nymphs in the text are usually fairly brief and do not necessarily include information given in the keys or in the section introducing each genus. Given at greater length are generic and species descriptions for the three genera whose distribution lies mainly in Canada and which the first author has studied closely, namely: *Leucorrhinia* (Walker, 1913, 1914, 1916b, 1927), *Sympetrum* (Walker, 1914, 1917b) and *Somatochlora* (Walker, 1925, 1941c); and in one such instance (*Leucorrhinia glacialis*) the nymph is described here for the first time.

Habitat and range

This section contains brief statements of the characteristic habitat where breeding takes place and of the geographical range of the species. Distribution is given by its known limits and in terms of provinces, territories and states rather than specific localities. Abbreviations used will be found on pages xi and xii.

Source references or collectors' names are only given when they are of special importance or interest. All source references are given in the Bibliography.

Distribution in Canada and Alaska

In the three provinces where this is possible (Nova Scotia, New Brunswick and Ontario), distribution is given by counties, specific localities being included only if they are of exceptional interest. In the other provinces and territories and in the state of Alaska individual localities are listed. Unless otherwise designated (by quotation marks and a comment) all place names given can be found in the official Gazeteer of Canada (Queen's Printer, Ottawa) or, for provinces that had no gazeteer at the time of writing (Newfoundland, Labrador, Prince Edward Island and Quebec), in current atlases or road maps. The abbreviations used for provinces and territories are listed on p. xi.

Localities and counties are listed approximately from south to north, except in Nova Scotia and New Brunswick where counties are placed in alphabetical order.

---

*These comprise three species of *Somatochlora* and one species each of *Williamsonia*, *Celithemis*, *Libellula*, *Sympetrum* and *Leucorrhinia*.

Source references or collectors' names are given only when they are of special importance or interest, when they post-date the first author's published provincial or regional lists or when the original locality is unknown. All published source references are listed in the Bibliography. Unpublished records come from the first author's notebooks. All records that appeared to me ambiguous or of doubtful authenticity have been omitted.

Different localities are separated by a semicolon; parts of the same one by a comma.

All localities listed in the Northwest Territories lie in the Mackenzie District. The Hudson Bay Railway (Manitoba) runs from The Pas (mile 0) to Churchill on Hudson Bay (mile 510).

Field notes

This section of the text contains information relating to flight period (and seasonal life history), to habitat (of adults and nymphs) and to behaviour. When I assumed responsibility for this volume, this section required supplying in full for 31 species and in part for 12 more.

In completing this part of the text, I have given priority to Canadian sources, drawing primarily on Walker's own publications and on those by Whitehouse (mainly for British Columbia and Alberta) and Robert (for Quebec). In addition I have been able to make use of the work of other entomologists in Canada (notably Fernet, Pilon, Pritchard and Trottier) that has appeared since the publication of Volume II.

For species that occur in Canada only occasionally, or for which relevant field observations in Canada were unavailable, I have called on published accounts of their biology in the United States (preferably from states close to Canada) or in other countries where conditions are ecologically similar to those in Canada. This approach conforms with that adopted by Walker in this and the two preceding volumes.

Among the source material for this book that was made available to me were about 30 of the first author's field notebooks and a quantity of his unbound handwritten notes. This material had been prepared by him for his own use and was, for the most part, difficult for someone else to interpret and collate. It has, however, been gone through for unpublished records of date and locality, an operation that yielded a small number of records which would otherwise have escaped notice. It did not prove logistically feasible to extract observations on habitat or behaviour from this handwritten material. The notebooks and unbound material have been deposited in the Royal Ontario Museum, Toronto.

The source of statements that derive from Walker's own (already published) observations has not necessarily been identified in the text; the source of observations published by other authors has always been given in the conventional manner, with one exception: the publications of Whitehouse (1941, 1948) have been used frequently for determining the limits of the flight period (particularly as for many species this tends to be more protracted in western Canada) but not necessarily cited in this connection.

To obtain information relating to the flight period and habitat I have also used the data accompanying specimens in the Royal Ontario Museum and the Canadian National Collection of Insects in Ottawa.

When interpreting the term "flight period" it should be remembered that this is not equivalent to the term "flying season" in the restricted sense of the reproductive period (see Corbet, 1962). As understood here, the "flight period" includes both prereproductive and reproductive phases of adult life, since the records available do not always specify whether the specimens were teneral or mature when caught. When inferring the limits of the flying season *sensu stricto* it is useful to remember that normally about one to two weeks separate the first emergence and the appearance of sexually mature adults at the breeding site; this is the time required for completion of the maturation period, which is typically spent away from water and during which adults progressively lose their teneral appearance. For the comparative treatment or amplification of this and other aspects of dragonfly biology and behaviour, the reader is referred to the appropriate section (pp. 32–57) of Volume I, and to books by P.-A. Robert (1958) and Corbet (1962).

ILLUSTRATIONS

No plates had been composed, or the material for them assembled, when I assumed responsibility for the completion of this book. Available to me were numerous unpublished drawings by the first author of species destined for inclusion in Volume III, but relatively few of these were completed or obviously definitive. In preparing the illustrations for this volume I have, wherever possible, called upon material by Walker: as drawings already published elsewhere (indicated by the expression "from Walker"); as sketches existing in his files ("by Walker"); and as redrawn copies of previously published illustrations the originals of which were no longer available ("after Walker"). The source of illustrations so derived (whether from Walker or other authors) is indicated at the end of the caption to the plate. Unless otherwise mentioned, the other (original) drawings are based on at least one specimen, usually about six, and by preference on Canadian material; and all were completed under my personal supervision.

All wings illustrated (to show venation or pattern) are from males unless otherwise specified.

SUPERFAMILY LIBELLULOIDEA TILLYARD

We are employing the superfamily name Libelluloidea to include all of those groups of Anisoptera which have been collectively regarded by many workers as a single family: the Libellulidae. This we have done in the belief that, although several groups of family rank are included, they have too much in common not to be recognized as a larger assemblage of related families, which should be typified by the original family Libellulidae and should therefore be given the superfamily name Libelluloidea.

This superfamily may be characterized as follows. The head is subhemispherical with the eyes meeting mid-dorsally in an eye-seam, which is generally shorter than in the Aeshnidae. The frons is concave above and the vertex is reduced to a cushion-like prominence, which is elevated and arched forward, thus concealing partly or wholly the large median ocellus, when viewed directly from above, but leaving fully exposed the much smaller lateral ocelli, which are situated near the bases of the antennae. The occiput is reduced to a triangular area of variable size. The median lobe of the labium is small and entire, the labial palpi (lateral lobes) being large and meeting in the median line; each may bear a minute distal movable hook.

In the males, only the posterior hamuli are functional, the anterior pair being vestigial or absent. The original or zygopterous ovipositor is absent, the female genitalia being represented by the vulvar lamina, an extension of the sternum of the eighth abdominal segment. It is usually short, flat and bilobed, and extends rearward over the sternum of segment 9, but it sometimes takes the form of a scoop and may stand out from the abdomen at an angle, as in some species of *Somatochlora* (Corduliidae) and *Sympetrum* (Libellulidae). In the Macromiidae it is almost vestigial. Oviposition is always exophytic.

The venation has certain very characteristic features, which reach their evolutionary climax in the family Libellulidae. These are as follows: The costal and subcostal series of antenodal cross-veins are in almost perfect alignment and the primary antenodals are no longer distinguishable. The triangles of the fore wings are no longer like those of the hind wings, being elongated anteroposteriorly and placed well beyond the arculus, whereas those of the hind wings are elongated laterally, or parallel to the long axis of the wing, and are retracted toward the arculus, reaching the same level in the Corduliidae and Libellulidae but only about halfway toward this level in the Macromiidae. With the anteroposterior elongation of the fore-wing triangle the subtriangle is swung mesocaudad and its primitively posterior side is more or less strongly broken or bent so that the subtriangle is actually four-sided and is usually divided into two or three cells, or more, three being the common number. In the three families represented in our fauna these characters are least marked in the Macromiidae and most marked in the Libellulidae.

The subtriangle of the hind wing has followed a different path. It has kept its original position but tends to remain small and weak or (by the loss of its basal cross-vein) to disappear. It is present in the North American Macromiidae but is more often absent in the Corduliidae. It is nearly always absent in the Libellulidae.

The anal loop remains short and broad in the Macromiidae but in all our Corduliidae and nearly all our Libellulidae the loop is elongate with two rows of cells divided by a "midrib" and terminating in an expansion, which culminates in the Libellulidae in a foot-like figure having both "toe" and "heel" (pl. 1: 2, p. 6). There are also other libellulid genera in which the anal loop is imperfectly developed but the only one of these known in Canada is *Nannothemis*, which contains our smallest species of Anisoptera: *N. bella*. In this tiny dragonfly the imperfect anal loop probably reflects the general decrease in the number of cross-veins that accompanies the marked reduction in size in all very small species.

The sectors of the arculus are sometimes fused for a short distance beyond their bases; when the sectors are separate this represents the more primitive state. They are separate in the Corduliidae but more frequently united at the base in the Libellulidae. In the Macromiidae they are also united.

Another character in which the Macromiidae stand apart from the other two families is the absence of development of the radial, median and apical supplements, or planates. These are strikingly developed in the Libellulidae, moderately well defined in the Corduliidae and feebly represented or scarcely noticeable in the Macromiidae.

The nymphs of the Libelluloidea are crawlers or climbers, and never deep-burrowers. They are usually not elongate but tend to be short-bodied and not much depressed. The head is generally blunt anteriorly but sometimes, in the Libellulidae, it projects in front of the eyes, owing to the prominence of the labium. The eyes are often very prominent. The face appears to be deep, owing to the broad labial palpi, which meet one another almost vertically along the middle line, thus covering the labrum and mandibles.

The chief diagnostic characters of the nymphs are found in the labium and the abdomen.

The labium is of the spoon-shaped type. The prementum is narrow at the base but widens distally up to the palpal articulations. Its concave dorsal surface bears the premental setae (sometimes known as mental setae) in two lateral groups. Beyond the palpal articulations, the prementum narrows to an angular apex but lacks a recognizable ligula. The palpi are broadly triangular, concave within, each with a greatly reduced movable hook, followed along the lateral margin (dorsal in the position of rest) with a series of about 6 to 12 palpal setae. (The latter have also been known as lateral or raptorial setae.)

The distal margins of the palpi are regularly scalloped, the crenations each bearing a short graded series of spine-like setae on the curved surface. The crenations vary greatly in length from base to apex, usually being longest in the Macromiidae and shortest in the Libellulidae, in some of which they are indicated only by the deeper notches which separate them and by the regular distribution of groups of setae close to the notches.

As none of the Libelluloidea are deep-burrowers, the legs show no special modifications of structure. Nymphs that live on a silted bottom are usually hairy, the legs being especially so. The sprawling macromiids have the longest legs, and some of the libellulids the shortest. The highest and most cultriform dorsal abdominal hooks are found in the Macromiidae.

KEY TO THE FAMILIES OF LIBELLULOIDEA

ADULTS

1 Anal loop short and broad without a midrib; triangle of hind wing about half as far beyond the arculus as triangle of fore wing; radial and median supplements (or planates) not distinctly developed (pl. 2, p. 18); abdomen without ventrolateral carinae; tarsal claws bifid, with branches nearly equal   *Macromiidae* (p. 16)

Anal loop long, composed of two rows of cells separated by a midrib, except towards the distal end, where it is generally widened; triangle of hind wings opposite the arculus, that of fore wing much farther distad; radial and median supplements distinct (pl. 20, p. 134; pl. 22, p. 148); abdomen with ventrolateral carinae on segs. 4 to 7 (pl. 21: 1, p. 138); tarsal claws unequally bifid with a long terminal branch and a very short inferior branch   2

2 Coloration usually somewhat metallic; posterior margin of eyes with a low tubercle; terminal expansion of anal loop not distinctly foot-shaped, without a distinct "heel" and "toe"*; sectors of the arculus separate at base; males with anal border of hind wings angular and an anal triangle present (pl. 1: 1, p. 6); mesopleural sutures more or less distinctly sinuate; fore and hind tibiae of males keeled on flexor surfaces; abdominal seg. 2 of males with auricles (pl. 21: 1, p. 138)
                                                            Corduliidae (p. 33)

Coloration rarely metallic (in none of our species), posterior margin of eyes without a tubercle; terminal expansion of anal loop nearly always more or less distinctly foot-shaped with "heel" and "toe"*; sectors of arculus only rarely separate, generally united at base, forming a short stalk; both sexes with anal border of hind wings rounded; anal triangle absent (pl. 1: 2, p. 6); none of the tibiae keeled; auricles absent; mesopleural sutures not distinctly sinuate   *Libellulidae* (p. 144)

NYMPHS

There appear to be no good external characters for separating the nymphs of Corduliidae and Libellulidae without breaking them up into

---

*Anal loop open below in *Nannothemis*.

artificial groups. In the following key it is therefore necessary to subdivide the Libellulidae so that certain genera of this family appear twice.

1 Head bearing a prominent frontal horn (pl. 38; 1, p. 254); abdomen three-fourths to four-fifths as broad as it is long; legs very long, the hind femora extending as far back as seg. 8; labial palpi with crenations (on distal margin) deep and obliquely incised _Macromiidae_ (p. 17)

Head without a frontal horn; abdomen not more than about two-thirds as wide as it is long; legs shorter, the hind femora usually extending as far back as some part of seg. 6; labial palpi with crenations moderately deep to very shallow 2

2 Crenations of labial palpi usually one-fourth to one-half as long as they are broad, or longer, having the form of somewhat rounded scoops (e.g. pl. 38: 7, p. 254) 3

Crenations of labial palpi much shallower, usually one-tenth to one-sixth as long as they are wide, in some genera indicated only by the spaces between the minute notches that separate them, each space bearing a group of short setae (e.g. pl. 38: 8, p. 254) _Libellulidae_ (p. 144)

3 Lateral spines of seg. 8 as long as mid-dorsal length of seg. 9 or longer (pl. 45: 5, 6, p. 276) _Libellulidae_, genus _Pantala_ (p. 275)

Lateral spines of seg. 8, when present, shorter than mid-dorsal length of seg. 9 (e.g. pl. 18: 1, p. 124) _Corduliidae_ (p. 33)

FAMILY MACROMIIDAE

These are large dragonflies of the size and proportions of an _Aeshna_, easily known by the oblique, belt-like yellow band which engirdles the thorax between the two pairs of wings, and appears on each side as a single oblique stripe. The head is peculiar in the large dorsal extent of the frons, which is delimited by a sharp ridge from the anterior frontal area. It is concave, with the sides sloping to a median furrow marked with a dark stripe. The eyes meet for a short distance on top of the head, the vertex and occiput being thus separated. The vertex is high and bent forward so that the median ocellus is concealed in direct dorsal view. The occiput is small and somewhat rounded. Behind the rear margin of each eye is a rounded tubercle. The wings are hyaline or brownish and the legs long, the femora bearing numerous short tooth-like spines and the tibiae two rows of long spine-like setae. The first and second tibiae or all three pairs bear on the flexor surface a slender keel which extends from a point beyond the base to the apex. The keel is longest on the hind tibiae where it is often fully five-sixths of the tibial length, whereas on the fore and middle tibiae it is often less than half the tibial length. The tarsal claws are bifid, the two branches being nearly equal in length.

The abdomen of the male is somewhat elevated, like that of *Cordulegaster*, which the species of *Macromia* resemble while in flight, because of their similar size and form and the dark brown and clear yellow colour pattern. There is, however, a metallic lustre in the coloration of *Macromia* that is absent in *Cordulegaster*. Our only other genus of Macromiidae, *Didymops*, also lacks the metallic lustre. A ventrolateral longitudinal carina is entirely absent.

In their venation the Macromiidae possess some relatively primitive features and some that are specialized. Primitive characters are: (1) the position of the triangle of the hind wing, which, instead of being close to the arculus, is about half as far distad as is the triangle of the fore wing; (2) the short, broad form of the anal loop which is not divided into two rows of cells by a midrib, as in the Corduliidae and Libellulidae; and (3) the near absence or absence of the radial supplement and absence of the median supplement. A specialized feature is the union of the sectors of the arculus for some distance beyond the latter. Other characters are the lack of a brace vein for the pterostigma, the undulating course of veins R_{4+5} and MA, and the size of the paranal cells of the fore wing, these being much larger than the accompanying marginal cells.

The main characters of the nymphs have been given in the key. The head is pentagonal in outline; the frontal horn is nearly erect and pyramidal; the eyes are very prominent, capping the anterodorsal angles. The antennae are slender and seven-segmented. The prementum of the labium is very broadly triangular; the palpi are also very wide and their distal margins are deeply cut into two series of crenations, each bearing a usually larger group of setae. The movable hook is small as in other Libelluloidea. The pronotum bears a pair of small erect tubercles, adjacent to those of the head at the posterolateral angles. The pterothorax is much depressed and, at the level of the middle and hind legs, decidedly wider than the head. The legs are long, slender, and spreading, the tarsal claws being very long and simple. The abdomen is very broadly oval or subcircular, being somewhat more than two-thirds as wide as it is long. It bears a segmental series of high cultriform dorsal hooks, and lateral spines on segments 8 and 9. The anal pyramid is short and broad.

The Macromiidae are insects of swift flight that develop in large streams, channels or lakes where there is considerable water movement. They are absent from ponds or other stagnant waters. They are widely distributed in the warmer parts of the world, but very few species are found in Canada and none has been recorded in the prairie provinces* or the far north. The two genera represented in our fauna may be separated by the key that follows.

Gloyd (1959) has reviewed in detail the status of this family.

---

*Except for the record of *Macromia illinoiensis* from Berens River, Manitoba.

PLATE 2

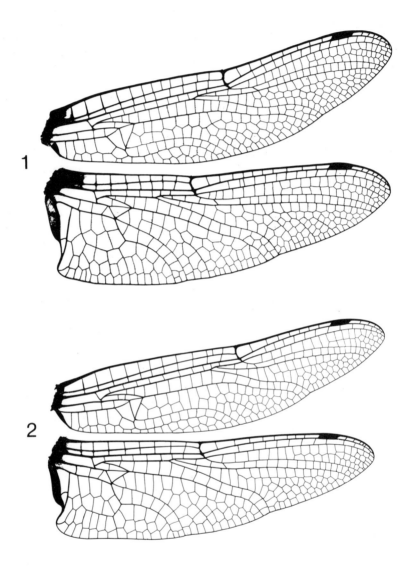

1

2

Key to the Genera of Macromiidae

ADULTS

1 Vertex simple, rounded, smaller than occiput, which has a bulbous swelling behind; nodus of fore wing about midway between base and apex of wing (pl. 2: 1, p. 18); coloration light brown and yellow, non-metallic   *Didymops* (p. 19)

Vertex bilobed, larger than occiput, which lacks a bulbous swelling behind; nodus of fore wing distinctly beyond middle of wing (pl. 2: 2, p. 18); coloration dark, with a metallic lustre and bright yellow markings   *Macromia* (p. 22)

NYMPHS

1 Lateral spines of seg. 9 reaching back as far as, or beyond, tip of epiproct; width of head across the eyes about equal to the greatest width behind the eyes; seg. 10 without a dorsal hook   *Didymops* (p. 19)

Lateral spines of seg. 9 not reaching back as far as tip of epiproct; width of head greatest across the eyes and decreasing posteriorly; seg. 10 with a small dorsal hook   *Macromia* (p. 22)

Genus **Didymops** Rambur

Ins. Neur., p. 142, 1842. Type species *Libellula transversa* Say (*D. servillei* Rambur).

A North American genus of two species: *D. transversa* (Say), widely distributed in eastern North America, and *D. floridensis* Davis, a similar but larger species known only from Florida. They are dull brownish insects marked with yellow and differ greatly in appearance from species of *Macromia*, lacking the metallic lustre and rich dark ground colour that are hallmarks of that genus.

As compared with *Macromia*, the line of contact of the eyes is shorter, being less than half a millimetre long, the vertex is arcuate, as viewed from in front, instead of biconical, as in *Macromia*, and the occiput is very large and swollen both above and behind.

The nymphs are readily distinguished from those of *Macromia* by the characters given in the key. In other respects they are extremely similar.

One species, *D. transversa* (Say), is known from the eastern provinces of Canada.

PLATE 2

Wings of Macromiidae: (1) *Didymops transversa*; (2) *Macromia magnifica*.
(1, 2 from Needham and Westfall, 1955.)

Didymops transversa (Say). (Pl. 2: 1, p. 18)

Libellula transversa Say, J. Acad. Phila., 8: 19, 1839.
Didymops servillei Rambur, Ins. Neur., p. 142, 1842.
Didymops transversa: Hagen, Syn. Neur. N. Amer., p. 135, 1861.

Easily recognized by the single yellow stripe on the sides of the thorax together with the dull brownish coloration, lacking any metallic lustre; appears like a dull *Macromia*, having none of the contrasting colours of the latter.

Male. Face light yellow with mouth-parts marked with brown and with darker brown cross-stripes on the anteclypeus and antefrons; postfrons with median dark streak between two pale spots, the dark streak forming the stem of a diffuse T-spot; occiput obscure olivaceous, with pale gray hairs especially behind; rear of head ochraceous. Pterothorax light to medium brown, densely clothed above with pale hair; median carina and ante-alar sinuses pale yellow; sides less hairy with a single oblique pale yellow stripe extending from middle coxae to base of fore wing immediately in front of the inter-alar space. Legs brown, darkening along sides of tibiae, tarsi nearly black; wings hyaline or brownish, with costal edges yellow and a dark brown spot at the base of the costal and subcostal spaces; pt. medium yellowish brown, venation otherwise dark brown; membranule pale gray, darkening towards the posterior apices. Abdomen dull brown, marked with light yellow basal annuli, which cover each segment from base to transverse carina, except for a mid-dorsal brown interval on segs. 2 to 6, this being broadest on 2 and narrowing caudad; annulus on 7 larger, unbroken; small paired basal yellow spots on 8.

Labial palpi very large, labrum also relatively large, postclypeus narrow and projecting far laterad; frons prominent, furrowed in front, bounded above by sharp ridge, smooth above, the median furrow shallower than in *Macromia*; vertex transversely arched, cushion-like, much smaller than the large triangular occiput, which is convex above and behind. Legs long, the hind femora being about a third longer than width of head. Wings seven-ninths of the length of the abdomen, including appendages; pt. short, surmounting two cells. Abdomen slightly thickened at base, constricted on seg. 3 to about one-half its basal width; seg. 2 with rounded auricles bearing no teeth, and small genital lobes; middle segs. subcylindrical, expanding on 7 to 10, forming a slight club, which is widest at apex of 9.

Anal apps. about as long as 9, the sup. apps. and inf. app. about equally long; sup. apps. separated by more than their own width, bowed, apices bluntly pointed but not turned outward in lateral view, horizontal; inf. app. bluntly triangular, sides convex, width at base about three-fourths of the length.

Female. Marked similarly to the male, except that the yellow abdominal spots are larger, owing to the greater width of the segments, the large spot on 7 nearly as large as those on 6, 5, 4, etc., but not divided as these are. Abd. segs. almost parallel-sided, narrowing gradually on 8 to 10. Anal apps. very short, only about as long as 10; vulvar lamina undeveloped, the two genital apertures raised on distinct papillae; wings usually tinged with brown.

Venation. Anx ♂♀ 11–13/8–9; pnx ♂ 6–10/8–10, ♀ 8–11/9–11; cells in triangle and st. 1, in spt. ♂♀ 2–3/2; paranal cells ♂ 6 – 7/5, ♀ 7/6; cells in anal triangle 2; cells in anal loop ♂ 6–9, ♀ 9–13; pt. subtending 2 or 2+ cells.

Measurements. Total length ♂ 52–57, ♀ 56–57; abd. ♂ 35–42, ♀ 38.5–40.0; h.w. ♂ 34–36, ♀ 35–38; h.f. ♂♀ 10–12; w.hd. ♂♀ 8.0–8.5; pt. ♂ 1.5–2.0/2.0–2.2, ♀ 2.0–2.4/2.2–3.0.

Nymph. Of the usual macromiid form, depressed, with broadly rounded abdomen and long sprawling legs. Head about five-eighths as wide as thorax; eyes small, prominent, capping the anterolateral angles of the head; frontal horn with base as wide as the space between the antennae, somewhat convex, lengthened in front, concave behind, sides of head bulging behind the eyes, the greatest width of this bulging region about equal to the width across the eyes, the posterolateral angles about in line with inner border of eyes. Prementum of labium wider than long (ratio 11:10), narrowed at base, the distal width nearly three times the proximal width; premental setae 5 or 6; palpal setae 5; crenations as in *Macromia*. Pronotum with lateral edges raised almost into a tubercle; pterothorax wider than prothorax, the meso- and metathorax nearly equally wide; wing-sheaths extending to proximal third of seg. 6. Abdomen widest at seg. 5, ratio of width to length, 11:15; lateral margins well rounded, a little more abruptly curved proximally than distally, lateral spines on 8 and 9, those on 8 about one-third the lateral length of that segment (excluding the spine), nearly parallel; spines on 9 slightly shorter than that segment length, extending behind to about tip of epiproct or sometimes beyond, slightly divergent; dorsal hooks on seg 1 to 9; hook on 1 very short, on 2 and 3 tall, slender, curved rearward, on 4 to 7 increasingly long, becoming dorsally horizontal with apices directed backward, highest on 6, somewhat lower on 8 and 9, 10 with only a minute median ridge, which becomes larger on the epiproct.

Colour pattern often nearly absent except on the legs, which bear three dark femoral annuli, a sub-basal, a median and an ante-apical annulus; tibiae darkening below, tarsi dark. On distinctly marked individuals there is a broad dark stripe along side of head, which may continue along the thorax, where it breaks up. Abdomen with numerous dark spots, the largest close to the lateral margins, segmentally arranged, tending to occupy front half of segment; also large median blotches on the more posterior segments.

Length 24–29; abd. 16–17; w. abd. 11.5–12.0; h.w. 6.0–6.5; h.f. 11.0–11.5; w. hd. 5.5–6.0.

Habitat and range. Forest streams with a gentle current, and lakes with moderate wave action; not normally in stagnant or very weedy waters. N.S. to Minn., s. to Fla., Ala., Okla. and Tex.

Distribution in Canada. N.S.—Halifax and Hants counties. *N.B.*—Charlotte, St. John, Sunbury and York counties. *Que.*—Hemmingford; St. John's; Fulford; Knowlton; Vaudreuil; Chateauguay; Montreal; La Trappe; St. Hyacinthe; Berthierville; Lac St. Pierre; Hull, Wakefield, Kazabazua and other stations in the Gatineau Valley; Mont Tremblant Park; Nominingue; Saguenay River (Cap Jaseux) (Fernet and Pilon, 1969). *Ont.*—Norfolk, Haldimand, York, Simcoe, Victoria, Peterborough, Hastings, Lennox and Addington, Frontenac, Prescott, Carleton, Lanark and Renfrew counties; Muskoka, Parry Sound, Nipissing (incl. Algonquin Park), Thunder Bay and Kenora districts.

Field notes. This is one of the first of the Anisoptera to appear on the wing in Ontario, our earliest record being on May 23, 1949, when an exuvia was found at Mud Lake, Victoria county. We have also seen males flying over Lake St. George, York county, on May 30, 1941, an observation that suggests a still earlier emergence. At Go Home Bay, Georgian Bay, where the season begins somewhat later, we found an exuvia on May 29, and observed individuals emerging up to June 16. June is the month of greatest abundance of this species in Ontario and Quebec and perhaps throughout its Canadian range. Its flight period is generally about over by the middle of July, although we have a record for July 30, 1926, from Dr. F. P. Ide, of a specimen taken at Kearney, Ontario, and two from De Grassi Point, for July 31, and

August 3, 1939, each of a single individual. The last is our only August record.

The young imagos usually appear first in clearings or roads in open woods or in bushy fields bordering woods. They fly low, 18 to 30 inches above the ground or slightly above the grass or herbage on which they often come to rest, perching in an oblique position. They fly with the abdomen raised and slightly arched, the apex in the males being depressed. Even during this wandering period their flight is swift.

When they return to the stream or lake shore to mate and oviposit, the males fly up and down, following a definite beat. They resemble *Basiaeschna janata* on the wing but fly more swiftly. During these patrols they fly rather low, seldom more than two feet above the water.

Mating, which we have observed in late June, exhibited nothing out of the ordinary. We have seen an ovipositing female at least once. She dipped the tip of the abdomen in the water rather rapidly around a group of dead tree trunks, which stood in the water close to the bank.

The nymphs were found in the vicinity of Go Home Bay, sprawling on the sand near the shore in clear, well aerated water. They are also found on muddy bottoms. The full-grown nymphs transform on bushes, tree trunks or buildings, sometimes at distances of forty or fifty feet from the water. At De Grassi Point in front of our cottage, the banks of the lake are about seven feet high. Nymphs of *Didymops* generally climb the bank, either over large boulders and a few feet of steep clay or up the railing of a staircase. Close to the edge of the cliff is a large oak tree and not infrequently the nymphs will climb five feet or more up the trunk of this tree, in addition to the bank of the lake, before transforming.

Genus **Macromia** Rambur

Ins. Neur., p. 137, 1842. Type species *M. cingulata* Rambur.

This is a genus of large dragonflies of dark, somewhat metallic coloration, rather sparingly marked with yellow. The thorax may or may not have dorsal longitudinal stripes, but there are always a single lateral yellow stripe and yellow antealar sinuses. The legs are black and the wings (pl. 2: 2, p. 18) hyaline or tinged with brown. Wings of immature females may be quite strongly amber-coloured. The abdomen is black with a series of median dorsal yellow spots, which may be very inconspicuous except in certain segments.

Distinctive features are the biconical vertex, which is larger than the relatively small occiput, the latter being also nearly flat and not bulging posteriorly as it does in *Didymops*.

The nymphs are as described for the family Macromiidae (p. 17), their distinctive features being, in our species at least: the presence of a reduced dorsal hook on abdominal segment 10 in the form of a ridge, besides well developed ones on all the other abdominal segments, and lateral spines on segment 9 that scarcely surpass segment 10, not nearly reaching the end of the anal pyramid. In one species of *Macromia* the sides of the head behind the eyes are distinctly convergent and the head is thus narrowed posteriorly. The genus is widely distributed in both the Old and New World.

Key to the Species of Macromia
Adults

1 With dorsal yellow thoracic stripes; vertex marked with yellow; largest yellow abdominal spot on seg. 8; males without a mesotibial keel; far western species   2

   Without dorsal thoracic stripes; vertex wholly black; largest yellow abdominal spot on seg. 7 (pl. 3: 1, p. 24); males with a mesotibial keel on the distal half; eastern species   *illinoiensis* (p. 23)

2 Yellow of labrum an unbroken transverse stripe; dorsal yellow spots of segs. 3 to 6 joined across mid-line, not or little diminishing in size from 4 to 6 (pl. 3: 3, p. 24)   *magnifica* (p. 27)

   Yellow of labrum divided into two spots; dorsal yellow spots of segs. 3 to 6 separated across mid-line and diminishing in size caudad, becoming very small on 6 (pl. 3: 2, p. 24)   *rickeri* (p. 30)

Nymphs

1 Dorsal hooks on segs. 5 and 6 resembling a bird's head when viewed laterally (pl. 3: 6, p. 24). Angle between posterior surface of frontal horn and dorsum of head about 45°   *illinoiensis* (p. 25)

   Dorsal hooks on segs. 5 and 6 gently falcate, not resembling a bird's head, when viewed laterally (pl. 3: 7, p. 24). Angle between posterior surface of frontal horn and dorsum of head about 90°   *magnifica* (p. 28)
                                                                        or *rickeri* (p. 31)

Macromia illinoiensis Walsh. (Pl. 3: 1, 4–6, p. 24)

Macromia illinoiensis Walsh, Proc. Acad. Phila., 1862: 397, 1862.

A large dark dragonfly of *Aeshna* size and proportions, dark brown with a slight metallic lustre and bright yellow markings. The large yellow spot (pl. 3: 1, p. 24) on abdominal segment 7 of the male shows conspicuously during flight.

Male. Labium yellow with brown margins; labrum yellow, brown at base and with brown hairs; anteclypeus very dark brown, postclypeus pale yellow, frons dark brown with a metallic lustre; with a pale spot on each side and smaller pale spots above in the dorsal concavity; vertex and occiput dark with metallic lustre; eyes in life bright green with a yellow spot on the posterior projection of each eye; rear of head black. Thorax dark brown with a rather dull green metallic lustre; without antehumeral markings but with clear yellow spots in the ante-alar sinuses and a lateral oblique stripe that passes over the metaspiracle occupying the metepisternum. Legs black with tibial keels light yellow. Wings hyaline, venation dark brown, including pt.; mem-

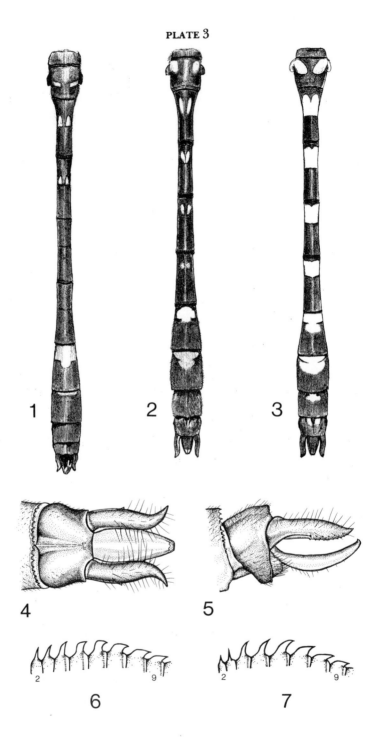

PLATE 3

branule gray, darkening to blackish on apex of anal triangle. Abdomen nearly black, with yellow spots (pl. 3: 1, p. 24) as follows: dorsal and lateral on seg. 2 in front of transverse carina; similar but smaller spots on 4 and 5, nearly or quite disappearing on 6; a large undivided dorsal spot on 7, a smaller basal spot on 8 and also small basal ventrolateral spots on 7 to 9; 10 wholly dark.

Vertex obtusely biconical, occiput about as long as the eye-seam, nearly flat behind; tibial keel of fore legs nearly one-half the tibial length, that of middle legs four-ninths of the tibial length; keel of hind legs about 1.5 mm. shorter than the tibia. Hind wing slightly more than four-fifths as long as abdomen (excluding apps.). Abdomen scarcely swollen at base; constricted on 3 at the transverse carina, thereafter slender and segs. nearly equal in width, rapidly enlarging on 7 to twice the diameter of 5 and narrowing again beyond the middle of 8; segs. 7 to 10 forming a club, which is distinctly tectate with a well developed median carina on 10.

Sup. apps. (pl. 3: 4, 5, p. 24) about as long as seg. 9, slightly bowed in proximal two-thirds, apices divaricate, acute; sup. apps. in profile nearly straight and horizontal, slightly tapering, apices acute, slightly upturned; inf. app. (pl. 3: 4, 5, p. 24) slightly longer than sup. apps., bluntly triangular with somewhat convex sides, the base being about one-third its length, in profile longitudinally concave.

Female. Somewhat larger than male, with stouter abdomen which is not clavate but is hairier than in male. Dorsal yellow spots larger than in male, conspicuous in tenerals, the paired dorsal spots of segs. 3 to 6 one-half to three-fourths as long as the distance from base of segment to transverse carinae. Spot on 7 as large as in male, extending a little beyond the transverse carina. Median carina of 10 as in male, anal apps. barely as long as 10; vulvar carina very short, obtusely bifid.

Venation. Anx ♂ 14–17/9–12, ♀ 15–18/10–12; pnx ♂ 7–10/9–13, ♀ 8–11/10–12; cells in triangle of fore wing ♂ ♀ 2(65%) or 1, and of hind wing ♂ 2(70%) or 1; cells in st. (♂) of fore wing 2 (30% ♂, 80% ♀) or 1.

Measurements. Total length ♂ 62.0–72.5, ♀ 61–72; abd. ♂ 45–50, ♀ 43–52; h.w. ♂ 39–43, ♀ 42–46 (NW: ♂ ♀ 40–49); h.f. ♂ ♀ 11–12; w.hd. ♂ 9–10, ♀ 9.5–10.5; pt. ♂ 2.7–3.0, ♀ 2.8–3.3.

Nymph. Head pentagonal; frontal horn with base filling the space between the antennae, concave behind; width of head across eyes decidedly greater than width behind eyes; sides of head not bulging as in *Didymops*, though very slightly convex, narrowing to the posterolateral angles, each of which bears a slight tubercle. Eyes capping the anterolateral angles, larger and more recurved than those of *Didymops transversa*. Prementum of labium much wider than long, ratio of width:length about 14:11; crenations on distal margin of palpus 7, rectangular about the middle; premental setae 5 plus 1 or more small inner ones; palpal setae 6. Pronotum with lateral edges strongly upturned, forming a tubercle; pterothorax widest at metathorax. Legs long and slender, the hind femur a little more than twice as long as the head is wide; wing-sheaths reaching nearly to base of seg. 7. Abdomen broadly ovate, two-thirds as wide as long, widest at seg. 5; lateral margins evenly rounded; lateral spines on 8 and 9, slender, directed straight back; those of 8 about one-third as wide as long, those of 9 reaching to posterolateral edge of 10 or a little beyond base of anal pyramid; dorsal hooks (pl. 3: 6, p. 24) on segs. 2 to 10, upright and slender on 2 and 3 becoming gradually more horizontal with apices directed caudad; cultriform on 5 and 6, highest on 6, decreasing on 7 to 9, on 10 very low but with a distinct backward point and continued upon the epiproct as a sharp median ridge.

PLATE 3

Macromia—dorsal abdominal patterns: (1) *M. illinoiensis*; (2) *M. rickeri*; (3) *M. magnifica*.
M. illinoiensis—anal appendages of male: (4) dorsal view; (5) left lateral view.
Macromia nymphs—dorsal hooks of abdomen, left lateral view: (6) *M. illinoiensis*; (7) *M. rickeri*. (6, 7 after Walker, 1937.) (Abbreviations: 2, 9—abdominal segments.)

Colour pattern often obscure but sometimes very distinct. Body, especially abdomen, pale drab (exuv.) with numerous irregular dots and blotches of a darker colour, tending to accumulate along the lateral margins and around the bases of the dorsal hooks; femora pale with three dark annuli, a sub-basal, a median and an ante-apical; tibiae irregularly and more minutely spotted; third tarsal segment darker than the others.

Length 28–30; abd. 16–19; w. abd. 10–12; h.w. 7–8; h.f. 10.8–11.5; w. hd. 7–8 (exuv.).

Habitat and range. Large somewhat rapid streams, and lakes exposed to moderate wave action. N.S. to w. Ont. (Lake of the Woods) and Man., s. to Ga., Ala., Miss., Iowa and Kans.

Distribution in Canada. N.S.—Annapolis, Antigonish and Hants counties. *N.B.*—Northumberland and Queen's counties. *Que.*—Hemmingford; St. John's; Fulford; Vaudreuil; Chateauguay; Montreal; St. Hyacinthe; Berthierville; Lac St. Pierre; Hull, Ironside, Wakefield and other stations in the Gatineau Valley; Mont Tremblant; Nominingue. *Ont.*—Kent, Middlesex, Norfolk, Simcoe, Hastings, Carleton and Renfrew counties; Muskoka, Parry Sound, Manitoulin, Nipissing (incl. Algonquin Park), Thunder Bay and Kenora districts. *Man.*—Berens River (52°25′N., the most northerly record; July 1, 1938, 1 ♀, W.J. Brown).

Field notes. This large handsome dragonfly is distributed over much the same territory as is *Didymops transversa* in Ontario and, although their habitats overlap to a considerable extent, *M. illinoiensis* prefers more rapid streams and more exposed lake shores.

We have observed this species to best advantage at the former Great Lakes Biological Station, Go Home Bay, Georgian Bay, Ontario. This was situated at the mouth of the Go Home River, in a typical section of the Canadian Shield where the rocky coast of Georgian Bay is extremely irregular, the innumerable headlands and islands being separated by inlets and tortuous channels in which the water may move sometimes in one direction, sometimes in another, depending on the direction of the wind. *M. illinoiensis* is common everywhere along these rockbound channels, except in the more sheltered bays, and in the many ponds, which occupy depressions in the rocky terrain.

"The long-legged, spider-like nymphs are not infrequently seen sprawling on the surface of the sand or mud bottoms or on the stones. The exuviae are often more or less muddy differing in this respect from those of *D. transversa*, which are always clean. Like *transversa* the nymphs often travel a considerable distance from the water prior to the emergence of the imago. I have found exuviae on the verandah of the dwelling-house, and under the eaves of the boathouse, on rocks along the shore a few feet or several yards from the water's edge or on tree trunks 3–6 feet from the ground" (Walker, 1915).

At Go Home Bay emergence began about the end of June, our earliest observation of this having been on June 28 and 31 in different years. In about a week's time after emergence "they were common about the island, flying rather low and frequently resting on the branches of trees. Both sexes appeared in about equal numbers and were easy to capture. In about a fortnight they had spread over the country and were no longer so easily obtained."

"During the latter half of July and throughout most of August they may be found in sunny weather flying back and forth along the edges of woods or in small open places partly enclosed by trees. They fly swiftly but, as a rule not beyond reach of the net, and as they follow a more or less regular beat, they are not very difficult to capture. Flight ceases at sundown and during dull weather" (Walker, 1915).

At De Grassi Point, Lake Simcoe, where we first met with this species in 1901, the males flew up and down a road through the woods, the distance covered in a beat being much longer than that observed at Go Home Bay. However, they flew in a rather restricted area (a few hundred yards at most) and the exuviae were discovered only along the part of the lake shore that was strewn with boulders and more than ordinarily exposed to wind and waves. *M. illinoiensis* was fairly common at the turn of the century but later almost vanished for many years. Then, more recently, we have seen it again but in smaller numbers. It is occasionally seen flying low over the water and we have sometimes seen a dragonfly that appears to be *M. illinoiensis* flying over the lake and striking the water with the end of the abdomen, presumably ovipositing.

Kellicott (1899) reported that he had "taken the female flying over ripples of our larger streams, and a pair in copulation resting on a low bush not over two feet from the ground."

Williamson (1900) quotes R. S. Osborn as saying that "at Sandusky, Ohio, this species may often be found congregated in great numbers in quiet nooks among the bushes near the marshes. They rest on the under side of twigs with the abdomen hanging down and several individuals may occupy the same twig."

The flight period, as observed in Ontario, ranges from June 15 (Normandale) to August 28 (Dunrobin).

Variations. Although ordinarily there is little variation in size in *M. illinoiensis* in any one locality, the few specimens we have seen from the Maritime Provinces are smaller than those from Ontario, and those from western Ontario are larger than those from southern Ontario. The variability of many venational characters of the male has been documented by Williamson (1909b).

Macromia magnifica MacLachlan. (Pl. 2: 2, p. 18; pl. 3: 3, p. 24)

Macromia magnifica MacLachlan, in Selys, Bull. Acad. Belg., (2) 37: 22 (reprint p. 11), 1874.

Macromia magnifica: Kennedy, Proc. U.S. Nat. Mus., 49: 313, 1915b.

A brightly coloured species of the Pacific Coast, conspicuous owing to the broad yellow annuli of the abdomen (pl. 3: 3, p. 24).

Male. Labium yellow, stained with reddish brown, especially in the palpi; face pale yellow with two broad dark-brown cross-stripes, one across basal half of labrum and anteclypeus, the other across frons; dorsum and sides of frons yellow divided by a median dark stripe that is widest behind the middle; vertex yellow except for a narrow black margin, eye-tubercles yellow, occiput dark-brownish; rear of head black. Thorax medium brown with a green metallic lustre, largely obscured by a dense, pale-brownish gray hair; median carina and ante-humeral stripes light yellow, the latter extending dorso-caudad more than half-way to the ante-alar ridges; ante-alar sinuses pale yellow. Sides of thorax somewhat brighter than dorsum owing to their having less hair; lateral stripe widening a little above the spiracle, metepimera with a post-eroventral yellow stripe. Legs black, except tibial keels which are pale yellow. Wings hyaline, costal margins pale yellow; venation with pt. blackish, membranule pale gray. Abdomen black, broadly annulate with yellow (pl. 3: 3, p. 24); segs. 1 and 2 brown, clothed with pale hairs, 1 without yellow markings, 2 with a pair of large yellow dorsal spots approaching one another mesally; yellow spots confluent across mid-line on 3 to 10, those on 3 to 6 occupying most of the dorsum in front of transverse carina, truncate behind and bilobed in front on 3, and on the other segments nearly reaching the base; largest on seg. 8; spots on 7 and 8 extending beyond transverse carina, with a somewhat rounded hind margin; 9 and 10 with much smaller basal spots.

Tibial keel of fore legs slightly more than half as long as the tibia; middle tibia without trace of a keel; that of hind leg about one-sixth shorter than tibia. Abdomen narrowest from the transverse carina of 3 to that of 6; thence widening gradually, more rapidly on 7, to attain a maximum width at the posterior end of this segment; 8 with parallel sides, 9 and 10 successively narrower.

Anal. apps. slightly longer than 10. Sup. apps. slightly sinuate, being proximately bowed and distally divaricate, with no external angle; apices somewhat bluntly pointed, in lateral view straight and horizontal. Inf. app. barely or not quite as long as sup. apps. and about one-half as wide as long, sides slightly convex, apex blunt, upward curvature as usual.

According to Needham and Westfall (1955) "Old specimens show a marked pruinosity of sides of thorax."

Female. Similar in pattern to male, except that the abdominal spots are more nearly alike, owing to the almost complete lack of constriction of seg. 3, and very little expansion of 7 and 8, as viewed dorsally. Dorsal yellow spots on 3 to 6 nearly alike, bounded behind and truncated by the transverse carina, spots on 7 and 8 extending beyond transverse carina on middle, 9 with a small basal spot and 10 without one. Vulvar lamina extremely short and deeply excavated.

Venation. Anx ♂ 10–13/7–9; pnx ♂ 6–8/7–9; cells in triangle and st. 1.

Measurements. Total length ♂ 69, ♀ *ca* 71*; abd. ♀ 47–48, ♂ *ca* 50*; h.w. ♂ 42–44*, ♀ 46*; h.f. ♂ 12–12.5*; w.hd. ♂ 9; pt. ♂ 2.6–2.8/2.6–2.8.

*These measurements are derived partly or wholly from Williamson (1909b).

Nymph (description from Kennedy (1915b)). Head pentagonal, five-eighths as long as wide, with short appressed hairs on the frontal horn, on bases of antennae and on sides below and behind the eyes. Horn acute, densely hairy and more erect than in *illinoiensis*; eyes very prominent; dorsolateral tubercules on hind angles of head a little more sharply defined than in *illinoiensis*. Antennae 7-segmented, first two segments stout and hairy, fourth the longest, slender; the whole length of the antennae slightly greater than that of *illinoiensis*. Sides of prementum and posterolateral angles less rounded than in *illinoiensis*; premental setae usually 4 in a short row on each side with a single detached seta at inner end of row, between which and the middle line are 2 to 4 shorter irregularly placed setae; in *illinoiensis* each main row of premental setae usually contains 5 setae with a single detached seta at inner end and other short irregularly placed setae between it and the median line. Abdomen very similar to that of *illinoiensis* but differing in three respects: (1) dorsal hooks of segs. 2 and 3 straight, whereas they arc slightly hooked in *illinoiensis*; (2) lateral spines on seg. 9 do not reach the posterior margin of seg. 10, whereas in *illinoiensis* they surpass it; (3) posterior edge of sternum of seg. 9 is fringed with hairs, whereas in *illinoiensis* there are seldom more than 2 to 4 short hairs to be found on this edge.

General colour of live nymphs dirty olive. Antennae pale, without markings. Frontal horn dark. Eyes and triangular areas between them, which represent the adult eyes, dark. Parts of head posterior to eyes pale. Dorsal surfaces of thorax and wing-sheaths dark brown, these being the darkest parts of the nymph, excepting the eyes. Legs with three femoral annuli; tibiae dotted, with hind leg the darkest. Dorsal surface of abdomen mottled and speckled with olive brown, with a more or less well defined spot at the outer edge of each segment. All ventral surfaces of the larva pale.

Length 31; abd. 20; w. abd. 11.5; h.w. 9.7; h.f. 13.5 (exuv.); w. hd. 8.

Habitat and range. In rapid streams, the nymphs being amongst matted tree-roots suspended in the water, or sometimes under stones. Calif., Ariz. and Nev., to Wash. and B.C.

Distribution in Canada. B.C.—Christina Lake; Okanagan River between Penticton and Dog Lake; Peachland; Vernon.

The distribution in Canada thus appears to be confined to the south-central or semi-arid regions of the south, chiefly in the Okanagan and Columbia valleys.

Field notes. M. magnifica was first taken in British Columbia by J.B. Wallis at Peachland, Okanagan Lake. Next we observed it on the Okanagan River at Penticton on July 22 and 23, 1926. All we saw were males and they "flew very swiftly, usually two or three feet above the water and a few yards from the shore, but were so scarce that their capture was practically hopeless" (Walker, 1927). At Christina Lake, Whitehouse (1941) captured them only on roads, but on the Okanagan River he caught a number from a rowboat close to the shore. These were all males and generally flew in pairs and when one was "safely netted, it paid to remain stationary for its companion to come along" (Whitehouse, 1941).

Kennedy (1915b) observed this dragonfly carefully on Satus Creek, Yakima county, Washington. This stream, after emerging from Satus Canyon and spreading out over the Yakima River flats, is cool and shallow and meanders from pool to pool over long gravel ripples. These pools vary greatly in size, sometimes being 200 feet long but contain neither aquatic vegetation nor brush. Exuviae of *Macromia* found on the trunks, branches and even twigs of alder trees growing along the bank, gave the clue to the habitat of the nymphs. The stream had undercut the banks around the roots of some of the trees, forming pools that were half full of an intricate mass of fibrous alder roots, and it was in this mass of roots that a *Macromia* nymph was found and it was on the trunks and branches of such trees that the exuviae were located.

Male *Macromia* were patrolling the pools, especially the larger ones. The patrols were long though rarely exceeding 300 feet. The flight of males over the water is apparently controlled largely by the presence of ovipositing females, which resort to water to oviposit early in the day, or in calm weather. Hence males are most often seen over water between 7 and 10 a.m. Few are seen in the afternoon or on windy days.

"In ovipositing the female would fly several times back and forth over a short beat of 40 or 50 feet, striking her abdomen on the surface of the water at 3 to 5 foot intervals. This beating back and forth generally lasted until a male discovered her, when she would be taken away in copulation." "The copulatory flight was ordinarily away from water over the surrounding trees, but ended in a long period of copulation while resting on some bush or tree" (Kennedy, 1915b).

At times when mating was not in progress, both males and females might be found patrolling glades and barnyards until late twilight and as much as half a mile from the water. "Here the flight varied from close over the ground to as high as the trees. One pair was seen in copulation at noon a half mile from the creek" (Kennedy, 1915b).

As to the flight period, there are not enough records from British Columbia to determine or even estimate it. All the specimens recorded were taken in July, the range of dates being from the 3rd to the 31st. Kennedy (1915b) estimates that in Washington they had probably been on the wing for about three weeks when he arrived on July 27. At this time they were at their height and "they rapidly disappeared after August 7, and none were seen after August 24." In California, this species has been recorded as early as June 6.

Variations. The variability of many venational characters of the male has been documented by Williamson (1909b).

Macromia rickeri Walker. (Pl. 3: 2, 7, p. 24)

Macromia rickeri Walker, Can. Ent., 69: 6–13, 1937.

Very closely related to *M. magnifica*, perhaps not specifically distinct but differing much in general appearance, owing to the darker ground colour and reduction in size of the yellow spots (pl. 3: 2, p. 24).

Male. Head very dark brown with slight bluish lustre. The following parts yellow: a pair of spots on the lower half of the labrum, sometimes narrowly connected, and another on the mandibles and postclypeus, except for a depressed spot on each side at upper margin and median part of lower margin; a pale dorsal spot on frons, narrowly separated from lateral spots by the marginal ridge; a small bilobate spot on summit of vertex. Median lobe of labium pale yellow, its distal margin and the palpi dark brownish. Pile of face blackish, not very dense. Eyes in an alcoholic specimen green with tubercle yellow. Rear of head black. Thorax chocolate brown with dull greenish reflections, clothed with pale-brownish gray hairs which are longest at front of dorsum. Yellow antehumeral stripe very short, much less than half the length of dorsum, ante-alar sinuses and a spot on mesinfraepisternum also yellow; lateral oblique stripe narrow, not quite attaining wing-base, widest in upper third, the hind edge somewhat excavate above and below the spiracle, lateroventral metathoracic margin narrowly yellow. Legs black with hind surfaces of pro- and mesocoxae pale yellowish. Wings hyaline, costal margins very narrowly yellow; venation otherwise blackish, including pt.; membranule pale gray. Abdomen dark greenish brown with yellow spots as follows: on seg. 2 widely separated, prolonged laterad to include the auricles, on 3 to 6 narrowly divided, successively smaller, those on 3 separated dorsally by a narrow V-shaped space, on 6 very small or even absent, on 7 and 8 large, undivided basal, the spot on 7 subtriangular, extending in middle a little beyond the transverse carina; spot on 8 largest, extending two-fifths of distance towards hind margin, broadly angular or rounded.

Vertex and occiput as in *magnifica*; front tibiae keeled on slightly more than their distal third; middle tibiae without keel; hind tibiae with a keel on all but about the proximal sixth. Abdomen with form as in *magnifica* but a little narrower on segs. 7 and 8 than in that species; it is also hairier at base and on terminal segments.

Anal apps. and genitalia as in *magnifica*.

Female. Similar in form to male, except for the usual differences in the abdomen. No widening such as that which forms the "club" in the male.

Colour pattern much as in male. Yellow spots on 2 not extended lateroventrally; those on 3 to 6 all mesally divided and of a more or less rounded form; decreasing in size caudad but usually to a much slighter extent than in the male. On 7 and 8 the yellow spots are of about the same size as in the male and are undivided mesally. The remaining segments are without yellow markings. The wings are hyaline or faintly yellow in mature individuals though decidedly flavescent in tenerals. Anal apps. about as long as 10, tapering to acute apices. Vulvar lamina represented by a slight thickening of the posterior edge of the sternum of 8, which is produced into an arcuate extension.

Venation. Anx ♂♀ 11–13/8–9; pnx ♂ 6–8/7–9, ♀ 7–8/7–9; cells in triangle and spt. 1.

Measurements. Total length ♂ 64–68, ♀ 65–70; abd. ♂♀ 46–50; h.w. ♂ 40–42, ♀ 41.0–43.5; h.f. ♂ 12.0–12.5, ♀ 12–13; w.hd. ♂♀ 9.5; pt. ♂♀ 2.5–3.0.

Nymph. Head pentagonal, five-eighths as long as broad, with short hairs, sparsely distributed except on the frontal horn and basal segments of antennae, where they are dense; frontal horn erect, acute; eyes more prominent than in *illinoiensis*; tubercles on posterolateral angles of head prominent like those of *magnifica*, more sharply defined than in *illinoiensis*. Antennae with segs. 1 and 2 thickened, the others tapering, the last very slender, 1 + 2 slightly shorter than 3 + 4; 4 + 5 slightly longer than 3 and slightly shorter than 6 + 7. Folded labium reaching posterior surface of mid-coxae; prementum constricted in its proximal fourth, slightly flaring distally, its greatest width slightly exceeding the length of one of the sides; premental setae usually 6 or 7, the 4 outermost closely crowded; the others more widely separated, there being usually 1 to 4 shorter setae, more regularly placed, on each side of the middle line; palpal setae 6 or 7; one or two smaller setae (setellae) at base of each palpus. Pronotum with lateral margin strongly bent upwards, more angulate than in *illinoiensis*. Abdomen as in *magnifica*: as compared with *illinoiensis*, the lateral spines of 8 and 9 are shorter and more incurved, those of 9 extending back beyond middle of 10, but not to level of posteroventral edge of 10, as they do in *illinoiensis*; dorsal hooks (pl. 3: 7, p. 24) on 2 and 3 straight, slender and erect, shorter than those of the following segments, which are successively wider (anteroposteriorly) at base and more sharply bent backward; they differ from those of *illinoiensis* in that the dorsal margins, particularly in the middle segments are not so prominently convex; in *illinoiensis* the hooks of 5 and 6, seen in profile, suggest a bird's head, whereas in *rickeri* and *magnifica* they are merely gently falcate; both seg. 10 and epiproct are carinate.

The colour pattern seems to be identical with that of *magnifica*, as described and figured by Kennedy (1915b).

Length 30; abd. 19–20; w.abd. 12.5 (11.5 exuv.); h.w. 9.5; h.f. 13; w.hd. 8.

Habitat and range. Forest lakes and rapid streams; s.w. B.C.

Distribution in Canada. B.C.—Cultus Lake; Chilliwack River; Kawakawa Lake; Salmon Arm.

Field notes. Cultus Lake, where most of the material of this species has been collected, lies in the southwestern mainland of British Columbia between the Fraser River and the international boundary, near the town of Chilliwack. The climate of the lowland part of this region is moist and equable and it belongs to the Vancouveran faunal area as defined by Klugh and McDougall (1924). The lake has been described by Ricker (Walker and Ricker, 1938) as follows: "Cultus Lake lies in a glaciated U-Valley between two mountain ranges, but itself is only about 160 feet above sea level. It is three miles long and one-half to one mile broad; mostly deep clear and with the epilimnion quite warm in summer (20–22° C.). The shores are of stones or bed-rock with small patches of sand. Emergent vegetation consists of *Scirpus validus* in several places among the stones and small clumps of *Typha latifolia* or even *Nymphaea polysepala* in the most sheltered locations; but most of the shore line is bare." Four odonates are typical of the lake proper: *Argia emma*, *Enallagma cyathigerum*, *M. rickeri* and *Aeshna umbrosa*.

Macromia rickeri flies over both the lake and its outlet, Sweltzer Creek, which is a warm stream of moderate current (volume of flow about one cubic foot per second in summer) but whether it is resident here or merely a wanderer has not been determined. All the nymphs and exuviae were collected from the lake itself or its shore. Full-grown nymphs were collected by Ricker from May 28 (1935) until June 12 (1936).

Whitehouse (1941) also collected the nymphs in Cultus Lake: "At Cultus Lake the mature nymphs live under stones and, when I was staying there eleven were 'pailed' June 17–23rd. Of this total five emerged safely, four being ♀s." "The first emerged about 8 a.m. June 23rd, and by 11 a.m. had spread her wings. The other four (and three more that died in the process) climbed up the sticks provided in the early evening, remaining with head and shoulders out of the water until dark. Transformation took place between 10 and 11 p.m."

Emergence or tenerals have been observed from May 1933 (one teneral, exact date not recorded) and June 22 (1936) to July 3 (1935), most of the emergence taking place during the last week of June.

The general flight period seems thus to extend from about the third week in June to the middle of August, although Ricker has one record for September 10, 1937, when two males were taken. There is also the record of a teneral female taken in May, 1933, but the exact date was not recorded.

Whitehouse (1941) also gives the following notes on the adults of *M. rickeri*. "June 25th a pair taken in cop., my earliest date of sexual maturity."

"At Kawakawa Lake, where the shore line is not stony, the ovipositing ♀ flew quickly some eight or ten inches above the water, swooping down every five or six feet to touch the tip of her abdomen. The water was five to ten feet deep, with a muddy bottom and reeds well below the surface."

FAMILY CORDULIIDAE

These are mostly middle-sized dragonflies of slender form and generally dark coloration, in which there is usually a metallic green or brassy lustre. A few genera, however, are composed entirely of species without this metallic lustre. Most of the characters by which the Corduliidae may be distinguished from the Libellulidae have been given in the key but, with the exception of the eye tubercle and the form of the anal loop, such characters are confined to the males. As a rule the abdomen is relatively longer in the Corduliidae than in our native Libellulidae, and a more or less clavate or "corduliform" abdomen is the rule in this family but exceptional in our Libellulidae. The males have large genital lobes and a pair of rather small toothless auricles. A longitudinal ventrolateral carina that bears numerous spines is present in both sexes, usually on abdominal segments 4 to 8.

The most striking differences between these two families on the one hand and the Macromiidae on the other have also been given in the key, namely the short broad anal loop with its absence of a midrib and the lack of development of supplements in the Macromiidae, and the marked difference in the tarsal claws.

The inclusion of the Macromiidae with the Corduliidae by many authors is based mainly on the presence in both families of the eye tubercle, the metallic coloration, the angular anal borders of the hind wing in the males, and the presence of tibial keels. Since all of these characters are shared with other families of Anisoptera they appear to be archaic features, probably being derived from some common ancestral group, and so do not represent any recent evolutionary trend. On the other hand, the similarity of the Corduliidae and the Libellulidae is not only close in the nymphal stages, as we have already noted (p. 15), but also in the imago, as seen in the highly specialized form of the anal loop. Although the foot-like shape of the loop is not as evident in the Corduliidae as in most of the Libellulidae, this is obviously a point of little significance.

These conclusions based on adult characters appear to be confirmed by the larval features since the nymphs of Macromiidae stand apart from those of the other two groups (see p. 16), whereas the nymphs of the Corduliidae and Libellulidae are so much alike that we have been unable to find a single reliable character on which a key can be based. As a general rule the crenations of the labial palpi are longer in the Corduliidae than in the Libellulidae but this character is insufficient for separating genera without leaving exceptions.

The Corduliidae of our territory are chiefly inhabitants of streams and lakes in which the oxygen content is high. They are less frequent in ponds, and perhaps this is due partly to competition with Libellulidae. In the far north, where libellulids play a smaller role and the corduliids a larger one, the latter appear to resort more generally to ponds, particularly muskeg or sphagnum pools.

KEY TO THE GENERA OF CORDULIIDAE

ADULTS

1 Veins MA and CuP of fore wing diverging towards wing margin (pl. 4: 1, p. 34; pl. 7: 2, p. 58)   2

  Veins MA and CuP of fore wing converging towards wing margin, often being parallel for most of the distance (e.g. pl. 1: 1, p. 6)   3

2 Medium-sized dragonflies (abd. and h.w. each more than 30 mm.); without metallic lustre; triangle of fore wing 3-celled, of hind wing 2-celled; hind wing with a large basal amber spot   *Neurocordulia* (p. 36)

  Small dragonflies (abd. and h.w. each less than 25 mm.); top of head metallic green; triangles of fore wing clear; hind wing without a basal spot   *Williamsonia* (p. 62)

3 Body without or almost without metallic lustre; tibiae of males with well developed keels on all three pairs of legs; vulvar lamina elongate, deeply bilobed and generally flexible, normally longer than sternum of seg. 9   4

  Body with a metallic green or blue lustre, mesotibiae of males without a keel or with only a vestige at the distal end; vulvar lamina bilobed or entire, sometimes shorter than sternum of seg. 9, and when as long more or less rigid   5

4 Abdomen of male widest at segs. 6 to 7; sup. apps. of ♂ converging from base, where they are separated by a space scarcely or not wider than one of the apps.; vulvar lamina bilobed to the base, the lobes divergent, longer than sternum of seg. 9 and flexible; one cross-vein behind pt.   *Epitheca** (p. 39)

  Abdomen of ♂ and mature ♀ widest at seg. 8; sup. apps. of ♂ diverging from base, where they are separated by a space wider than one of the apps.; vulvar lamina bilobed but not from base, much shorter than sternum of seg. 9, the lobes nearly parallel in our species and rigid; wings usually with 2 cross-veins behind pt.   *Helocordulia* (p. 57)

*Includes *Epicordulia* and *Tetragoneuria*.

PLATE 4

Wings of Corduliidae: (1) *Neurocordulia yamaskanensis*; (2) *Epitheca princeps*.

PLATE 4

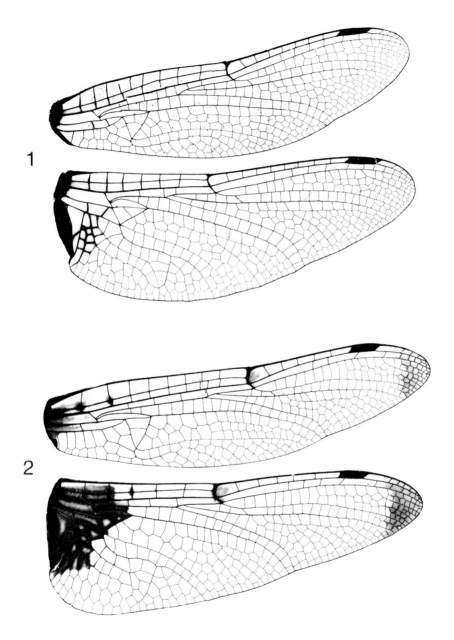

1

2

5 Hind wing with a second cubito-anal cross-vein forming a subtriangle
   (pl. 7: 3, p. 58); mesotibiae of males with no trace of a keel; inf. app. of
   ♂ usually subtriangular; rarely divided once   *Somatochlora* (p. 65)

   Hind wing without a second cubito-anal cross-vein (pl. 1: 1, p. 6; pl.
   20: 2, p. 134); mesotibiae of males with a vestigial keel at the distal end;
   inf. app. of ♂ triangular or bifurcate   6

6 Sup. apps. of ♂ each tapering to an upturned point; inf. app. of ♂
   triangular   *Dorocordulia* (p. 137)

   Sup. apps. of ♂ cylindrical without an upturned pointed apex; inf. app.
   of ♂ bifurcate, the two rami each being bifid   *Cordulia* (p. 133)

NYMPHS

1 Dorsal hooks present and well developed on some of the abdominal
   segments   2
   Dorsal hooks absent or reduced to low knobs   6

2 Lateral spines on seg. 8   3
   No lateral spines on seg. 8   *Williamsonia** (p. 62)

3 Crenations on distal margin of labial palpi nearly semicircular or even
   more deeply cut; lateral spines on seg. 8 divergent   *Neurocordulia* (p. 36)

   Crenations of labial palpi much shallower than a semicircle; lateral
   spines on segs. 8 and 9 about equal in length, those of 8 not or scarcely
   divergent   4

4 Lateral spines on seg. 9 more than three times as long as those on seg. 8,
   being at least as long as the mid-dorsal length of seg. 9 and usually much
   longer   *Epitheca* (p. 41)

   Lateral spines on segs. 8 and 9 short and about equal in length, or
   absent, those of seg. 9 never as much as twice as long as those of seg. 8   5

5 Dorsal hooks on segs. 6 to 9 only, minute on seg. 6, on the other
   segments as long as the segment or longer   *Helocordulia* (p. 59)

   Dorsal hooks on segs. 3 or 4 to 9 spine-like and usually curved
                                                     Somatochlora (p. 66)

6 Sides of thorax with a broad dark longitudinal stripe   7
   Sides of thorax uniformly coloured   *Somatochlora* (p. 66)

7 Total length 23 mm.; dorsal hooks absent or vestigial   *Cordulia* (p. 133)
   Total length 19 mm.; dorsal hooks stubby but distinct
                                                     Dorocordulia (p. 139)

---

*This couplet is constructed on the assumption that the nymphs of *W. fletcheri* (at present
unknown) and *W. lintneri* (described by White and Raff, 1970) both lack lateral spines on
segment 8.

Genus **Neurocordulia** Selys

Bull. Acad. Belg., (2) 31: 278, 1871. Type species *Libellula obsoleta* Say.

These are medium-sized, rather stocky corduliids of non-metallic coloration, usually brownish or yellowish without conspicuous spots or stripes. All three pairs of legs are keeled. The wings (pl. 4: 1, p. 34) are subhyaline or dusky, usually marked with amber or brown spots and along the borders from base to nodus or even farther, the spots sometimes being situated over each antenodal vein. The most distinctive venational character is the divergence of MA and CuP in the fore wing. Cross-veins are also somewhat more numerous than usual, both the triangle and subtriangle being nearly always three-celled in the fore wing, whereas in the hind wing the triangle is normally crossed and the subtriangle is clear. The anal loop is somewhat squarely truncate, being broader than a right angle.

The vertex is nearly erect, forming a rounded transverse arch over the middle ocellus, which is quite concealed from above.

The superior anal appendages are subcylindrical and relatively slender in the proximal half, but are somewhat suddenly enlarged immediately beyond the middle, the enlargement being chiefly lateral. The apices are immediately acute and slightly elevated. The inferior appendage is triangular and usually three-fourths to five-sixths as long as the superiors.

The females are similar to the males in general form and coloration. The anal appendages are short and slender and the vulvar lamina in our only species is very short.

The nymphs are very compact, of a yellow or olive colour, the legs sometimes strongly banded with dark brown. The cuticle is unusually firm and thick. The head is wide and pentagonal but the eyes are not very prominent. The frontal ridge extends forward a little beyond the inter-antennal space and is covered with very short setae. The vertex is slightly raised, bearing a number of still smaller setae. The sides of the head behind the eyes are very oblique; yet the hind angles are prominent and the hind margin of the head is concave. The antennae are shorter than the head and are seven-segmented. The labium is short and wide with 8 to 11 premental setae and 5 to 6 palpal setae. The crenations of the palpi are deeply and widely separated, each crenation being rounded at the end. The abdomen is more than usually convex, both longitudinally and transversely. It bears dorsal hooks on segments 2 to 9, those on 2 to 5 being upright and slender, and the others much stouter, becoming progressively lower and their tops somewhat recurved and bluntly rounded. The only lateral spines are on segments 8 and 9; they are broad at the base and slender at the apex, those on 8 being divergent, and those on 9 being only slightly longer than those on 8.

Neurocordulia is a Nearctic genus of six species found chiefly in the southern United States. Only one species is known from Canada.

Neurocordulia yamaskanensis (Provancher). (Pl. 4: 1, p. 34)

Aeshna yamaskanensis Provancher, Nat. Canad., 7: 248, 1875.
Epitheca yamaskanensis Provancher, Nat. Canad., 9: 86, 1877b.
Epitheca jamascarnensis (Hagen): Selys, Bull. Acad. Belg., (2) 45: 191, 1878.
Neurocordulia yamaskanensis: Walker, Can. Ent., 45: 164–166, 1913.

An inconspicuous light brown dragonfly of medium size, with an amber spot at the base of each hind wing and with the venation darkened to brown. The adults fly only for a short time after dusk.

Male. Eyes brown; labium and labrum pale to lemon yellow, darker towards the outer margins. Face above labrum olivaceous, without dark cross bands, dorsum of frons and vertex clear brown, of nearly the same depth; occiput reddish brown. Rear of head dull ochraceous with a transverse fringe of pale brownish hair a little below the postero-dorsal eye margin. Pterothorax moderately dark brown above, the median carina paler, hair pale brownish on dorsum, not very dense, more so on the sides; sides with a small yellowish spot around the spiracle; legs moderately dark brown, coxae and trochanters paler, tarsi dark brown. Wings subhyaline or tinged with brown in old individuals. Costal veins pale ivory yellow; pt. dark brown. Hind wings with an amber spot at base between membranule and triangle, veins darkened, the median space in anterior cell of anal triangle nearly clear. Abdomen, viewed from above, with a broad median dark stripe, sides and seg. 10 paler.

Median furrow rather shallow; eye-seam short, about one-fourth as long as occiput, which is attenuated in front and slightly convex behind. Tibial keels on fore legs for about one-third of their length, on middle legs nearly one-half, and on hind legs three-fifths of their length. Abdomen very deep between segs. 2 and 3, at base of 4 about half as deep as at base of 3, narrowest at the transverse carina of 4, thence widening and becoming depressed, widest at rear end of 6 to 8.

Sup. apps. suddenly narrowed a little beyond base, thence in proximal half narrow, cylindrical, and slightly converging; beyond the middle stouter, somewhat divergent, but with pointed apices nearly parallel, lateral margins slightly convex and mesal margins slightly concave; in lateral view slightly arched; inf. apps. three-fourths as long as sup. apps. elongate-triangular, gently upcurved.

Female. Similar to male in size, form and colour, except for the usual differences between the sexes; lat. apps. twice as long as 10 or three-fourths as long as 9, cylindrical, with acute apices. Vulvar lamina of female very short with a wide V-shaped emargination, each lateral lobe rounded but with a somewhat prominent edge.

Venation. Anx ♂ 9–10/6(7), ♀ 9–10/6; pnx ♂ 7/7–8, ♀ 8–9/9–10; cells in triangle 2–4, in spt. 1; cubito-anal cross-veins in hind wing 2, rarely 1; pt. subtending 2 or 3 cells in fore and hind wings.

Measurements. Total length ♂ 51–55, ♀ 51–52; abd. ♂ 35–37, ♀ 37.0–37.5; h.w. ♂ 31–33, ♀ 34–35; h.f. ♂ ♀ 7; w.hd. ♂ ♀ 8; pt. ♂ 2.5–2.9/2.9–3.6, ♀ 2.4–3.6/2.7–3.0.

Nymph. Head broadly convex above and on the sides eyes not very prominent; frontal ridge with numerous minute and very short hairs. Hind angles of head prominent, distance between them a little greater than half the greatest width of head; hind margin excavate. Folded labium extending very slightly behind fore coxae; prementum about four-fifths as long as wide; the labial suture abruptly deflexed, bluntly obtusangulate; premental setae 9 to 11, the mesal 3 or 4 much smaller than the others; palpi triangular; palpal setae 6; movable hook about as long as one of the palpal setae; crenations of the distal margin of palpus very deeply cut, each bearing a group of about 2 to 5 stiff spiniform setae. Supra-coxal processes very prominent. Legs short, length of hind femur being slightly less than width of head. Abdomen ovate, widest at seg. 6 or

7, slightly greater than two-thirds of the length; curve of lateral margins somewhat greater in distal than in proximal half, lateral spines on 8 and 9 of nearly the same width, those on 8 strongly divergent, those on 9 parallel and extending caudad scarcely or not at all beyond tip of anal pyramid, which is very short; dorsal hooks on segs. 2 to 9, very slender and erect at base but becoming successively stouter, and after seg. 5 increasingly bent backward, being represented on seg. 8 by little more than a ridge; epiproct about equilateral, the paraprocts a little longer; cerci very short.

Colour yellowish to orange brown, variegated with dark brown. Head dark brown above, generally paler in the centre and on frontal ridge; femora and tibiae dark with two pale annuli, a median and an ante-apical. Abdomen yellowish brown, more or less distinctly blotched with dark brown, especially on the dorsal hooks, lateral margins, spines and muscle scars.

Length 22.0–24.5; abd. 13.0–13.5; w. abd. 9–10; h.w. 6–7; h.f. 5.5–6.0; w. hd. 7.0–7.5.

Habitat and range. Rocky shores of lakes and rivers, somewhat exposed to currents or wave action. Me. to central Ont. and Mich., s. to W.Va., Ky. and Mo.

Distribution in Canada. Que.—Mt. Yamaska, near St. Hyacinthe (type locality); Ironside; Choisy; Ste. Anne de Bellevue. *Ont.*—Essex (Pelee I.; Leamington), Carleton and Lanark (Mississippi Lake at Innisville) counties; Muskoka, Parry Sound, Nipissing (Algonquin Park) and Sudbury districts.

Field notes. This species is abundant in the Muskoka and Parry Sound districts, in Sturgeon Falls and doubtless throughout the more southern parts of the Canadian Shield or the Precambrian regions of Ontario and Quebec.

The nymphs frequent rocky shores of large lakes, channels or rivers, showing a preference, in the case of lakes, for the more exposed, wave-washed shores and, in the case of streams, for the vicinity of falls and rapids. In such situations they cling to the undersides of boulders or other loosened rock masses. Since the exuviae are most commonly found on steep rocky shores, rising almost perpendicularly from the water, it seems probable that nymphs may emerge at times from water of considerable depth, probably eight or ten feet, although we have often taken nymphs of several stages in water less than two feet deep. The important factor is probably the oxygen concentration, as their occurrence near falls or rapids suggests.

The nymphs are usually associated with various nymphs of Ephemeroptera (especially Heptageniidae) or of the damselfly *Argia moesta*, on which the *Neurocordulia* nymphs probably feed.

Emergence was frequently seen during the summers that we spent at Go Home Bay, Georgian Bay, Ontario. Our first observation was on June 28, 1907, but exuviae were seen at least a day earlier, but in 1912 adults emerged in the laboratory from June 23 to July 8. Most of them transformed between 7 and 8 a.m. (EST), but a few were taken in the evening. For some days, imagos could be obtained only in this way, but it was at last discovered that they were flying about the island at dusk. It was further ascertained that their time of flight was limited to about half an hour daily commencing soon after sundown (a little after 8 p.m.), and continuing until shortly after 8.30 p.m., after which they retired to the shelter of the trees. This flight period is nearly coincident with that of the mayflies *Ephemera*, *Hexagenia* and others, upon which they appeared to feed predominantly.

"The majority of individuals thus engaged are females. The males are found at the same time flying over the water within a few inches of the water, close to the shore, which they follow very closely. They fly back and forth on a regular beat and with extraordinary swiftness. During these flights the males apparently do not feed, but seem to be on the watch for females, for now and then a male is seen to pounce upon a female, the pair then sailing away over the water or up into the trees where copulation takes place at rest. Except when thus seized by the males, no females were observed close to the water, though plenty of them could always be seen over the rocks nearby" (Walker, 1915).

The flight period as observed at Go Home Bay and other localities on the Precambrian Shield is June 23 to July 30. Exuviae, however, were found abundantly on the shore of Mississippi Lake, Lanark county, on June 7, 1940.

Genus **Epitheca** Burmeister

Handb. Ent., 2: 845, 1839. Type species *E. bimaculata* Charpentier.
Epicordulia Selys, Bull. Acad. Belg., (2) 31: 259, 1871.
Tetragoneuria Hagen, Syn. Neur. N. Amer., p. 140, 1861.

The type species of *Epitheca* is *E. bimaculata* Charpentier. We have recently (Walker, 1966b) stated our reasons for uniting the genera *Epicordulia* and *Tetragoneuria* with the Old World genus *Epitheca* Charpentier.

These are medium-sized to large Corduliidae apparently without metallic lustre though sometimes showing slight brassy reflections on the thorax through the somewhat dense hair.

The frons is somewhat tumid and wrinkled as in most Corduliidae, the eye-seam much shorter than the occiput which is smooth and rounded above and behind. The face in all our species is yellowish, rather thinly beset with short black hairs, and with a darker stripe covering the clypeus and extending dorsally along the eye margins, and sometimes reaching the dorsal surface. The thorax in North American species is unstriped, being of a nearly uniform reddish brown, the colour dulled by a heavy covering of pale grayish brown hair, which is thickest above, especially in front and behind, and thinner on the sides. The median carina, sutures and lower margin of the mesepisternum are darkened and there is generally an orange-brown spot divided, or partly so, by the dark brown spiracle.

The legs are all furnished with well-developed tibial keels, which are longest on the hind legs. The wings (pl. 4: 2, p. 34) are hyaline or flavescent, usually with dark brown spots at the base of the hind wing and often with

PLATE 5

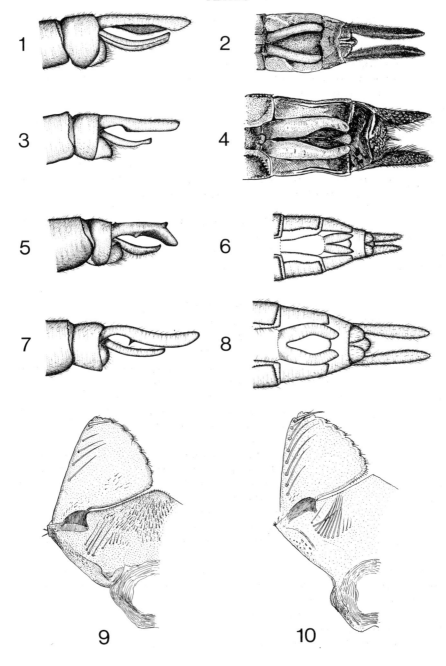

traces at the base of the fore wing and/or on the antenodal cross-veins; in one species nodal and apical spots are generally present. The abdomen is black, without metallic reflections, sparsely hairy, except on the first two segments and towards the posterior end. It is slightly swollen at the base, a little constricted at segment 3, with the middle segments depressed, widest at 6, tapering distad and, in the male, very slender at 9 and 10. The vulvar lamina is bilobed to the base, and the two rami are long, divergent and flexible, and extend to the end of segment 9. This form of the vulvar lamina is apparently correlated functionally with the habit of laying eggs in gelatinous strands which is a distinctive feature of members of this genus.

The antenodal cross-veins of the hind wing are often only four but may be five or six, often in the same species. There is only one cross-vein behind the pterostigma. The sectors of the arculus are separate. In Eurasian species the hind wing has two cubito-anal cross-veins and hence a subtriangle. In North American species (*Tetragoneuria* and *Epicordulia*) there is usually only one cubito-anal cross-vein (the anal crossing) and hence the subtriangle is absent. But in *Epicordulia*, there is occasionally a second cubito-anal cross-vein and hence a subtriangle on one or both sides. It is evident, therefore, that the presence or absence of a subtriangle on the hind wings is of no value as a generic character.

The nymphs are smooth, broadly depressed and have large laterally flattened dorsal hooks on abdominal segments 2 to 9 and prominent lateral spines, those on segment 9 being more than three times as long as those on segment 8 and usually much longer than the mid-dorsal length of segment 9.

The head may or may not bear dorsal tubercles; it appears trapezoidal in dorsal view, being narrowed straight back to rounded hind angles, separated by a straight posterior border. Palpal setae (on each side) 5 to 8.

Kormondy (1959) has described morphological changes that take place during nymphal development of *Epitheca canis*, *E. cynosura* and *E. spinigera*.

Key to the Species of Epitheca
Adults

1 Wings with apical and usually also nodal dark spots (pl. 4: 2, p. 34)

princeps (p. 42)

  Wings without apical or nodal spots   2

PLATE 5

Epitheca—anal appendages of males, left lateral view: (1) *E. princeps*; (3) *E. cynosura*; (5) *E. canis*; (7) *E. spinigera*.
Epitheca—vulvar lamina, ventral view: (2) *E. princeps*; (4) *E. cynosura*; (6) *E. canis*; (8) *E. spinigera*.
Epitheca nymphs—prementum (left side) and palpus of labium, dorsal view when flattened: (9) *E. princeps*; (10) *E. spinigera*
(1, 7 modified after Walker, unpublished; 3, 5, 6, 8 after Kormondy, 1959.)

2 T-spot on dorsal surface of frons clearly defined (pl. 6: 2, p. 50); sup.
  apps. of male (in lateral view) with a sharp ventral spine near the base
  (pl. 5: 7, p. 40); vulvar lamina as in pl. 5: 8 (p. 40)   *spinigera* (p. 51)
  T-spot on dorsal surface of frons indistinct or incomplete (pl. 6: 1, p.
  50); sup. apps. of male without a sharp ventral spine; vulvar lamina
  not as in pl. 5: 8 (p. 40)   3
3 Sup. apps. of male (in lateral view) with a well-defined dorsal tubercle (pl.
  5: 5, p. 40); vulvar lamina with rami sub-cylindrical and extending about
  to posterior margin of seg. 9 (pl. 5: 6, p. 40)   *canis* (p. 54)
  Sup. apps. of male without a well-defined dorsal tubercle (pl. 5: 3, p.
  40); vulvar lamina with rami sub-conical and extending well beyond
  the posterior margin of seg. 9 (pl. 5: 4, p. 40)   *cynosura* (p. 45)

Nymphs

1 A pair of small tubercles on top of head; distal half of dorsal surface
  of prementum heavily setose; palpal setae 4 or 5 (pl. 5: 9, p. 40)
                                                                princeps (p. 43)
  No tubercles on top of head; distal half of dorsal surface of prementum
  with few, or usually no, setae; palpal setae 6 to 8   2
2 Lateral spine of seg. 9 short, barely attaining the level of the tips of the
  anal appendages (epiproct and paraprocts)   *canis* (p. 55)
  Lateral spine of seg. 9 long, extending beyond the level of the tips of the
  anal appendages, usually by twice its length   3
3 Premental setae usually not more than 10   *cynosura* (p. 46)
  Premental setae usually not less than 11   *spinigera* (p. 52)

Epitheca princeps Hagen. (Pl. 4: 2, p. 34; pl. 5: 1, 2, 9, p. 40)

Epitheca princeps Hagen, Syn. Neur. N. Amer., p. 134, 1861.
Epicordulia princeps: Selys, Bull. Acad. Belg., (2) 31: 275, 1871.

This is our largest corduliid, having the size of a small *Aeshna*. It is peculiar
in being spotted after the manner of *Libellula pulchella* or *L. quadrimaculata*.
Towards the northern boundary of its range, however, the nodal spots may
disappear and little may remain of the pattern seen in more southern
individuals, except for the large basal spot of the hind wing.

Male. Labium and labrum pale yellow to buff; face ochraceous to brown, moderately beset
with brownish hairs which are longest on the frons at the anterolateral angles near the dorsum;
frons with an anterior creamy spot encroaching a little on the dorsum; vertex brown; occiput
dull brownish with pale hairs. Pterothorax moderately dark brown, rather densely clothed with
pale hair, which conceals a slight metallic lustre, darkest on the dorsum except for a black spot
on the anteroventral angle of the mesepimeron, which lies between two yellow spots; legs brown
at bases, palest on coxae, darkening to black towards distal ends of femora; tibiae and tarsi black.
Wings hyaline, usually with basal, nodal and apical dark-brown spots, of which the largest is the
triangular basal spot of the hind wing. Nodal spots very variable (see *Variations*), sometimes

absent; the apical spot is a marginal area beyond pt. Abdomen dark brown, nearly black, segs. 3 to 7 with a lighter brown streak above the lateral carina; 2 with large brown areas behind the auricles; intersegmental membranes also paler than segmental dorsa. Venation, pt. and many spots all dark brown.

Frontal furrow very shallow in front, deepening only immediately before the vertex, which is moderately elevated and slightly emarginate, as viewed dorsally; median ocellus deep set; lateral ocelli scarcely larger in diameter than the basal segments of the antennae. Occiput with hind margin rounded. Proportions of tibial keels to the tibiae which bear them as follows: fore tibia, 4/7; middle tibia 2/3, hind tibia 5/6. Abdomen slender, somewhat swollen on seg. 2 and base of 3; 2 with small auricles; 3 somewhat constricted about the middle, widest at 6 and 7; 2 to 9 with strong lateral keels, minutely serrate.

Anal apps. as in pl. 5: 1 (p. 40).

Female. Averaging slightly larger than male, with the wing spots a little larger than those of males found at a similar latitude (see *Variations*). Females always with at least a trace of the nodal and apical spots. Vulvar lamina (pl. 5: 2, p. 40) deeply bilobed, extending about to end of seg. 9; sup. apps. about 6 mm., longer than segs. 9 + 10, flexible.

Venation. Anx ♂♀ 7–8/5–6; pnx ♂ 5–7/6–7, ♀ 5–6/5–8; paranal cells ♂ 4(5), ♀ 5(6).

Measurements. Total length ♂ 59.0–61.5, ♀ 63–68; abd. ♂ 39–42, ♀ 42–48; h.w. ♂ 40–42, ♀ 40.5–48.0 (NW: ♂♀ 38–43); h.f. ♂ 9.0–9.5, ♀ 9–10; w.hd. ♂♀ 9.0–9.5; pt. ♂ 3.0–3.2/3.0–3.5, ♀ 3.0–3.7/3.0–3.5.

Nymph. Eyes moderately prominent, sides of head straight, each about three-fourths as long as the hind margin, which is very slightly concave; posterolateral corners bluntly obtusangulate; dorsal surface with a pair of small tubercles each behind the mesal angle of one of the compound eyes and each directed rearward. Antennae 7-segmented, the third segment slightly longer than the first or second but much more slender; seg. 4 half as long as 3. Folded labium (pl. 5: 9, p. 40) reaching to the fore coxae; length of prementum equal to about three-sevenths of its width, distal half of dorsal surface heavily setose; premental setae usually 8 on each side, in two distinct groups, comprising 4 long outer setae and about 4 short inner ones; palpal setae 4 or 5 and an additional smaller one at base of the movable hook which is very short. Legs moderately long, the length of the hind femur being slightly greater than the width of the head across the eyes; hind tibiae slightly longer; hind wing-sheaths about three-fourths as long as hind femora, reaching a little beyond base of seg. 6 (exuv.). Abdomen broadly ovate, widest at seg. 6, moderately convex; lateral spines of seg. 8 about one-fourth the length of the margin of that segment (including the spine), decidedly curved; lateral spines of seg. 9 slightly divergent, about as long as that segment's dorsal length immediately mesad of the spine; dorsal hooks on segs. 2 to 9, these being slender and erect on 2, 3 and 4, high and strongly cultriform on middle segments, highest on 6, and thence successively lower to 9, directed caudad and decidedly arcuate above; paraprocts half as long as lateral spine of 9, slightly longer than epiproct and twice as long as cerci.

Colour of exuviae light brownish, marked with paler gray and darker brown; posterolateral surfaces of head dark brown mottled with pale gray, the dark area extending as a widening stripe across the side of the thorax and over the wing-sheaths; legs pale gray with two dark brown femoral annuli and less distinct tibial annuli, the hind tibiae being not clearly annulate. Abdomen indistinctly marked except on the dorsal hooks, which are darker.

Length 26–27; abd. 14–17; w. abd. 11–12; h.w. 6.5–7.6; h.f. 7–8.5; w. hd. 7.0–7.5.

Habitat and range. Lakes and rivers in quiet waters. Me. and s. Que., s. Ont., Mich. and Minn., s. to Ga., Ala., Miss., La. and Tex.

Distribution in Canada. Que.—Beauharnois; Vaudreuil; Hull; Wakefield; Berthierville. *Ont.*—Essex, Kent, Elgin, Norfolk, Oxford, Welland, Wentworth, Waterloo, Peel, York, Prince Edward, Simcoe, Peterborough, Lennox and Addington, Leeds, Prescott, Carleton and Lanark counties; Muskoka and Nipissing (Algonquin Park) districts.

Field notes. This large corduliid is common in southern Ontario and south-western Quebec. Small lakes, large quiet streams and the channels among the islands of Georgian Bay are favourable haunts for the nymphs.

During our three summers at the former Great Lakes Biological Station, on Georgian Bay, we found *E. princeps* to be a common insect. It is the dragonfly most often seen flying far out over the open water. Of its flight Needham (1901) says "The males at least prefer the surface of still water over which they will sweep back and forth in zig-zag lines and broad curves hour after hour."

In the Georgian Bay region we found the nymphs in debris on the bottom of shallow bays and inlets and the larger ponds. In the words of Needham (1901) they "sprawl on the bottom amid fallen reeds, or clamber over submerged logs." They are often also found clinging to the undersides of stones.

Our earliest date for emergence in Ontario is June 7 (Erindale, Peel county), but at Go Home Bay, Muskoka district, June 16 was the first date on which we observed tenerals on each of the two years on which we made our observations, and it was on June 26 that our first imago emerged in the laboratory.

The nymphs prefer a broad surface, such as a stump or rock or boathouse for support in transforming but sometimes they do so on a clump of grass or sedge stems. They may travel several yards before finding a satisfactory support. Transformation takes place early in the morning according to Needham (1901).

The tenerals of *E. princeps* fly in a leisurely fashion and rest frequently and so are easily captured, but later, adults wander afar over the water or land, often at considerable heights. In fine weather they seem to be in constant flight from early morning until after dusk. We observed them often at Go Home Bay during the evening flying rather higher in pursuit of the larger Ephemeroptera (chiefly *Hexagenia*).

Males may also be often seen patrolling the lake shore in a short beat of only a few yards. They generally keep close to the shore and fly at a height of about four to seven feet, so that they are not very difficult to net.

On a windy day at Berthierville, Quebec, Brother Robert (1963) saw hundreds of individuals perched on dry bushes at the edge of a clearing.

Regarding oviposition, Needham (1901) states that "the eggs are dropped by the female while flying alone, dips being made far out in open water and widely distributed." An egg mass similar to that of *E. cynosura*, described by Needham (1901) and Williamson (1905) is formed by *E. princeps* (Montgomery, 1929) in which the vulvar lamina is also long and deeply bilobed. Two egg masses examined by Montgomery (1929) were spherical, and about six or seven millimetres in diameter; when placed in water they absorbed water rapidly, and soon assumed a flat disc-like shape, about 25 millimetres in diameter and three thick.

The flight period of *E. princeps* at Go Home Bay, as recorded by us, ranged from June 26 to August 5, or approximately six weeks. Our earliest and latest dates for Ontario, however, are June 6 and August 20, the first being from Erindale (Peel country) and the second from Lyn (Leeds county). This makes a period of nearly ten weeks. The flight period in Essex and Kent counties is doubtless somewhat longer than this but must be considerably shorter than it is in the southern part of the range of *E. princeps*. Needham and Westfall (1955) give May 12 (Tennessee) as the earliest record and September 14 (Ohio) as the latest.

Variations. Variation in the size of the wing spots is marked in this species and appears to be correlated with latitudinal and altitudinal distribution. In the extreme south of Ontario, e.g., at Rondeau Park and Point Pelee, both of which are on the shore of Lake Erie, the spots reach their maximum size for Canada. The nodal spot on specimens from these localities may be 10 mm. long and the apical spot may cover the entire wing-tips beyond the pterostigmata. In the vicinity of Toronto, and in Peel county, individuals are sometimes seen with equally large spots, but usually they are smaller. The small glacial lakes on the watershed between Lake Simcoe and Lake Ontario produce a variety with still smaller spots, and this is the variety that is met with over the greater part of Ontario and southern Quebec. Northward on the east coast of Georgian Bay, in the Precambrian region, *E. princeps* is abundant and, for the most part the spots are still smaller, the nodal spots in the male often disappearing entirely while the apical spots are reduced to a narrow margin. The basal spots of the hind wings remain almost unchanged.

North of Georgian Bay, *E. princeps* is unknown. Since intermediate or fairly large-spotted specimens are sometimes found, even in the Georgian Bay region, we have found it impossible to divide *E. princeps* into subspecies, although it is clear that in general the large spotted form is Carolinian and the small spotted form is Alleghenian. The latter form tends to be very slightly smaller than the more southern form.

Epitheca cynosura (Say). (Pl. 5: 3, 4, p. 40)

Libellula cynosura Say, J. Acad. Phila., 8: 30, 1839.
Epophthalmia lateralis Burmeister, Handb. Ent., 2: 847, 1839.
Cordulia cynosura: Selys, Bull. Acad. Belg., (2) 31: 270 (reprint p. 36), 1871.
Tetragoneuria cynosura: Calvert, Trans. Amer. Ent. Soc., 20: 252, 1893.
Tetragoneuria morio Muttkowski: Kormondy, Misc., Publ. Mus. Zool. Univ. Mich., 107: 68, 1959.

An average-sized species whose only wing spots, except the pterostigmata, are the basal spots of the hind wings which are very variable in size. The males have superior appendages lacking a ventral spine and a dorsal tubercle. Found in Canada only east of the Great Plains.

Male. Front of head light yellow with a broad olivaceous cross-bar covering clypeus, lower margin and sides of frons, dorsum of frons brownish with a dark cross-bar in front of vertex, from which an ill-defined T-spot extends forward; eyes of living insect blue; facial hair dense, blackish; vertex and occiput dull, olivaceous; rear of head black. Pterothorax light orange brown, obscured by a dense pale buff hair. Mesepimeron with a wide dark posteroventral stripe, behind which is a bright orange spot extending to rear of hind coxae, and another orange spot behind the spiracle; metepimeron pale orange brown, obscured by the growth of hair. Legs, except coxae, dark reddish brown. Wings hyaline except base of hind wing, where there is a reddish brown spot of very variable size, sometimes confined to vicinity of veins about the first two or three anal veins and base of C and Sc, sometimes extending along costal border, last antenodal and including the anal loop, except the "foot" (see also *Variations*). Abdomen blackish with a brighter orange brown marginal stripe or series of elongate spots on each side from seg. 2 to 9, interrupted segmentally, rather thinly clothed with buff-gray hairs.

Frons concave above, appearing squarish in dorsal view, the anterolateral corners angular; vertex hairy, convex above and in front, projecting upward and forward well over median ocellus; occiput large, longer than eye-seam; hind surface convex, hairy. Abdomen depressed, widening from base to seg. 2 beyond transverse carina; somewhat constricted at the transverse carina of 3, then widening again to the end of 5, and gradually but considerably narrowing to the end.

Sup. apps. as in pl. 5: 3 (p. 40) narrowing and converging in the proximal third, thence diverging at first gradually then more rapidly, subcylindrical, bluntly rounded at distal extremities, in profile showing an abrupt angulation at the basal third. Genital lobes long and slender.

Female. Similar in markings to male; abdomen with lateral reddish brown marginal stripe distinctly divided with a row of spots. Abdomen wider and more depressed than in male, less tapering caudad, but always somewhat narrowed from 6 to the end. Vulvar lamina (pl. 5: 4, p. 40) extending nearly to the end of the 10th sternum, bifurcated to the base, V-shaped, the two lobes flattened and flexible; cerci straight, nearly twice as long as midlength of seg. 10.

Venation. Anx ♂ ♀ 6–7 (5–8)/4 (5); pnx ♂ ♀ 5–6/5–6 (4–7); paranal cells ♂ ♀ 4.

Measurements. Total length ♂ 39–44, ♀ 37–41; abd. ♂ 26–28, ♀ 25–28; h.w. ♂ 26–29, ♀ 28.5–29.0 (NW: ♂♀ 26–31); h.f. ♂♀ 7.0–7.5; w.hd. ♂♀ 7.0–7.5; pt. ♂ 2.0–2.5/2.0–2.5, ♀ 1.7–2.0/2.0–2.2.

Nymph. Head two-thirds as long as wide, widest across the moderately prominent eyes, which are a little anterior to the middle; postocular margins sloping rapidly mesocaudad to the rounded posterolateral angles; posterior margin barely concave; dorsal surface of head behind eyes smooth or with at most only a trace of a pair of tubercles. Relative lengths of 7 antennal segments approximately 10:10:11:8:10:8:7. Prementum a little more than three-fourths as long as wide, somewhat flaring distally; length equal to about nine-tenths of the width; premental setae 9 or 10, those at both ends a little shorter; palpal setae usually 6, with sometimes a smaller basal 7th; crenations of distal margin of palpus rather deep, most of them bearing 5 or 6 coarse spiniform setae of various sizes. Thorax barely wider than head between hind coxae. Legs rather long and slender, the length of the hind femur exceeding the width of the head by about one-fifth. Hind wing-sheaths reaching to about three-fourths of the length of seg. 6. Abdomen broad, ovate, depressed, widest between segs. 6 and 7, then narrowing slightly to base of 8, and rapidly on 9; lateral spines on 8 and 9, that on 8 about one-sixth as long as that segment's lateral margin (excluding the spine), somewhat curved mesad; the spine on 9 about two-thirds as long as that segment similarly measured, nearly straight and very slightly divergent or almost parallel, extending rearward about 1.5 times the length of seg. 10 and the anal pyramid; dorsal hooks on segs. 1 to 9 increasing in height on 1 to 6, thereafter decreasing in height but increasing in length backward, the first 3 slender but straight, the following ones cultriform; the last an arch-like crest with the apex projecting over the base of seg. 10.

Colour pattern often inconspicuous; thorax with a dark stripe on each side extending back from the head and dividing above the hind coxae into two branches, one terminating at base of hind wing-sheath, the other at the first abdominal segment. Femora uniform or with two annuli and apex dark; distal half of tibiae and a sub-basal annulus also sometimes darkened. Abdomen with a series of segmentally arranged pairs of pale spots or blotches, each spot being halfway between the middle line and the lateral margin. Between the spots is a broad median area, which in strongly marked individuals is somewhat darker than the lateral fields and is bounded by a dark stripe next to the pale spots.

Length 21–22 (Garman, 1927: 16); abd. 12.5 (11.3)*; w. abd. 8.5; h.w. 5.5 (6.06)*; h.f. 7.5; w. hd. 6.5 (5.3)*.

*These are averages recorded by Kormondy (1959) from Michigan material reared in captivity.

Habitat and range. Marsh-bordered lakes, bays and mouths of slow streams with submerged and emergent vegetation. "Essentially an inhabitant of marl lakes with eutrophic tendencies . . ." (Kormondy, 1959). Fla. to Tex. and Kans., n. to Me., N.S., N.B., s. Que., s. Ont., Mich. and Minn.

Distribution in Canada. N.S.—"Nova Scotia" (Hagen, 1875, as *Cordulia semiaquea*). N.B.—Charlotte and Sunbury counties. *Que.*—Beauharnois; Vaudreuil; Knowlton; Lanoraie; Hull, Ironside, Wakefield, Alcove, Kazabazua and other stations in the Gatineau Valley; Lac Mercier; Mont Tremblant. *Ont.*—Essex, Kent, Norfolk, Oxford, Brant, Haldimand, Welland, Lincoln, Peel, York, Ontario, Prince Edward, Bruce, Simcoe, Hastings, Frontenac, Leeds, Prescott, Carleton, Lanark and Renfrew counties; Muskoka, Parry Sound and Nipissing (incl. Algonquin Park and Timagami) districts.

Field notes. E. cynosura is one of the most abundant of midsummer dragonflies in southern Ontario, although its numbers fluctuate widely. At De Grassi Point in some years it outnumbers all other Anisoptera during the second half of June.

At Go Home Bay, Georgian Bay, and vicinity the nymphs are abundant in sheltered bays and channels among the islands, where small marshy coves along the rocky shores are favourite haunts, the water never becoming so nearly static as it does in the large marshes and bog-margined ponds. When the time for emergence approaches the full-grown nymphs may be found lurking under stones or drift-wood close to shore and when ready for transformation they climb up the slender rushes or sedges in considerable numbers. Sometimes half a dozen or more exuviae may be seen on a single rush. Or the nymphs may climb wharves or boathouses, often five or six feet above the water, and transform there.

Most of the imagos emerge within a few days. In 1912 we collected full-grown nymphs on May 29, and the first imago emerged on June 13. The last day for emergence was June 19 and the last adult recorded was seen on July 31. At De Grassi Point, where the season begins somewhat earlier than at Go Home Bay, June 5 (in 1938) is our earliest record for observing the imagos, although we have one record for May for St. George Lake, some 30 miles farther south, when a few adults were already flying over the lake, having completed the teneral stage.

After emergence the tenerals disperse widely over the countryside. They fly chiefly in sunshine, frequenting the borders of woods, small clearings, glades and forest roads, not very far from the water.

Kormondy (1959) recognizes four styles of flight in the males of *E. cynosura*, two of them involving also females. In brief these are as follows:

Patrol flight is exhibited during territorial behaviour of the male over water. It features extended periods of hovering, considerable manoeuvring and no alighting; some feeding occurs but is of secondary importance.

Feeding flight is exhibited away from water. It features little or no hovering, being moderately fast with both vertical, and considerable horizontal, displacement. It usually occurs six or more feet above ground, and in mid-morning to early afternoon. In the feeding flight, activity is primarily directed toward procuring food. While there is usually area localization, no territorial behaviour is exhibited. In this flight type the pattern consists of a series of larger and smaller loops, followed in no fixed order or direction.

Copulatory flight involves both sexes and occurs about three feet above the surface of water. It features no hovering, and the flight speed is about half that of the male in feeding flight. Direction of flight is more or less linear for considerable distances, a quarter of a mile or more being not unusual. Unlike the other flights listed the copulatory flight is quite noisy.

Swarming flight involves both sexes and one or more species. No "swarming" of *Tetragoneuria* was observed by Kormondy that did not also involve other species either of *Tetragoneuria* or other genera. Corbet (1962) has termed this activity "swarm-feeding," in order to distinguish it from the reproductively oriented swarming behaviour of many other insect groups, and because in dragonflies this is essentially a feeding flight in which both prey and predator are aggregated.

A type of flight that involves the female only is described by Kormondy as the *preoviposition flight*.

The process of oviposition is unusual and, as far as we know, is peculiar to the genus *Epitheca* in the sense in which we are now using this name (see Walker, 1966b). It was first described by Williamson (1905), whose curiosity was aroused while observing the process in *E. cynosura*. The individual he was watching as it skirted the shores of an Indiana gravel pit, seemed to be carrying a pellet of some kind attached beneath the end of the abdomen. This individual proved to be a female and the pellet, which was placed in water, slowly unwound itself into a strand of eggs about three inches long. At first the eggs were compact with little gelatinous material apparent, but this material increased rapidly until the diameter of the strand was about a quarter of an inch.

Kormondy (1959) describes the behaviour of the female, in searching for a suitable place for oviposition. First she makes a general survey of the area (preoviposition flight), and then several less casual flights, leading to a final choice of a suitable site. During these flights the extruding eggs are held by the long vulvar lamina which acts as a flexible lower member of a vice. While the abdomen is pulled through the water in this position, the eggs are released in a gelatinous string or strand until the abdomen touches some more or less solid object. When this takes place the end of the string adheres to the object, the rest of the string floating on or just below the surface (Kormondy, 1959).

The season of flight at Lake Simcoe, Ontario reaches its height at about the end of June when we have sometimes seen the imagos flying in swarms. The year 1947 was one of particular abundance of *E. cynosura*. The height of numbers was apparently reached on June 29, during a hot spell. They flew in swarms along roads and in the open fields along the edges of woods until dusk. During July their numbers fell off rapidly although they were still fairly common on the 13th and 14th. We saw no more of them after this last date, which appears to be our latest record for the Lake Simcoe region.

A female taken near Joe Lake, Algonquin Park, on July 18, 1942, is our latest record. Others were seen on the same day; so the end of the flight period in this region was probably at least some days later.

Variations. E. cynosura exhibits a wide range of variation in the size of the pigmented area at the base of the hind wings. Some individuals have very little pigment at the extreme base of the wings, whereas others have a conspicuous patch that may extend distally beyond the triangle even to the last antenodal vein along the costal border.

Prior to Muttkowski's studies of *Tetragoneuria* (1911b), specimens of *E. cynosura* with large basal wing spots had been confused with a similar southeastern variant (*E. semiaquea* Burmeister). Muttkowski separated the true *semiaquea* from the heavily marked specimens of *cynosura*, and applied to the latter the trinomial "*T. cynosura simulans*," although he admitted that there was a complete gradation from typical *cynosura* to *simulans*. He also commented on the fact that *simulans* is found only in the northern United States and Canada and that the gradation is the opposite to that of *E. princeps* in which it is the southern specimens that are heavily marked. We follow Needham and Westfall (1955) in placing the name *simulans* in the synonymy of *cynosura*. In Ontario specimens with a minimum of wing-pigmentation are found along the north shore of Lake Erie and sometimes in the vicinity of Toronto. Generally, however, specimens from the Toronto district have

PLATE 6

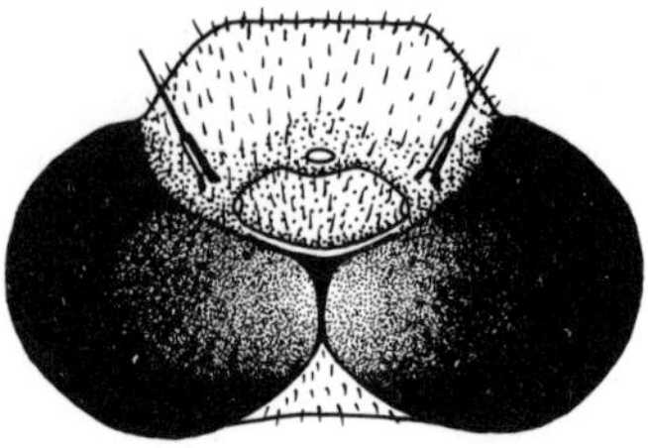

1

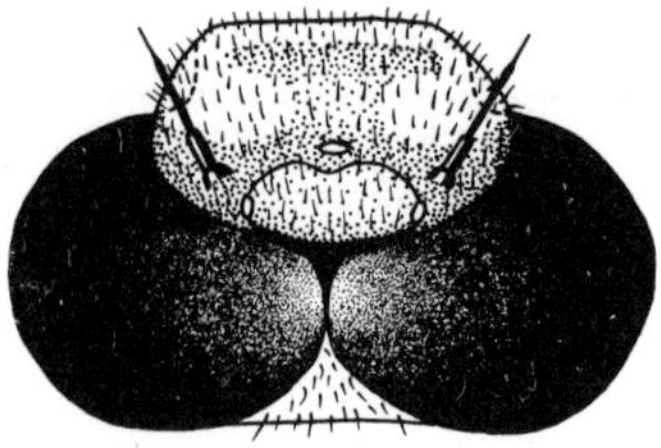

2

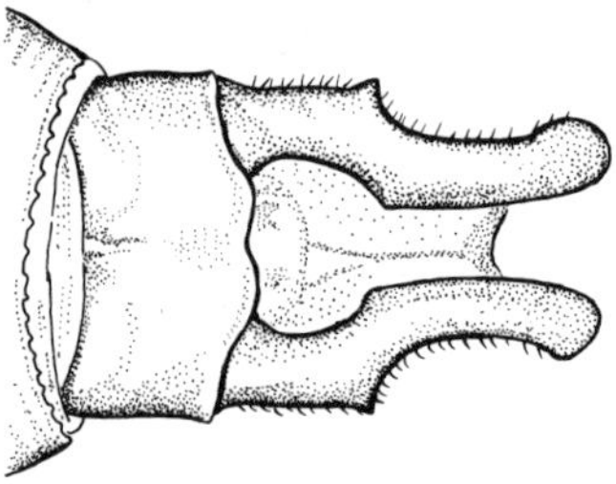

3

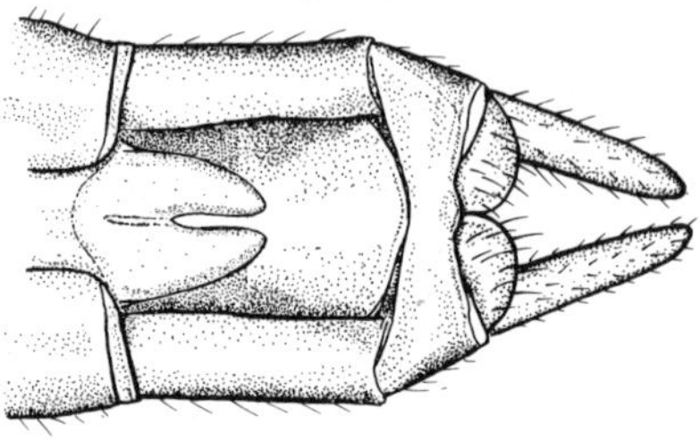

4

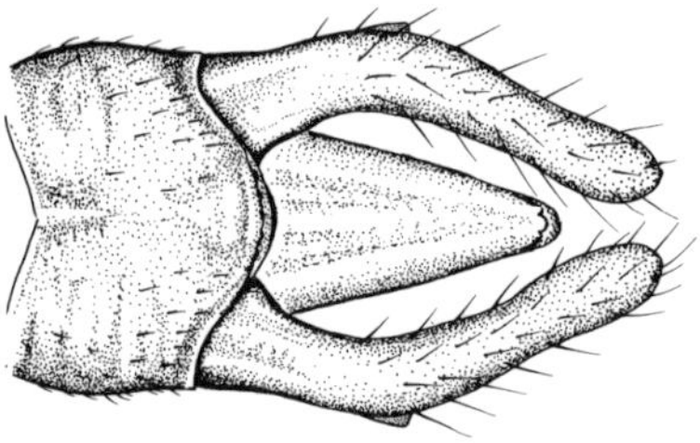

5

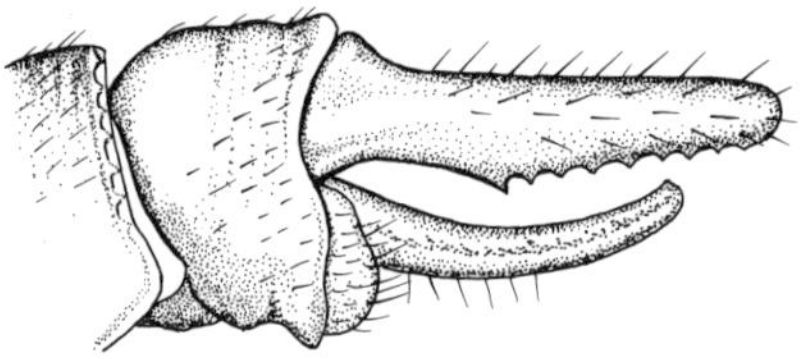

6

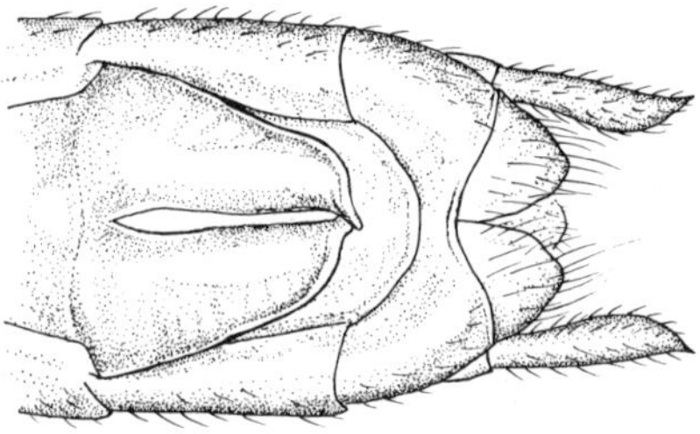

7

spots reaching the apices of the triangle of the hind wings but the spots may be larger or smaller. At Lake Simcoe the spots average a little larger than at Toronto but in one recent year the specimens had decidedly less than the usual pigmentation, suggesting an influx from farther south. In northern Muskoka district an extreme degree of wing pigmentation is found; and there every individual would be classed as *simulans*.

Epitheca spinigera (Selys). (Pl. 5: 7, 8, 10, p. 40; pl. 6: 2, p. 50)

> *Tetragoneuria spinigera* Selys, Bull. Acad. Belg., (2) 31: 269 (reprint p. 35), 1871.
>
> *Tetragoneuria morio* Muttkowski (females only): Kormondy, Misc. Pub. Mus. Zool. Univ. Mich., 107: 68, 1959.

A very dark species of variable size, usually larger than *cynosura* and sometimes larger than any other Canadian species of *Epitheca* except *princeps*, the males being easily recognized by the spine on the underside of the superior anal appendages (pl. 5: 7, p. 40).

Male. Labium pale dull yellow, labrum and a large anterior spot on frons orange yellow, face elsewhere dark olivaceous; frons above with a clearly defined T-spot (pl. 6: 2, p. 50) having a narrow cross-bar and a wide stem, flanked by orange brown eyes (blue in life); vertex and occiput dull olivaceous, rear of head black. Thorax dark reddish brown with a dense cover of grayish brown hair; pleural sclerites margined with blackish; anteroventral margin of the mesepimeron and ante-alar ridge of mesepisternum with a dense black fringe. Legs black, except rear surfaces of fore femora, which are reddish brown. Wings with a slight yellowish tinge, pt. and venation dark yellow-brown; membranule smoky except a small marginal area in front of anal triangle. Hind wing with basal dark spot not reaching the triangle, consisting of a broad streak on the costal and subcostal spaces to about as far as arculus and an irregular open blotch behind it, covering the anal triangle and edging a few cells beyond to about the first row of the anal loop, keeping this form with very little variation; fore wing with one or two small basal spots. Abdomen mostly black with sides of segs. 1, 2 and most of 3 reddish brown, and with elongate lateral brown spots, becoming successively shorter; anal appendages black, venter largely brown.

Frons viewed from above with rounded anterolateral angles, vertex slightly concave in front; eye-seam as in *cynosura*, much shorter than occiput. Abdomen with segs. 1 and 2 somewhat swollen, constricted at middle of 3, widening behind to end of 5, where abdomen is relatively wider than in *cynosura*, narrowing again to 10; greatest width: 3.6 mm. in smallest specimens, 5.3 mm. in largest.

Sup. apps. (pl. 5: 7, p. 40) scarcely shorter than segs. 9 + 10, converging to end of basal third, subparallel in middle third, diverging distally to blunt round apices, an apparent swelling just before middle, formed by slight inferior carina in lateral view and showing an inferior spine at the proximal third; no angulation near the middle such as appears in *cynosura*. Inf. app. two-thirds as long as sup. apps., very slightly concave above, apex bifid.

PLATE 6

Epitheca—head, dorsal view: (1) *E. canis*; (2) *E. spinigera*.

Helocordulia uhleri: (3) anal appendages of male, dorsal view; (4) vulvar lamina, ventral view.

Williamsonia fletcheri: (5, 6) anal appendages of male: (5) dorsal view; (6) left lateral view; (7) vulvar lamina, ventral view.

(1, 2 after Robert, 1963; 3 after Needham and Westfall, 1955.)

Female. Similar to male in size and coloration. Abdomen more depressed especially on basal segments, lateral margins subparallel and much more abruptly narrowed towards the posterior end; anal apps. shorter than 9 + 10, vulvar lamina (pl. 5: 8, p. 40) similar to that of *cynosura*, the two lobes diverging more than one-half their length, then curving mesad and converging slightly towards apices, reaching almost to distal margin of 10th sternum.

Venation. Anx ♂ ♀ 7–8 (6)/4–5; pnx ♂ ♀ 6–8/5; paranal cells ♂ ♀ 6–7 (8).

Measurements. Total length ♂ 42–53, ♀ 45–52; abd. ♂ 29–35, ♀ 30–36; h.w. ♂ 29–36, ♀ 31.0–37.0; h.f. ♂ 7.3–8.2, ♀ 8–9; w.hd. ♂ 6.8–8.0, ♀ 7.5–8.0; pt. ♂ 2.0–2.5/2.0–2.7, ♀ 2.3–3.0/2.3–3.0.

Nymph. On the basis of characters now available to us, it may not always be possible to distinguish this species from *E. cynosura.* The relative length and direction of the lateral spines on abd. seg. 9, used by Needham (1901) and Needham and Westfall (1955), has not been found reliable as a diagnostic character. Kormondy (1959) records 11 or 12 premental setae (*vs.* 9 or 10 in *cynosura*) and 7 or 8 palpal setae (*vs.* usually 6 in *cynosura*) but overlap can be expected (pl. 5: 10, p. 40). *E. spinigera* attains a greater size in part of its range, and where *cynosura* is apparently not present. *E. spinigera* reaches its greatest size along the north shore of Lake Superior, where it is the only species of *Epitheca* so far recorded. Measurements of exuviae from Silver Islet, Ontario in this region are as follows: length 25–26; abd. 16; w. abd. 10; h.w. 6; h.f. 8; w. hd. 6. Specimens from Shakespeare Island, Lake Nipigon are, on average, slightly smaller: length 23.5–24.0; abd. 14–15; w. abd. 8–9; h.w. 5.5–6.0. Those from Go Home Bay, Georgian Bay, Ontario, are of the usual size, and are of about the same size as are exuviae of *E. cynosura*: length 22.0–23.5; abd. 13–14; w. abd. 8–9; h.w. 5.0–5.5; h.f. 7–8. Specimens from the Prairie Provinces and British Columbia are of about the same size as are those from Georgian Bay and eastward. Average measurements recorded by Kormondy (1959) from Michigan material were: abd. 13.6; h.w. 6.77; w. hd. 5.8.

Habitat and range. Marshy borders of lakes and slow streams but not in muskeg or other stagnant waters. N.S. to the Laurentides, Que., w. to B.C. and s. to N.J., W.Va., Ind. and Calif. Also recorded from La. but this isolated, southerly record of an essentially northern species requires confirmation (see Kormondy, 1959, p. 33).

Distribution in Canada. N.S.—Annapolis, Antigonish, Halifax and Queen's counties; Cape Breton I. (Victoria county). *N.B.*—Sunbury and York counties. *Que.*—Knowlton; La Trappe; St. Bruno; Aylmer; Hull; Ironside; Wakefield; Mont Tremblant Park; Laurentides Park (Lac Reine Elizabeth); Saguenay River (Cap Jaseux); Abitibi county. *Ont.*—Waterloo, York, Ontario, Bruce, Simcoe, Victoria, Peterborough, Lennox and Addington, Leeds, Glengarry, Russell, Carleton, Lanark and Renfrew counties; Muskoka, Parry Sound, Nipissing (incl. Algonquin Park), Sudbury, Thunder Bay, Cochrane and Kenora (incl. Patricia Portion) districts. *Man.*—Stockton; Treesbank; Aweme; Onah; Winnipeg; Victoria Beach; Lake Winnipeg; The Pas. *Sask.*—Deep Lake; Waskesiu Lake; Heart Lake; La Ronge. *Alta.*—Edmonton; "Billy" (presumably Bilby); near Flatbush. *B.C.*—Cultus Lake; Christina Lake; Kootenay Lake; Kaslo; Vernon; Squilax; Vancouver I. (Sooke; Langford Lake; Goldstream; Victoria; Forbes Landing; Alberni; Nanaimo district).

Field notes. In southern Ontario, *E. spinigera* appears on the wing about ten days after *E. canis* and two weeks before *E. cynosura*. It is a northern species and, like some other northern dragonflies, has a short flight period in the southern part of its range.

At Go Home Bay, Georgian Bay, in 1912 (Walker, 1915) we found the first exuvia on May 29, and on June 1 we found large numbers of them clinging to the rushes and floating in the water in the open marsh at the outlet of one of the small lakes. No imagos were found except a single crippled teneral with its exuvia; it was a male and readily identified. Much search was made for nymphs here but without success. A number of *Epitheca* nymphs, however, were found under stones along the shore of an open channel where the water was doubtless colder and two of them yielded female nymphs of *E. spinigera*, which emerged on June 2. The others proved to be *E. cynosura*. Thus, although the nymphs of these two species may be associated with each other, we are inclined to the opinion that *spinigera* is most at home in somewhat more marshy stations than those preferred by *cynosura*.

The earliest record we have of *spinigera* is May 22, when two specimens were taken at Glen Morris (Brant county) and shown to me. On May 24, 1941 a few were flying about Kelly Lake (York county) but were all teneral, having doubtless emerged on that day. On May 31, 1936 *spinigera* was abundant at the De Grassi Point, flying along the borders of woods in sunny sheltered glades. They were more than half a mile from the nearest marshy water.

When at the height of their numbers during the first week in June in southern Ontario, adults sometimes appear in very large aggregations.

The decline in numbers, following this period of abundance, is very rapid. In 1942 we noticed that they were already scarce at De Grassi Point on June 13 and that the latest date on which they were observed during the summer at nearly the same locality was June 21. Farther north or east they may be still common at this date, e.g., in the Gatineau Valley, Quebec, from June 19 to 23, 1933, and at Joe Lake, Algonquin Park, until July 3. Near Fredericton, New Brunswick, we found *spinigera* still flying on July 10, 1932 and at Nyanza, Cape Breton Island, Nova Scotia, we took a single male on August 2, 1948. The flight period in British Columbia is also later than in southern Ontario and Quebec, the flight period for the Province as a whole being June 6 to July 30 (Whitehouse, 1941). "At Thetis Lake this species was emerging in numbers July 16th, the nymphs crawling over the grassy slope and gravel pathway and transforming in the seats and fencing. The distance traversed was 20 feet."

Whitehouse found the egg-masses of *spinigera*. He found females flying along the shore sometimes with an inch of the egg ribbon hanging. "The lateral view of females in flight gives the illusion of miniature sea-planes, due, apparently to the tip of the abdomen being canted upward and a glimpse obtained of the forked vulvar lamina; but, once netted, the peculiarity is gone" (Whitehouse, 1941). This position we have frequently noted in *E.*

canis, but not in any other species. Whitehouse found egg clusters from June 25 to July 7 "too large to be contained in cupped hands, and to which scores of females must have contributed."

Variations. Unlike *E. cynosura*, in which the wing pigmentation increases significantly northward, that of *spinigera* appears to be unaffected, except as an apparently rare recessive. According to Kormondy (1959) size varies significantly in all three species from Michigan, which are the same as those from Canada. Size evidently increases northward in *spinigera* and *canis* but we have not observed this variation in *E. cynosura*, perhaps owing to scarcity of material in *cynosura* from northern Ontario. For instance, specimens from New Brunswick, southern Quebec, southern Ontario, southern Manitoba and British Columbia are of similar size, without much variation, the usual length being about 42–45 mm., whereas from the north shore of Lake Superior and northward the length runs to 50–53 mm. We have occasionally seen specimens from other northern localities in Ontario and Manitoba that are above average size but we have not enough material to speculate on what factor influences the dimensions of this species. It may be noted, however, that it is only in relatively cool regions that the large individuals have been found.

Epitheca canis MacLachlan. (Pl. 5: 5, 6, p. 40; pl. 6: 1, p. 50)

Epitheca canis MacLachlan, Ent. Mon. Mag. 23: 104, 1886.

This is a medium-sized species easily known by the complete absence of the transverse mark which forms the cross-bar of the T-spot on the top of the frons and in the male, by the shape of the superior anal appendages.

Male. Labium, labrum and frons pale yellow to orange-yellow, clypeus and sides of frons dull olivaceous, T-spot (pl. 6: 1, p. 50) represented by the stem, this being only a short clearly defined black stripe on the median furrow. Vertex medium brown, occiput dull olivaceous. Pterothorax light brown, appearing paler owing to the dense covering of pale gray-brown hair, which covers also the sides but is thinner there; darker brown hairs on ante-alar ridges and sinus; a dark-brown border on interpleural suture in front and an elongate orange spot behind, partly divided by the dark spiracle; legs blackish or very dark brown, the hind surface of profemora orange brown. Wings hyaline in young adults, becoming brownish later; venation and pt. dark brown; membranule whitish on basal third or fourth. Dark spot at base of hind wing consisting of an anterior spot extending from base to first antenodal often followed by several much smaller and decreasing spots on the next two or three antenodals and a posterior spot on hind part of anal triangle and continued along veins to nearest cells of anal loop. Abdomen dark brown above with whitish hair which is very scanty except on the venter of segs. 1 to 3; and with elongate, somewhat ill defined, orange-brown spots along the lateral margins; intersegmental membranes rather pale.

Abdomen narrower than in *spinigera*, slightly constricted on 3 and widening to about the end of 6, narrowing again to the end, the width of 10 being less than half of 6.

Sup. apps. (pl. 5: 5, p. 40) convergent in proximal two-thirds, apical third somewhat divergent and steeply declivent, with a dorsal knob at the beginning of the declivity, also a ventral tubercle at about two-fifths of the length; inf. app. very bluntly triangular, about three-fourths as long as the sup. apps.

Female. Similar to male in colour pattern except in the following respects: the posterior basal spot of the hind wing is absent; the membrane of fully mature females is more deeply stained with yellow-brown from base to about the distal end of pt., where it becomes more or less suddenly paler; the lateral orange brown spots of the abdomen form an almost continuous stripe to the end of 9. Abdomen barely constricted at 3, segs. 4 to 7 almost parallel-sided, 8 to 10 rapidly narrowed. Anal apps. about as long as 9. Vulvar lamina (pl. 5: 6, p. 40) about the same length, the two rami diverging at an angle of about 40°.

Venation. Anx ♂ 8 (9)/5–6, ♀ 7–8 (9)/5–6; pnx ♂ ♀ 6–7(4–8); paranal cells ♂ 6/4, ♀ 6–7/4 (5).

Measurements. Total length ♂ 44–45, ♀ 40–47; abd. ♂ 28–30, ♀ 28.5–32.0; h.w. ♂ 28.0–31.5, ♀ 28.5–32.0; h.f. ♂ 7–8, ♀ 7.0–7.5; w.hd. ♂ 6.5–6.0; pt. ♂ 2.0/2.2–2.5, ♀ 2.0–2.5/2.2–2.8.

Nymph. Easily distinguished from those of *cynosura* and *spinigera* by the much shorter lateral spines of abd. seg. 9. Head with posterolateral angles more prominent and hind margin more deeply concave. Prementum of labium three-fourths as long as wide; premental setae 11 or 12, palpal setae 7 or 8. Legs a little shorter than in *cynosura* (relative lengths of hind femur 7 *vs*. 7.5 mm.). Lateral spines of abd. seg. 9 not reaching quite as far back as tips of paraprocts, about as long as the length of the segment in mid-dorsal line; lateral spines of seg. 8 about one-fourth as long as those of 9; dorsal hooks as in *cynosura*.

Colour pattern similar to that of *cynosura* or *spinigera*; the dark femoral annuli, however, are less definite and more spotty in well-marked individuals of both species; abdominal pattern showing a similar segmental series of pale spots on each side, as in *cynosura*, bounded mesally by brown spots which merge into the somewhat paler broad median abdominal field, the dorsal hooks being paler; stripes of thorax similar to those of *cynosura*.

Length 21–23; abd. 13–14 (13.7)*; w. abd. 8.5–8.8; h.w. 6.0–6.5 (6.85)*; h.f. 7.0–7.7; w. hd. 5.5 (5.5)*.

*These are averages recorded by Kormondy (1959) from Michigan material reared in captivity.

Habitat and range. Bog ponds and discharges from acid waters. Maritime Provinces, Laurentian Hills, Que., Lake Simcoe, Ont. (occasional) and common northward (to 54°N. in Sask.); probably continuously distributed from eastern Provinces to B.C. where it occurs n. to 54°N. (Sumas Canal); in the United States from Wash. to Calif. The record from Okla. belongs to *E. spinosa* (Kormondy, 1959).

Distribution in Canada. *N.S.*—Colchester and Halifax counties; Cape Breton I. (Inverness county). *N.B.*—York county. *P.E.I.*—(July 8, 1944, 1 ♂, H. C. White). *Que.*—Hull; Wakefield; Kazabazua; La Trappe; Woburn; Mont Tremblant. *Ont.*—Peel, Dufferin, Ontario, Simcoe, Victoria, Peterborough, Leeds, Russell, Carleton, Lanark, Haliburton and Renfrew counties; Muskoka, Parry Sound, Nipissing (incl. Algonquin Park), Sudbury, Cochrane and Kenora districts. *Sask.*—Stanley Rapids, Churchill River (June 21, 1957, D. S. Rawson). *B.C.*—Sumas Canal; Burns Lake.

Field notes. *E. canis* is the earliest species of *Epitheca* to appear on the wing in Ontario—in fact it is probably the earliest of all the Anisoptera to emerge in eastern Canada, appearing very soon after the arrival of *Anax junius* from the south. At De Grassi Point we have observed it in flight on May 10, 1936, and we have a record from Petawawa Forest for May 6. More frequently emergence begins about the middle of May and continues until about May 25, since we have seen tenerals until this date.

In some years, however, at De Grassi Point, individuals are already flying about by May 22, in sunny glades, pastures or scattered along the edges of woods, within half a mile of the creek. At this time their flights are short and they frequently settle on trees or bushes. By the first week in June their numbers have reached their peak and the males may be commonly seen flying back and forth over the water.

The stream from which they emerge near De Grassi Point has a wide pond-like basin close to the lake into which it opens by a narrow mouth. Upstream the basin narrows and is partly shaded by alders. Both floating and emergent vegetation become rapidly scantier and, twenty yards farther up, it is in dense shade and is almost devoid of aquatic vegetation. *E. canis* flies neither over the open pond nor the shaded stream but in the intermediate section, where there is still considerable floating vegetation though much less than in the pond. It is here that *E. canis* may be seen when abundant, as the two preceding species often are, but it is present more consistently than they are. It also sometimes patrols a small drainage ditch along the main road that leads through the same area of woods at De Grassi Point.

A hundred miles farther north, in Algonquin Park, *E. canis* is very common and shows a distinct preference for small bog-margined lakes or ponds; and at Lake Abitibi, still farther north by 230 miles, we find it to be not so uncommon on the drowned portions of the Ghost River and along the railway near Low Bush. The specimens taken here in June have hyaline wings but in the females taken in July the wings are deeply suffused with amber-brown. This appears to be an invariable feature of the wings of old females (Walker, 1928).

Recently we received a few specimens of *E. canis* taken below Stanley Rapids, Churchill River, Saskatchewan, said to have been present in "thousands." In this northern region *E. canis* thus appears to be no exception to the two previous species in its habit of "swarming." During the time of maturity in early June, pairs in tandem may be occasionally seen. On June 3, 1933, as we were crossing a pasture near the De Grassi Point stream, a mature female was pounced upon by a male and fell to the ground. When picked up she appeared to be stunned but soon recovered. But before we had left this spot another female was knocked down by a male, perhaps the same one, and stunned. Apparently the seizure of the female's head by the anal appendages of the male was the cause of the stunning.

These mature females are conspicuous not only because of their wing coloration but also because of their peculiar flight (with the end of the abdomen turned up) and on several occasions we have seen them flying in this fashion with two egg strings, two or three inches long, trailing from the genital openings. We have not witnessed oviposition in *E. canis*, but on June

13, 1931 we found on the De Grassi Point stream two egg masses at the surface but caught on the submerged vegetation. This was on that same section of the stream where only *E. canis* breeds and we found a few exuviae of this species nearby. The egg masses, however, do not appear to be distinguishable from those of *E. cynosura* as far as we could see.

According to Kormondy (1959) such females trailing broken egg strings during flight have been disturbed by males and have made their escape.

The flight period in southern Ontario may extend from May 6 (Petawawa) to June 27 (Brockville) but, farther north and east, it may continue at least until July 18.

Genus **Helocordulia** Needham

Bull. N.Y. State Mus., 47: 496, 1901. Type species *Cordulia uhleri* Selys.

This genus was founded by Needham on several very slight venational characters (pl. 7: 1, p. 58), none of which is constant. One of these, the number of cubito-anal cross-veins in the hind wing, is stated by Needham (1901, p. 496) to be usually two, whereas we found in five males and five females, taken at random of our only species, *H. uhleri*, that none of the males had a single wing with two cubito-anal cross-veins, and that of the females only two had two cross-veins on each side. Two others had two on one side and one on the other, and the fifth had one on each side. Clearly there is no character here that is of even specific, much less generic, value. Another character used by Needham, the presence of two cross-veins behind the pterostigma, is usual but not constant in *H. uhleri*, whereas in *Epitheca*, its nearest relative, there is only one cross-vein, usually with a long space on each side of it. The narrowing of the trigonal interspace to a single row of cells, although usual, is also not constant (see also *Variations*).

Other features that separate *Helocordulia* from *Epitheca* are found in the vulvar lamina, the male genitalia, the anal appendages of the male and the form of the male abdomen.

The difference in the vulvar lamina seems to be the most important character, as it is associated with a difference in oviposition. In *Epitheca* the vulvar lamina is divided to the base and the two lobes or rami are long, divergent and flexible, extending to the end of segment 9 or beyond. In *Helocordulia* the vulvar lamina is rigid, and is only one-half or two-thirds as long as the ninth sternum, and its two lobes are much less divergent, if at all. In *Epitheca* these elongated lobes serve as part of the vice that holds the long egg-strand, constituting a special adaptation to hold the egg pellet in place before it is unwound, whereas those of *Helocordulia* appear not to function in this way. Oviposition has apparently not been described in *Helocordulia* but we have seen the eggs adhering singly to the under side of segment 10 in a dried female. There is almost certainly no gelatinous strand in which the eggs are embedded as is the case in *Epitheca*.

PLATE 7

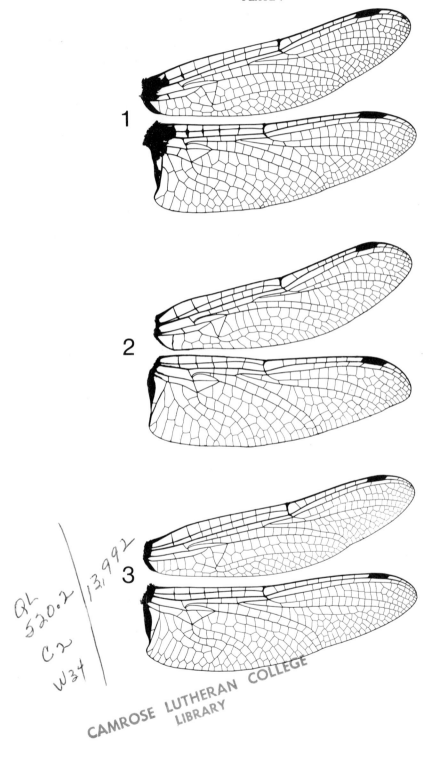

The superior appendages are bowed, with the apices directed rearward and with a lateral tooth near the middle, thus differing from those of *Epitheca* as described in the key.

The male genitalia are much the same as those of *Epitheca* but the hamuli do not terminate in a definite hook as do those of the North American species of *Epitheca*. The two Palaearctic species, however, also lack this hook, which, in any case is not of more value than a specific character.

The form of the male abdomen is more like that of *Cordulia* than *Epitheca*, since it widens gradually from the constriction on segment 3 to 8, whereas in our species of *Epitheca* the abdomen is widest at the end of 6 or between 6 and 7, becoming very narrow on segment 10.

The nymphs are small and stocky with inconspicuous eyes. There are 13 to 14 premental setae and 6 to 7 palpal setae, the outer 8 on each side being distinctly longer than the inner ones. Dorsal hooks are confined to abdominal segments 6 to 9, being minute on 6 and long on 7 to 9. The lateral spines on segments 8 and 9 are short and about equal in length. Kennedy (1924) has published a detailed description of the nymphs of this genus.

This genus contains two known species found only in eastern North America; one occurs in Canada.

Helocordulia uhleri Selys. (Pl. 6: 3, 4, p. 50; pl. 7: 1, p. 58)

Cordulia uhleri Selys, Bull. Acad. Belg., (2) 31: 274 (reprint p. 40), 1871.
Cordulia uhleri: Beutenmuller, Prelim. Cat., p. 164, 1890.
Helocordulia uhleri: Needham, Bull. N.Y. State Mus., 47: 496, 497, 1901.

A small dark corduliid with a somewhat clavate abdomen and clear wings, whose basal spots are broken and bicoloured with dark brown and clear yellow.

Male. Labium, labrum, anterior surface and part of postfrons light yellow; clypeus and sides of frons, including rear part of its dorsum, olivaceous or dull ochraceous; a short black median stripe in the median furrow; vertex and occiput olivaceous; rear of head laterally olivaceous along the eye margin, the hidden central area black and a fringe of pale hairs marking the mesal margin of the pale area. Pterothorax light brown with a slight metallic lustre, rather thinly clothed with pale brownish hair, especially thin on the sides, the median carina and sutures darker. Legs black below the coxae. Wings hyaline with basal spots, chiefly dark brown but partly golden yellow, continued along the costal and subcostal space in two or three of the antenodals. A larger spot on hind wing generally divided into three parts, the largest in front, the second and third parts at the base and near the apex respectively of anal triangle; membranule wholly pale or nearly so. Abdomen mostly black, thinly clothed with pale hairs which are denser on basal segments and towards posterior end than elsewhere. Segs. 1 and 2 dull ochraceous above, including muscles, clear orange-brown below auricles and in front of trans-

---

PLATE 7

Wings of Corduliidae: (1) *Helocordulia uhleri*; (2) *Williamsonia fletcheri*; (3) *Somatochlora tenebrosa*. (1–3 from Needham and Westfall, 1955.)

verse carina. Seg. 3 with two large dorsolateral orange-brown spots narrowly divided mid-dorsally at base; sides of these segments with smaller basal red-brown spots as far as 8; segs. 9 and 10 and apps. dark.

Head as in *Epitheca*, vertex convex in front, scarcely emarginate; eye-seam about one-half as long as occiput, which is rounded behind. Abdomen with segs. 1 and 2 moderately swollen, narrowed on 3 at transverse carina where it is about half as wide as 2, widening again from 4 to 8, width of 8 being slightly more than twice the minimum width of 3.

Sup. apps. (pl. 6: 3, p. 50) separated by more than the basal width of one app. diverging from base, then bowed, the blunt apices being approximated; sides with a rectangular tooth; inf. app. about four-fifths as long as sup. app., tapering distally, but terminating in a bifurcate distal margin, which is about half as wide at base and has also a minute median tooth.

Female. Similar to male in dimensions and colour pattern but with a wider abdomen, which is more depressed in young females and thicker and more impressed in mature ones; vulvar lamina (pl. 6: 4, p. 50) one-half to two-thirds as long as the sternum of 9, flat and rigid, bilobed from about the basal third, the lobes almost contiguous; anal apps. about 1.5 times as long as 10 or slightly shorter than 9.

Venation. Anx ♂ 7–9/5–6, ♀ 7–9/5–7; pnx ♂ 5–6/5–7, ♀ 6–7/6–7; cells in triangle ♂ ♀ usually 2; usually 2 cross-veins under pt. (see also *Variations*).

Measurements. Total length ♂ 40–44, ♀ 42–43; abd. ♂ 27–32, ♀ 28–30; h.w. ♂ 27–28, __ 28–29 (NW: ♂ ♀ 25–29); h.f. ♂ ♀ 6–8; w.hd. ♂ 6.0–6.5, ♀ 6.0–7.0; pt. ♂ 2.0–2.5, ♀ 2.3–3.0 (see also *Variations*).

Nymph. Similar in general form to a small *Epitheca*, but dark brown without distinct markings, except femoral annuli, and with well developed dorsal hooks on segs. 6 to 9 only. Head only about five-eighths as long as wide, eyes a little less prominent than in *Epitheca*; lateral margins and posterolateral angles more rounded; hind margin nearly straight. Antennae 7-segmented, the relative lengths of the segments being about 10:10:14:10:13:15:9; frontal disc rounded, set with numerous scurfy hairs. Folded labium 4.1 mm. long and extending to middle of fore coxae; prementum seven-eighths as long as wide (3.5 *vs*. 4.0 mm.); premental setae 10 or 11; palpal setae 6 or 7; crenations of distal margin rather deep and rounded, bearing groups of 4 or more relatively coarse spiniform setae. Lateral margins of pronotum produced into an angular, somewhat raised propleura with a smaller process over each procoxa; legs slender, sparsely hairy, femora with a dark ante-apical annulus and a dark apex, these markings usually being obscured by a coating of debris. Abdomen ovate, widest at seg. 6, the width here being two-thirds of the length, narrowing more rapidly posteriorly than anteriorly; dorsal hooks on segs. 6 to 9, smallest on 6, largest on 8, all but the first cultriform; lateral spines on 8 and 9, directed posteriorly, those on 9 somewhat longer, being almost one-third as long as the lateral margin of that segment (excluding the spine) and reaching to the level of the tip of the anal pyramid.

Length 19–20; abd. 9.5–10.0; w. abd. 7; h.w. 5.5; h.f. 6.5–7.0; w. hd. 5.0–5.5.

Habitat and range. Edges of rapid streams and sometimes lakes formed by an expansion in the course of a rocky stream. The Maritime Provinces, the Abitibi region, Que., and the n. shore of Lake Superior, Ont.; s. to N.C., Ala., Tenn. and La.

Distribution in Canada. *N.S.*—Annapolis, Halifax and Pictou counties. *N.B.*—Charlotte, Sunbury and York counties. *Que.*—Knowlton; Potton Springs; Mégantic; Rawdon; Hull; Ironside; Wakefield; Mont Tremblant Park; Saguenay River (Cap Jaseux); near Dolbeau (Lake St. John); Abitibi county. *Ont.*—Ontario, Hastings, Frontenac, Carleton and Renfrew counties; Muskoka, Parry Sound and Nipissing (incl. Algonquin Park) districts.

Field notes. From Ontario we have a few tenerals, taken on the Magnetawan River by Mr. Paul Hahn on May 25, 1918. This locality is only slightly farther south than Mont Tremblant Park.

We have never found *Helocordulia* in abundance, but given the proper conditions—a rocky stream, fairly rapid, and unpolluted, and the right time of the year (June in Ontario and Quebec; June to early July in the Maritimes)—this species is likely to be seen. It is inconspicuous and flies with great speed, within two or three feet of the water, following the shore line closely. It is not very difficult to net, however, if one strikes from behind and, if adults are not seen, the exuviae are likely to reveal the presence of the habitat of *Helocordulia*, for they are often found clinging to stumps, wharves and rocks, where they may remain until the season of flight is well over.

As we have not observed the habits of *H. uhleri* at all closely we give the following notes translated from Robert (1953). "This corduliid is one of the first dragonflies to make its appearance at the beginning of summer. At this time it may be seen along roads, clearings and even outlets of streams. This is its period of maturation; it is especially occupied in feeding and the females are not yet ovipositing. Toward the end of June they return to the river margins, over which they fly on their long journey. From strong winds they take refuge in the shrubbery of the river bank. At sundown they still show much activity, passing and repassing a locality, always in search of a mate or in pursuit of prey. At the old dam below Lake Monroe it abounds at the end of June and beginning of July. The females hide in the accumulation of debris, laying their eggs in the water." At dusk adults are easier to secure because at that time their flight-style is more regular (Robert, 1963).

Helocordulia evidently does not oviposit after the manner of *Epitheca*. The female lacks the long flexible vulvar lamina with its widely divergent arms, which is found in *Epitheca* and which keeps the egg pellet in place until it is unwound and deposited. We have a dried female of *Helocordulia* with a few eggs adhering to the underside of the 10th abdominal segment. It is evident that these were not embedded in a gelatinous strand, although each may have had its individual coat of jelly. We regard this difference and the vulvar lamina associated with a different mode of oviposition as a more significant generic distinction than the inconstant and trivial venational characters used by Needham (1901) to characterize *Helocordulia*.

"The nymphs live in the borders of the creek, mainly in the shallow places, filled with red-rotten vegetable debris—the haunts of the giant crane fly, *Tipula abdominalis*, and the phantom fly, *Bittacomorpha clavipes*, larvae" (Needham, 1901).

Variations. A detailed study of variability of size and several venational characters was undertaken by McMahan and Gray (1957) on a population of

H. uhleri near Durham, North Carolina. This showed that there is a slight sexual dimorphism in total and abdomen length and a distinct dimorphism in wing length; there are also sex differences in the number of cells in the triangles and in the number of antenodal, cubito-anal and brace veins. A respect in which the Durham population differed from material available to us is in having a predominance (65–70 per cent of specimens) of 1 cell in the triangle in both sexes. The ranges for lengths of abdomen and hind wing were, as expected, wider than those recorded above.

Genus **Williamsonia** Davis

Bull. Brooklyn Ent. Soc., 8: 93, 1913. Type species *Dorocordulia lintneri* Hagen.

These are small, dark corduliids, having the size and proportions of a medium-sized *Leucorrhinia*. They have few or no areas of bright metallic coloration. The thorax is brownish and without distinct markings and is rather thinly clad with soft hairs. The abdomen is black with paler apical rings on some of the segments. Tibial keels are present on the fore and hind legs of the males but not on the middle legs. The wings (pl. 7: 2, p. 58) are hyaline, the triangles, supratriangles and subtriangle of the fore wing are all clear. The hind wing lacks the second cubito-anal cross-vein and is thus without a subtriangle. The anal angle is nearly rounded. There is a single row of cells in the trigonal interspace for a variable distance in both the fore and hind wing. The veins enclosing this space (MA and CuP) are subparallel for the greater part of their course but diverge as they approach the wing margin. The genital lobes of the male are prolonged rearward and are pointed and somewhat hairy. The superior appendages are rather wide apart at the base, beyond which they are bowed and terminate bluntly; the inferior appendage is triangular. The vulvar lamina is as long as the sternum of segment 9, against which it usually lies closely; it is entire and its distal margin is broadly rounded. It has a median crease that may give it a bilobed appearance.

The two species of this genus are the smallest of North American Corduliidae; both are rare inhabitants of bogs, and appear early in the spring. They are probably glacial relicts.

The type species, *W. lintneri* (Hagen), which has a restricted distribution in the northeastern United States, differs from *W. fletcheri* in having pale orange markings on the apex of segments 2 to 9, whereas in *fletcheri* the pale areas are present only on the apices of segments 2 to 4 and are pale yellowish.

Williamsonia was the last among the genera of North American Anisoptera to be discovered in the nymphal stage. It was not until 1970 that Harold B. White and Rudolf A. Raff obtained the first known specimens (of *W. lintneri*) in early May from a bog in Norfolk county, Massachusetts. One of these emerged in captivity, thus confirming its identity and that of the twelve exuviae that White and Raff had obtained from the same site.

It is likely that the following characters of the nymph will prove to be generic and therefore apply also to *W. fletcheri*, the nymph of which has yet to be discovered. The description is taken from the account by White and Raff (1970).

The nymph somewhat resembles that of *Dorocordulia lepida* but differs from it, and from all other known North American corduliid nymphs, in the combined presence of dorsal hooks on abdominal segments 3 to 9 and lateral spines on segment 9 only. The dorsal hooks are small, lie close to the tergites and project posteriorly. The lateral spines on segment 9 form about one-third of the lateral margin of that segment (including the spine). Prominent setae on the posterior margin of segment 9 are about twice as long as the lateral spines. The head lacks dorsal horns. The labium has 8 & 8 palpal setae; premental setae can be 11 + 11 or 12 + 12; the distal margin of the palpus has seven or eight crenations which become progressively deeper towards the movable hook; each crenation except that nearest the movable hook bears two spiniform setae unequal in length. There is a pale mid-dorsal longitudinal line on the abdomen.

Williamsonia fletcheri Williamson. (Pl. 6: 5–7, p. 50; pl. 7: 2, p. 58)

> *Cordulia lintneri* Hagen: Selys, Bull. Acad. Belg., (2) 45: 187 (reprint p. 9), 1878.
> *Dorocordulia lintneri*: Martin, Coll. Zool. Selys, 17: 36, 1907.
> *Williamsonia fletcheri* Williamson, Can. Ent., 55: 96, 1923b.

This is our smallest corduliid and one of the rarest. It is very dark and scarcely metallic, except the frons which is brassy green. Its flight period is early and very short.

Male. Labium and labrum pale reddish brown; anteclypeus with a pale median spot, post-clypeus shining black, frons black below passing into metallic green above, its entire surface pitted; vertex more finely pitted, with a duller lustre; occiput nearly black, with a posterior tuft of long pale brownish gray hairs; rear of head black with a series of similar hairs in line with those of the occiput. Pronotum with a median pale yellow spot. Pterothorax brown, paler next to the ante-alar sinus, beset with long but not dense hairs, slightly brassy on the sides. Legs black. Wings hyaline, venation and pt. brown; membranule whitish in front, shading into gray behind. Abdomen mostly black, moderately hairy, the hairs longest, though thin, on 1 and 2, very short elsewhere; brown on 1 and 2, deepening on dorsum of 2 with a pale yellowish half-ring between 2 and 3, and similar but less distinct pale half-rings following 3 and 4; anal apps. black.

Frons tumid with median furrow very slight and short, vertex nearly flat above, slightly rounded in front, with surface closely and finely pitted. Legs of moderate length, fore tibiae of male with a keel in the distal half; hind tibiae with about distal three-fourths with a keel, middle tibiae without a keel. Abdomen slightly swollen at segs. 1 and 2 and a little constricted at middle of 3, widening to end of 6, sides parallel from 6 to 9, 5 longest, 7 nearly as long as 6, very slightly constricted between 8 and 9.

Sup. apps. (pl. 6: 5, 6, p. 50) rather wide apart at base, diverging in proximal two-fifths, then bent abruptly mesocaudad, the blunt apices meeting or slightly overlapping; in profile nearly straight, widened at the bend, where there is a small ventral tooth, followed by a row of still smaller ones. Inf. app. a little shorter than sup. apps., triangular, with base equal to nearly two-thirds of length. Genital lobes long and distally incurved. Hamuli very large with inferior margin indented before the middle; distal part elongate with apices hooked.

Female. Similar to male in coloration but with the metallic lustre of the frons somewhat less brilliant. Abdomen stouter but of similar proportions; segs. 1, 2, and base of 3 slightly swollen, 3 narrowing to midlength, 4 to end of 5 widening gradually, beyond 5 subequal to end of 9; 10 half as long as 9, apps. one-half longer again than 10. Vulvar lamina (pl. 6: 7, p. 50) as long as tergum of 9, a little shorter than sternum of 9, horizontal in usual position, appearing to be bilobed, somewhat transversely convex; the distolateral angles rounded, the distomesal angles nearly rectangular. A pair of slender spines projects from supra-anal plate between cerci.

Venation. Anx ♂ 7 (8)/5, ♀ 7–8/5; pnx ♂ 5–7/6–7, ♀ 6–7/6 (5); paranal cells ♂ ♀ 4–5/4; anal triangle 2-celled, wide; anal loop rather short, generally with 10–12 cells, rarely with a single ankle cell; sole 3-celled; one row of post-trigonal cells in fore wing as far as middle fork; usually 2 cross-veins under pt. and when only the second is nearly always just beyond the pt.; sectors of arculus well separated at origin; cells at anal angle elongated.

Measurements. Total length ♂ 32–34, ♀ 29–33; abd. ♂ 22–24, ♀ 20.0–22.5; h.w. ♂ 21.5–22.0, ♀ 21–23; h.f. ♂ 5.0–5.5, ♀ 4.5–5.5; w.hd. ♂ 4.5–5.0, ♀ 4.8–5.0; pt. ♂ 1.7–2.0/1.7–2.0, ♀ 1.7–2.0/1.8–2.3.

Nymph. Unknown (1972), but probably similar to that of *W. lintneri*, the characters of which are described in the introduction to this genus (p. 63).

Habitat and range. Sphagnum bog pools. N.B. to James Bay, Ont., w. to Lake Winnipeg, Man., and s. to Me., N.Y. and Mich.

Distribution in Canada. N.B.—York county (Fredericton). *Que.*—Lanoraie; Kazabazua. *Ont.*—Carleton county (Mer Bleue: Carlsbad Springs); Nipissing (Timagami) and Cochrane (Moose Factory, James Bay) districts. *Man.*—Lake Winnipeg (Hagen (1890b) as *Dorocordulia lintneri* Hagen).

Field notes. The only locality where this small corduliid is well-known is the Mer Bleue, an extensive peat bog or muskeg in Carleton county, Ontario, about twelve miles from Ottawa. Very little, however, has been found out about its habits, chiefly because of its very early and brief flight period, but also because of its scarcity and very local occurrence, even within this bog.

The dates of capture at the Mer Bleue range from May 19 (1941) to June 11 (1925) but none of the specimens, so far as we are aware, were tenerals and we therefore believe that the time of emergence must begin well before any of the recorded dates. On the few occasions when we have found this inconspicuous species, it was making short flights close to a few small dark pools in the more open part of an alder thicket close to a rocky island, which rises abruptly from the flat green expanse of the bog. Although they were not particularly active, sometimes coming to rest on dead branches, most of them escaped owing to the treacherous nature of the boggy terrain. On June 5, 1940, we captured a pair in copula.

On May 19, 1941, Professors J. S. Rogers and F. P. Ide visited a part of this bog with which we are not familiar. They captured several specimens of *W. fletcheri* including a pair in copula. They noted that some of the males were patrolling a small pool of open water. It is evident that, at this early date, *W. fletcheri* was already quite mature.

The single specimen from New Brunswick, a male, was taken near Fredericton on July 1, 1931, a date at which the flight period would be long over at the Mer Bleue.

We have made frequent efforts to discover the nymph of this elusive insect, in the pools of the Mer Bleue, but the only corduliid nymphs that we could find turned out to be *Somatochlora franklini* and *S. walshii*. The nymph of *W. fletcheri* remains unknown. Until recently this applied to nymphs of the genus as a whole; but White and Raff (1970) have now discovered and described the nymph of *Williamsonia lintneri* (Hagen).

Genus **Somatochlora** Selys

Bull. Acad. Belg., (2) 31: 279, 1871 (*Chlorosoma* Charpentier, preoccupied). Type species *S. metallica* Van der Linden.

These are dark corduliids with a metallic lustre, reduced pale markings and usually moderate pubescence, which is sometimes rather dense on the thoracic pleura. The eyes in mature individuals are brilliant green, the face partly yellowish including the labium, anteclypeus and parts of the antefrons. The remaining parts of the face as well as the postfrons and vertex are usually black with a metallic lustre.

The yellow markings of the pterothorax, if present, are restricted to one or two pairs of lateral spots, which are sometimes distinct, sometimes more or less obscure in mature individuals. The abdomen is chiefly dark metallic greenish to black, the pale markings being few and inconspicuous. Abdominal segment 2, which is inflated, usually has a pair of lateral yellowish spots and sometimes another pair of dorsal ones. Small basal lateral yellow spots may be present on segments 3 to 7 or 3 to 8, and pale intersegmental annuli are common features of several northern species. The legs are usually blackish and the wings are most frequently hyaline; but tenerals, especially females of some species, may be deeply yellowish or smoky.

Tibial keels are present only on the fore and hind legs, those of the fore legs occupying the distal half, and those of the hind legs extending from near the base to the apex. The middle tibiae have no trace of a keel.

The venation (pl. 7: 3, p. 58) is similar to that of *Epitheca*. There are usually five antenodals in the hind wing, more rarely four or six, and seven to eight or nine in the fore wing. Postnodals are generally seven to nine in the hind wing, and six to eight in the fore wing, but vary considerably in each species. Triangles are two-celled, the supratriangles being clear. The subtriangle of the fore wing is typically three-celled and libelluloid; that of the hind wing is constantly present although small and clear.

The male genitalia are associated with large genital lobes and small simple auricles. The hamuli are large flattened hooks directed caudad. They offer good specific characters in the details of their form.

The best specific characters are found in the anal appendages of the male and the vulvar lamina of the female. The latter is sometimes short and bilobed, as in *Cordulia* and other related genera, but is generally as long as the sternum of segment 9 or longer, and entire or only slightly notched. A specialized form, compressed and spout-like, and modified for alternately stabbing the eggs into the mud or sand of the stream bottom or bank and washing the lamina with water, is found in a number of species, including those of the typical or *metallica* group of the genus.

The nymphs of *Somatochlora* are of the average corduliid build and moderately to very hairy. The head is about twice as broad as long, being widest across the eyes, which are small and not very prominent. The lateral margins of the head meet the hind margin in a broad curve. The pubescence of the head is longest and thickest on the edge of the frons, on the basal joints of the antennae, in a transverse band across the eyes, and in two postocular bands, a lateral and a dorsolateral one. The hinge of the labium (the labial suture) reaches the base of the mesothorax; the prementum is about as wide as long; the premental setae (on each side) number from 8 to 15, usually 11 to 12, and the palpal setae 7 to 8. The movable hook is about one-fifth the length of the labial palpus. The crenations of the labial palpi number eight to ten and are rounded, one-third to one-half as high as long, and each bears a group of setae, arranged in a graded series of 3 to 7 or 3 to 8. The pronotum has the lateral margins somewhat elevated and projecting. The legs are rather short to moderately long. The hind femora vary in length from about the width of the head across the eyes to one-third longer. The hind-wing sheaths generally reach as far back as the base to the middle of the sixth abdominal segment.

The abdomen is more or less obovate, its greatest width being at segment 6 or segments 6 and 7, being equal to about three-fourths of its length and distinctly wider than the thorax. Abdominal segment 9 narrows rapidly so that its mid-dorsal length is only about one-half as long as the lateral margins.

Dorsal hooks may be present on segments 3 to 9 or 4 to 9. When well developed they are compressed, acute and somewhat falciform but slender and never cultriform. In some species they are reduced to mere knobs and in many they are entirely absent. Lateral spines may also be present or absent. When present, they are restricted to segments 8 and 9 or to 9 alone. Those on segment 9 are no longer than the lateral margin of the same segment (excluding the spine), and are usually much the stouter; those on 8, if present, are still shorter. The anal pyramid is about as long as segments 9 and 10 together. The epiproct is triangular, usually slightly longer than wide and slightly acuminate. The cerci are of nearly the same length. The paraprocts are generally a little longer than the other anal appendages and are slender and pointed. The pubescence of the abdomen is longest and thickest along the lateral and posterior margins of segment 9. It is elsewhere very variable according to the species.

The coloration is generally dark and nearly uniform, the thorax lacking the dark lateral longitudinal stripes of the nymphs of *Cordulia* and *Dorocordulia*, and the legs only rarely showing distinct pale and dark annuli.

This genus, the largest in the Corduliidae, is Holarctic and predominantly Canadian, being found as far north as Odonata exist. Of the twenty-four North American species, seventeen occur in Canada. This Canadian total does not include *S. linearis* (Say). A single female adult of this species, reportedly taken at Ste. Hyacinthe, Quebec is in the Provancher collection. This is mainly a species of the Carolinian zone, and until confirmation of this locality record is obtained, it seems wise to postpone formal inclusion of this dragonfly among the Canadian species (see Walker, 1925, pp. 97–98; Robert, 1963, p. 199). It has, however, been included in the keys to adults and nymphs.

Key to the Species of Somatochlora
Adult Males

1 Apices of sup. apps. bifid with a dorso-caudal and a ventral tooth (pl. 8: 9, 10, p. 68)　　　　　　　　　　　　　　　　　　　　　　　　2

　Apices of sup. apps. without a ventral tooth　　　　　　　　　　3

2 With yellow lateral thoracic spots large and well defined; sup apps. little enlarged beyond the middle (pl. 8: 9, p. 68)　　　*ensigera* (p. 88)

　Without yellow lateral thoracic spots; sup. apps. considerably enlarged beyond the middle　　　　　　　　　　　　　　(*linearis*, see p. 67)

3 Apices of sup. apps. slender and curved upward　　　　　　　4

　Apices of sup. apps. not curved upward　　　　　　　　　　14

4 Sup. apps. distally parallel or only slightly bent mesad; thorax with 2 yellow lateral spots; pt. 4 to 5 times longer than wide　　　　5

　Sup. apps. strongly bent mesad beyond the middle, forming a decided angle; metepimeral spot wanting; pt. 6 times longer than wide　　8

5 Abdomen distinctly shorter than hind wing, widest at posterior end of seg. 5, the width of which is about equal to its length　　　　6

　Abdomen as long as hind wing, widest at rear end of seg. 6, distal width of seg. 5 much less than its length　　　　　　　　　　7

6 Sup. apps. with a conspicuous tuft of hair, without a sub-basal ventro-mesal angle (pl. 8: 1, 2, p. 68); hamuli hooked, their ventral edges curved (pl. 11: 1, p. 86); mesepimeral yellow spot elongate
　　　　　　　　　　　　　　　　　　　　　　　walshii (p. 74)

　Sup. apps. without a conspicuous tuft of hair but with a sub-basal ventro-mesal angle (pl. 8: 4, p. 68); hamuli truncate, their ventral edges angular (pl. 11: 2, p. 86); mesepimeral spot rounded
　　　　　　　　　　　　　　　　　　　　　　　minor (p. 78)

PLATE 8

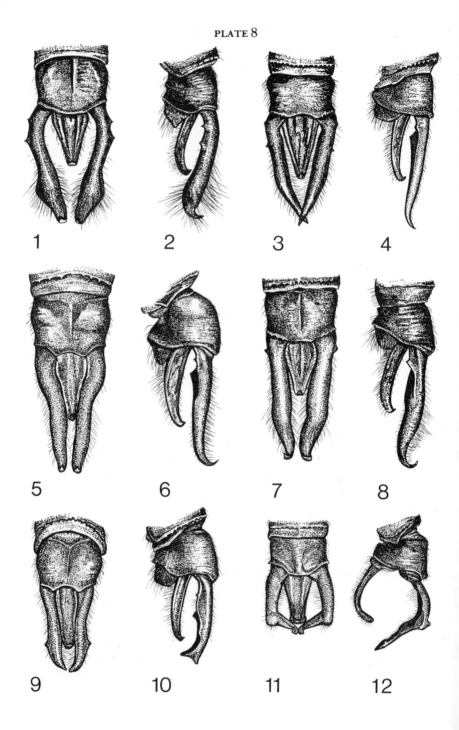

7 Lateral thoracic spots light yellow, conspicuous and well defined even in old individuals; sup. apps. distally parallel, with a sub-basal, ventro-mesal tooth (pl. 8: 5, 6, p. 68); hamuli with inferior edges angulate (pl. 11: 3, p. 86) *elongata* (p. 82)

Lateral thoracic spots in mature individuals dark yellow, inconspicuous; sup. apps. distally approximated, without a sub-basal ventro-mesal tooth (pl. 8: 7, 8, p. 68); hamuli with inferior margins rounded (pl. 11: 4, p. 86) *williamsoni* (p. 84)

8 Hind wing with a dark brown spot at base covering the anal triangle; abdomen widest at distal end of seg. 7; sup. apps. with a prominent distally-pointing sub-basal ventral spine 9

Hind wing without a brown basal spot; abdomen widest at seg. 5 or 6; sup. apps. without such a sub-basal ventral spine 10

9 Sup. apps. converging from base to middle, the lateral carinae bent ventrad, each with a spine arising from it at about the proximal fourth of the appendage's length (pl. 9: 11, 12, p. 76); hamuli with ventral margin obtusangulate, apices blunt (pl. 11: 11, p. 86) *whitehousei* (p. 110)

Sup. apps. subparallel from bases to distal bend, the lateral carina wholly lateral, the spine arising from it at about the proximal fifth (pl. 10: 1, 2, p. 80); hamuli with ventral margin rounded, apices acute (pl. 11: 12, p. 86) *septentrionalis* (p. 113)

10 Abdomen without pale apical annuli except on seg. 2 *sahlbergi* (p. 116)

Abdomen with pale apical annuli, entire or divided mid-dorsally 11

11 Inf. app. bifurcate with a broad apical margin; sup. apps. with a prominent sub-basal lateral spine (pl. 10: 11, p. 80) *cingulata* (p. 130)

Inf. app. triangular with apex truncate 12

12 Inferomesal carina of sup. apps., in lateral view, showing a prominent angle near the middle (pl. 10: 10, p. 80) *hudsonica* (p. 126)

Inferomesal carina of sup. apps., in lateral view, not angulate 13

13 Abd. seg. 8 slightly wider than 10; sup. apps. with a slight, rounded basal prominence but no sub-basal spine visible in dorsal view (pl. 10: 5, p. 80) *albicincta* (p. 119)

Abd. seg. 8 barely narrower than 10; sup. apps. with a small sub-basal lateral spine visible in dorsal view (pl. 10: 7, p. 80) *brevicincta* (p. 125)

PLATE 8

Somatochlora—anal appendages of males (odd nos., dorsal view; even nos., lateral view): (1, 2) *S. walshii*; (3, 4) *S. minor*; (5, 6) *S. elongata*; (7, 8) *S. williamsoni*; (9, 10) *S. ensigera*; (11, 12) *S. tenebrosa*.
(1–12 after Walker, 1925.)

14 Sup. apps. angularly bent near the middle, the distal parts directed meso-ventrad; inf. app. not triangular strongly arched (pl. 8: 11, 12, p. 68); labrum in part yellow   *tenebrosa* (p. 90)

Sup. apps. not angularly bent near the middle; distal third or fourth bent or curved mesad; labrum black   15

15 Hind wing about three-fourths as long as abdomen, with a dark brown spot at base, extending beyond anal triangle; abdomen very slender at base, widest beyond seg. 7; metepimeral spot absent   *franklini* (p. 94)

Hind wing more than three-fourths as long as abdomen, without a dark brown spot at base; abdomen widest at seg. 6 or rarely 7   16

16 Sup. apps. with the lateral ante-apical tubercle conspicuously visible in dorsal view, apices meeting one another in an arc (pl. 9: 9, p. 76); lateral lobes of postclypeus normally black; thorax with two roundish yellow lateral spots (the metepimeral one sometimes obscure)

semicircularis (p. 106)

Sup. apps. with the lateral ante-apical tubercle concealed or inconspicuous in dorsal view, the apices meeting or approaching one another at an angle; lateral lobes of postclypeus generally yellowish or brown   17

17 Abdomen without lateral yellow spots beyond seg. 3; lateral thoracic spots obscure, the metepimeral ones indistinct or absent; lateral ante-apical tubercle of sup. apps. visible from above (pl. 9: 3, p. 76); hind wing distinctly shorter than abdomen   *kennedyi* (p. 98)

Abdomen with lateral yellow spots at bases of segs. 4 or 5 to 7 or 8; both lateral thoracic spots present; lateral ante-apical tubercle of sup. apps. concealed in dorsal view; hind wing barely shorter than abdomen   18

18 Lateral thoracic spots dull and ill defined, elongate; sup. apps. only slightly arched, the lateral ante-apical tubercle not appearing in lateral view as an inferior prominence (pl. 9: 8, p. 76)   *incurvata* (p. 104)

Lateral thoracic spots pale yellow, distinct, ovate; sup. apps. strongly arched, the lateral ante-apical tubercle appearing in lateral view as an inferior prominence (pl. 9: 6, p. 76)   *forcipata* (p. 101)

Adult Females

1 Vulvar lamina subperpendicular to the long axis of the abdomen, compressed, triangular in lateral view, as long as the lateral margin of seg. 9 or longer   2

Vulvar lamina usually horizontal or inclined, not compressed or, if slightly so, shorter than the lateral edge of seg. 9   7

2 Labrum at least partly yellow, postclypeus wholly yellow   3

Labrum black, postclypeus or its middle part black   5

3 Lateral thoracic spots absent (*linearis*, see p. 67)

Lateral thoracic spots present 4

4 Lateral thoracic and basal abdominal yellow spots large and well defined; vulvar lamina longer than anal apps. (pl. 12: 5, p. 92)

ensigera (p. 88)

Lateral spots distinct but not well defined; vulvar lamina shorter than anal apps. (pl. 12: 6, p. 92) *tenebrosa* (p. 90)

5 Abd. less than 35 mm., scarcely as long as hind wings; both lateral thoracic spots ovate, clear yellow; vulvar lamina about as long as anal apps. (pl. 12: 2, p. 92) *minor* (p. 78)

Abd. more than 35 mm., slightly longer than hind wings; mesepimeral spot elongate; vulvar lamina shorter than anal apps. 6

6 Yellow markings of thorax and base of abdomen clear and well defined; no yellow spots posterior to seg. 3, vulvar lamina bluntly pointed, not tapering and little compressed (pl. 12: 3, p. 92) *elongata* (p. 82)

Yellow markings of thorax and base of abdomen dull and inconspicuous, except in tenerals; yellow basal lateral spots on segs. 5 to 7 or 5 to 8; vulvar lamina slender-pointed, somewhat tapering and strongly compressed (pl. 12: 4, p. 92) *williamsoni* (p. 84)

7 Vulvar lamina as long or nearly as long as sternum of seg. 9, entire, scoop-shaped 8

Vulvar lamina distinctly shorter than sternum of seg. 9 12

8 With two yellow lateral thoracic spots and yellow spots on abd. segs. 5 to 7 or 5 to 8; vulvar lamina as long as sternum of seg. 9 or longer 9

Only the mesepimeral spot present, this sometimes being obscure, and no yellow spots on abd. segs. posterior to 3; vulvar lamina a little shorter than sternum of seg. 9 11

9 Abdomen about as long as hind wings or slightly shorter; mesepimeral spot elongate, metepimeral spot subovate; vulvar lamina in natural position declined, largely black (pl. 12: 1, p. 92) *walshii* (p. 74)

Abdomen slightly longer than hind wings; vulvar lamina in natural position horizontal, yellowish 10

10 Lateral thoracic spots pale yellow, ovate, well defined; vulvar lamina about as long as sternum of seg. 9 or a little longer (pl. 13: 2, p. 96)

forcipata (p. 101)

Lateral thoracic spots dull yellow, obscure or ill defined; vulvar lamina longer than sternum of seg. 9 (pl. 13: 3, p. 96) *incurvata* (p. 104)

11 Fork of R$_2$ with 6 to 9 cells; hind wing with a brown spot at base (sometimes suffused over the entire area); lateral lobes of postclypeus black; mesepimeral spot usually distinct; anal apps. scarcely longer than vulvar lamina (pl. 12: 7, p. 92) *franklini* (p. 94)

Fork of R_2 with 11 to 20 cells; hind wings more or less yellow at base but without a distinct spot; lateral lobes of postclypeus brown; mesepimeral spot obscure; anal apps. nearly twice as long as vulvar lamina (pl. 13: 1, p. 96)   *kennedyi* (p. 98)

12 Thorax with two lateral yellow spots, the mesepimeral one ovate, well-defined, the metepimeral one much smaller, sometimes rather obscure; vulvar lamina half as long as sternum of seg. 9, apically notched (pl. 13: 4, p. 96)   *semicircularis* (p. 106)

Thorax without a metepimeral spot, the mesepimeral spot small or wanting   13

13 Hind wings with a dark brown spot at base   14

Hind wings without a dark brown basal spot   15

14 Vulvar lamina entire or only slightly notched (pl. 14: 4, p. 102), somewhat compressed and projecting (pl. 13: 5, p. 96)
                                                                  whitehousei (p. 110)

Vulvar lamina bilobed (pl. 14: 5, p. 102), flattened and not projecting (pl. 13: 6, p. 96)   *septentrionalis* (p. 113)

15 Abdomen without pale intersegmental annuli except between segs. 2 and 3; vulvar lamina less than half as long as sternum of seg. 9, emarginate or bilobed (pl. 13: 7, p. 96; pl. 14: 6, p. 102) *sahlbergi* (p. 116)

Abdomen with pale annuli, complete or partial in nearly all the inter-segmental membranes; vulvar lamina variable   16

16 Vulvar lamina more than half as long as sternum of seg. 9, entire or only slightly emarginate, often projecting   17

Vulvar lamina not more, usually less, than half as long as sternum of seg. 9, not projecting, more or less bilobed or emarginate   18

17 Vulvar lamina about three-fifths to two-thirds as long as sternum of seg. 9, more or less projecting (pl. 14: 2, p. 102); pale annuli of abdomen dorsally entire   *hudsonica* (p. 126)

Vulvar lamina four-fifths as long as sternum of seg. 9 (pl. 14: 8, p. 102); pale annuli of abdomen mesally broken except those following segs. 1 and 2   *brevicincta* (p. 125)

18 Total length less than 55 mm.*; thorax brassy green with a distinct pale mesepimeral spot; vulvar lamina nearly half as long as sternum of seg. 9, deeply emarginate or bilobed (pl. 14: 1, 7, p. 102)   *albicincta* (p. 119)

Total length more than 55 mm.; thorax very dark brassy, with a pale lateral spot; vulvar lamina one-third as long as sternum of seg. 9, obtusely excavated but scarcely bilobed (p. 14: 3, 9, p. 102)
                                                                  cingulata (p. 130)

---

*Except in subspecies *massettensis* Whitehouse, in which the total length is 58 mm.

Nymphs

Nymph unknown

ensigera (p. 89)
incurvata (p. 105)
brevicincta (p. 126)

1 Dorsal hooks present 2

Dorsal hooks absent 7

2 Dorsal hooks falciform, acute, the last one projecting beyond the middle of seg. 10 (e.g. pl. 19: 7, p. 128) 3

Dorsal hooks low knobs, not falciform, the last one not projecting to the middle of seg. 10 (pl. 19: 6, p. 128) *williamsoni* (p. 85)

3 Epiproct of male nearly flat above, ante-apical tubercles (male projection) not at all elevated 4

Epiproct of male longitudinally concave, ante-apical tubercle slightly elevated (pl. 19: 3, p. 128) *tenebrosa* (p. 93)

4 Dorsal hook on seg. 4 more than one-half as long as seg. 4 5

Dorsal hook on seg. 4 not more than one-fourth as long as seg. 4 6

5 Hind femora 6.75–7.0 mm.; hind tibiae 8 mm.; lateral spines on seg. 9 one-third or more as long as the lateral margin of seg. 9 (not including the spine) (pl. 19: 2, p. 128) *minor* (p. 79)

Hind femora 8 mm.; hind tibiae 9 mm.; lateral spines on seg. 9 less than one-fourth and usually one-sixth as long as the margin of seg. 9 (pl. 18: 1, p. 124) *elongata* (p. 82)

6 Lateral spines on seg. 9 more than twice as long as their basal width (pl. 19: 1, p. 128) *walshii* (p. 75)

Lateral spines on seg. 9 less than twice as long as their basal width

linearis (p. 67)

7 Lateral spines absent or, if present, on seg. 9 only 8

Lateral spines present on both segs. 8 and 9 13

8 Fringe of hair on hind margins of abdominal tergites 6 to 9 forming two rows of dorsolateral tufts (pl. 16: 3, p. 112); lateral spines on seg. 9 represented by very minute denticles, or absent *forcipata* (p. 103)

Fringe of hair on hind margins of abdominal tergites 6 to 9 not forming two rows of dorsolateral tufts 9

9 Lateral spines present on seg. 9 10

Lateral spines wholly absent 12

10 Fringe of hair on hind margins of abdominal tergites 7 and 8 forming a distinct median tuft (pl. 16: 2, p. 112) *kennedyi* (p. 99)

Fringe of hair on hind margins of abdominal tergites 7 and 8 not forming a distinct tuft (e.g. pl. 16: 1, 4, p. 112) 11

11 Fringe of hair on hind margins of abdominal tergites 8 and 9 as long or
    nearly as long as mid-dorsal length of these segments (pl. 16: 1, p. 112);
    total length 17 mm.   *franklini* (p. 95)
    Fringe of hair on hind margins of abdominal tergites 8 and 9 much
    shorter than mid-dorsal length of these segments (pl. 16: 4, p. 112);
    total length 21–22 mm.   *semicircularis* (p. 107)
12 Premental setae 9 or 10; palpal setae 6 or 7   *whitehousei* (p. 111)
    Premental setae 11 to 13, palpal setae 8   *septentrionalis* (p. 114)
13 Lateral spines on seg. 9 about one-half as long as the lateral margin of
    seg. 9 (not including the spine); middle crenations of labial palpi bear-
    ing usually 2 setae each, a longer and a shorter one   *sahlbergi* (p. 117)
    Lateral spines on seg. 9 usually less than one-third as long as the margin
    of seg. 9; middle crenations of labial palpi bearing typically 3 setae each,
    in a graded series   14
14 Length under 25 mm.; hind femora less than 7 mm.; abdomen with-
    out median dorsal prominences; epiproct of male without lateral knobs
    (e.g. pl. 19: 4, 5, p. 128); coloration uniform or nearly so   15
    Length over 25 mm.; hind femora over 7 mm.; abdomen with a median
    series of slightly elevated prominences; epiproct of male with a distinct
    knob on each side (pl. 18: 7, p. 124); coloration not uniform
                                                                cingulata (p. 131)
15 Lateral spines on seg. 9 one-fifth to one-third as long as the lateral
    margin of seg. 9 (not including the spine) (pl. 19: 4, p. 128)
                                                               albicincta (p. 121)
    Lateral spines on seg. 9 minute, one-tenth to one-sixth as long as the
    margin of seg. 9 (pl. 19: 5, p. 128)   *hudsonica* (p. 127)

Somatochlora walshii (Scudder). (Pl. 8: 1, 2, p. 68; pl. 11: 1, p. 86; pl. 12: 1,
    p. 92; pl. 19: 1, p. 128)

Cordulia walshii Scudder, Proc. Bost. Soc. Nat. Hist., 10: 217, 1866.
Epitheca walshii: Selys, Bull. Acad. Belg., (2) 31: 293, 1871.
Somatochlora walshii: Selys, Bull. Acad. Belg., (2) 45: (reprint pp. 27, 39),
    1878.

A middle-sized species with a short abdomen, especially in the male, and
two unequal pale spots on each side of the thorax.

Male. Labium and anteclypeus pale to brownish yellow; labrum black; post-clypeus and
antefrons brownish yellow to dark brown; postfrons and vertex black with a metallic lustre;
occiput brown to blue-black. Rear of head black. Thorax dark brassy varied with brown below,
lateral spots pale yellow, often obscured with brown. The mesepimeral spot an oblique streak,
metepimeral spot much shorter, subovate. Fore and middle legs from coxae to basal half or
more of outer face of tibiae, brown; distally black. Wings hyaline with the usual yellowish tinge,
which is sometimes deepened to amber in teneral females. Abdomen with seg. 1 dark grayish
brown, 2 blackish in front of transverse carina, dull metallic green above and behind, with pale

yellowish lateral spots above and behind, remaining segs. chiefly bronze black, seg. 2 with anterodorsal yellow and basal ventrolateral drab spots, often confluent; segs. 5, 6, and 7 with small basal lateral yellowish spots, 8 and 9 with whitish somewhat incomplete apical annuli. Pt. smoky brown; membranule smoky with the basal third or fourth whitish.

Abdomen, including anal apps., slightly shorter than hind wing, widest at posterior end of 5, the width here about equal to length of 5; narrowing to middle of 9, sides of 9 and 10 subparallel.

Sup. apps. (pl. 8: 1, 2, p. 68) widely separated at base, in proximal half strongly bowed, in distal half swollen and convergent, and the slender apices approaching or meeting one another. In lateral view the sup. apps. appear slender, without a ventral spine or angle, decurved with apices reflexed. Near the base are two small outer teeth, most clearly seen in dorsolateral view. Sup. apps. conspicuously hairy above. Inf. app. (pl. 8: 1, 2, p. 68) two-thirds as long as sup. apps., of the usual long triangular form with blunt apex. Hamuli (pl. 11: 1, p. 86) obtusely bent, the convex margin rounded, the distal limb acutangulate.

Female. Longer than male, the abdomen about as long as the hind wing. Seg. 2 with only one lateral yellow spot on each side and without distinct dorsal spots. Seg. 3 with yellow anterodorsal spots much larger than in male, anteroventral spots grayish or whitish, ill defined; yellow basal lateral spots on 5, 6, and 8 larger than in the male.

Vulvar lamina (pl. 12: 1, p. 92) dark brown, yellowish, laterally directed ventrocaudad, slightly longer than sternum of 9, strongly transversely convex but scarcely compressed, apex narrowly rounded or subangulate, ventral profile nearly straight. Apps. one-fourth longer than 9 + 10, bluntly pointed.

Venation. Anx ♂ ♀ 7–8/5–6; pnx ♂ ♀ 5–7/(5) 6–8; two rows of post-trigonal cells, sometimes increasing to 3 rows, especially in ♀; cells between fork of R_2 ♂ 8–15/9–17, ♀ 11–16/12–20; cells between IR_3 and Rspl usually 5 to 7; pt. 5 times as long as wide.

Measurements. Total length ♂ 41.5–46.0, ♀ 44–52; abd. ♂ 25–29, ♀ 29.5–35.0; h.w. ♂ 30–32, ♀ 31–34 (NW: ♂ ♀ 25–34); h.f. ♂ 7.1–7.9, ♀ 7.3–7.6; pt. ♂ 2.5–2.7, ♀ 2.5–2.8.

Nymph (material 1 male exuvia; 1 female nymph). Uniform dark brown, not very hairy. Head as described for the genus; hairs distributed as follows: a group of 15 to 20 rather long hairs over the posterolateral convexities and a few lateral ones below these; a thin fringe of somewhat shorter marginal hairs on the frons, and a row of very short hairs on the lateral half of the postocular suture. Folded labium reaching posteriorly to base of middle coxae; prementum about as broad as long, constricted near the base; premental setae 10 (9 on one side of female), the inner 3 much shorter than the others; palpal setae 7; crenations of distal margins of palpus with a group of spiniform setae in a graded series, usually 4 or 5, the longest about as long as one of the crenations. Hind femora about one-seventh longer than width of head, each with 2 to 4 scattered dorsal hairs and still fewer ventral hairs; length of hind tibiae about equal to width of abdomen; tibial setae numerous. Abdomen (pl. 19: 1, p. 128) widest between segs. 5 and 6; dorsal hairs nearly absent; lateral marginal hairs forming a fairly dense fringe which is short and inconspicuous, except on 8 and 9, on which it is increasingly long; dorsal hooks on segs. 4 to 9, on 4 very minute, on 5 one-half as long as its segment, longest on 8, slightly falciform on 7 to 9; lateral spines on segs. 8 and 9, on 9 slightly more than one-half the lateral length of the segment (excluding the spine), and 3 or 4 times as long as basal width, those of 8 about one-half as long as those of 9; anal appendages as long as mid-dorsal lengths of segs. 9 + 10, the epiproct somewhat acuminate, in profile barely concave in the male; cerci barely longer than epiproct; paraprocts slightly longer than the cerci.

Length 20.5; abd. 11.0; w.abd. 7.2; h.w. 6.5; h.f. 6.4; w.hd. 6.

Habitat and range. Small quiet streams running through bogs and marshes. Nfld. to Wash., Vancouver I., B.C., n. to Hudson Bay, and s. to Conn., Pa., Mich., Wis. and Minn.

PLATE 9

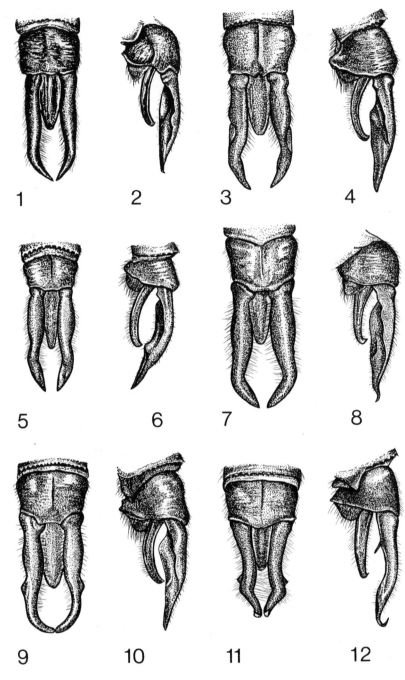

Distribution in Canada. Nfld.—Avalon Peninsula. *N.S.*—Colchester and Pictou counties. *N.B.*—Charlotte county. *Que.*—Covey Hill; Knowlton; St. Thérèse; Lanoraie; Ironside; Mont Tremblant Park; Nominingue; Saguenay River (Cap Jaseux); Eastmain (James Bay); Lake St. John; Lake Mistassini. *Ont.*—Brant, Wellington, Peel, York, Ontario, Prince Edward, Bruce, Simcoe, Hastings and Carleton counties; Muskoka, Nipissing, Thunder Bay and Cochrane districts. *Man.*—Dauphin. *Alta.*—Nordegg (summit of Coliseum Mt.). *B.C.*—Harrison Bay; Eagle River, near Revelstoke; Tête Jaune; Vancouver I. (Forbes Landing).

Field notes. The breeding places of *S. walshii* are small slow streams of clear water in boggy or marshy places. It avoids ponds of any sort and is also absent from streams with an easily perceptible flow. A typical habitat observed near St. Andrews, New Brunswick, July 6 and 9, 1923, was a limited portion of a brook, which flowed through an open treeless marsh. Although there was a definite channel through the marsh the water had spread over a much wider area and was almost everywhere thickly grown over with horsetails (*Equisetum fluviatile*) and various sedges.

Males were flying actively in all parts of the marsh but particularly over the main channel. They flew usually about waist high, hovering frequently in the air with the abdomen slightly flexed, but darting back and forth very rapidly and dodging the net well.

We have also taken *S. walshii* in a floating bog in the upper reaches of a small stream in the Temagami Forest Reserve, Ontario, in many other boggy streams and even a boggy ditch along a railway track, the ditch being fed by springs and choked with watercress.

Only once have we found a newly emerged individual, a male with its exuvia (Walker, 1941c). This was on June 5, 1940, at the Mer Bleue, a large open bog, near Ottawa. The insect was resting in the low sedges and grasses that were growing in the bog a little beyond the open water. The exuvia was soon found clinging to a sedge about a foot above the water.

Apart from the above date of emergence, June 5, our earliest records of capture are June 18, 1934, and June 19, 1919, both from De Grassi Point. The rarity of *S. walshii* in this region and the paucity of records of tenerals suggest that the flight period begins earlier in June than our dates of capture indicate.

The flight period thus begins no later than the first half of June but the

---

PLATE 9

Somatochlora—anal appendages of males (odd nos., dorsal view; even nos., lateral view): (1, 2) *S. franklini*; (3, 4) *S. kennedyi*; (5, 6) *S. forcipata*; (7, 8) *S. incurvata*; (9, 10) *S. semicircularis*; (11, 12) *S. whitehousei*.
(1–12 after Walker, 1925.)

majority of our dates of capture are in July, specimens may still be taken in August, and we have one record for September 1, 1957. In British Columbia Whitehouse (1941) gives a flight period from July 25 to August 18, but estimates it to be longer—from July 10 into September.

Oviposition has been observed by Kennedy (*in litt.*), Whitehouse (1941) and the writer. According to Kennedy (Walker, 1925, p. 62) "One female seen ovipositing in creek. She made several very nervous hasty darts over creek, finally making a single quick rush down to surface, tapping it once and so hard as almost to submerge herself. This performance was repeated three times with fast irregular flights between each time. Then scared away by collector." Our observations are more like those of Whitehouse (1941) who gives the following account. "Males, when not visiting the water, would fly together in nearby sunny glades at the edge of the conifers. A number of females were observed in process of ovipositing and were watched most carefully at the stagnant pot-holes of the drainage creek . . . The insect would fly down to the surface of the water amid the broad-leaved reeds, going from one small open space to the next, the dip, dip, dip, of her abdomen never ceasing except when she manoeuvred around the reed stems. Some scores of eggs must have been released at one laying and within an area of three or four feet." "The eggs disposed of, she would rise to just above the reeds, poise for an instant to look around her, and dart away."

Somatochlora minor Calvert. (Pl. 8: 3, 4, p. 68; pl. 11: 2, p. 86; pl. 12: 2, p. 92; pl. 15: 1, p. 108; pl. 19: 2, p. 128)

Somatochlora elongata var. *minor* Calvert, Ent. News, 9: 87 (footnote), 1898.
Somatochlora elongata: Martin, Coll. Zool. Selys, 17: fig. 21 (p. 23), 1906.
Somatochlora minor: Walker, Can. Ent., 39: 69–74, 1907.

A medium-sized species with a short abdomen, and two rounded yellow spots on each side of the thorax.

Male. Labium, anteclypeus, sides and lower margin of post-clypeus ochre yellow; other parts of face and dorsum of head metallic greenish black; rear of head black. Pterothorax steel blue with green reflections, the brilliance somewhat obscured by the coat of pale brownish pile, sides with two oval light yellow spots, the larger in front; ventral parts of thorax, most of coxae, fore trochanters and most of outer surface of fore femora brownish, legs otherwise black. Wings slightly washed with amber; pt. smoky-brown; membranule smoky, the basal two-fifths or less whitish. Abdomen: first two segs. dark brown, second with a dorsal and two ventrolateral yellow spots and a pale apical annulus, remaining segs. greenish black with a dull lustre; seg. 3 with a dorsolateral and a larger ventrolateral yellowish spot, often confluent; remaining segments without yellow spots.

Abdomen distinctly shorter than ind wing, widest at distal end of 5, narrowing again more gradually to base or middle of 9. Seg. 5 shaped as in *walshii*, the distal width being almost equal to the length.

Genital lobes large; hamuli (pl. 11: 2, p. 86) almost squarely truncate with a decided ventral angle. Sup. apps. (pl. 8: 3, 4, p. 68) about as long as 9 + 10, somewhat widely separated at base; where there is a short lateral angle beyond which they converge in a long curve, tapering distally to finely pointed apices, which meet; lateral margins bearing two minute spines or angles, one near the base, the other at the proximal fourth. In profile they appear very slender, slightly arcuate and upcurved at apices. The two lateral spines appear as projections of the lateral carina and there appears also a sub-basal ventromesal angle. Inf. app. (pl. 8: 3, 4, p. 68) two-thirds as long as sup. apps. with the usual triangular form. Hair of appendages rather scanty.

Female. Seg. 2 generally with only one lateral yellow spot, the posterior lateral and dorsal spots being small or absent. Dorsal spot on 3, however, is larger than in the male, whereas the ventral spot is smaller. Edge of 9th tergum and sides of vulvar lamina also yellowish. Wings sometimes deeply stained with amber.

Abdomen subcylindrical in mature females, scarcely shorter than hind wing. Vulvar lamina (pl. 12: 2, p. 92) about as long as 9 + 10, strongly compressed, triangular, spout-shaped, projecting almost perpendicularly to the abdominal axis, with a broad base, tapering to a slender apex.

Venation. Anx ♂ ♀ 6–8/5(4 or 6); pnx ♂ ♀ 7–8(6 or 9)/8–9(7 or 10); usually in ♂ two rows of post-trigonal cells in the fore wings, with a 3rd row frequently represented by only 1 cell, rarely 3 rows (in ♀ usually 3 rows); cells between fork of R₂ ♂ 11–17/13–18, ♀ 14–22/16–21; pt. 4.5 times as long as wide.

Measurements. Total length ♂ 42–44, ♀ 44–50; abd. ♂ 26.5–29.0, ♀ 31–33; h.w. ♂ 30.5–32.2, ♀ 32–34; h.f. ♂ ♀ 8–9; pt. ♂ 1.9–2.5, ♀ 2.5–2.7.

Nymph (pl. 15: 1, p. 108). Uniform brown, not very hairy; very like *S. walshii*, lateral and posterolateral region of head clothed with rather long hairs. Antennae also hairy, the two basal segments with rather close-set hairs. Prementum of folded labium reaching posteriorly to base of mesothorax, about as broad as long, less constricted at base than in *S. walshii*; premental setae 11 to 13, the outer 7 or 8 much longer and closer together than the others, the 5th or 6th from the outside longest; palpal setae 6 to 8; crenations of distal margin of palpus each (except some of the smallest) with groups of 5 or 6 spiniform setae, the hindmost seta almost as long as the crenation itself, the others in general successively shorter. Hind wing-sheaths reaching middle of seg. 6; hind femora one-sixth or one-seventh longer than width of head across eyes; hind tibiae as long as width of abdomen, with numerous setae. Abdomen (pl. 19: 2, p. 128) widest at seg. 6 or 7; dorsal setae very sparse; lateral marginal setae, moderately dense, longer and denser on 8 and 9. Dorsal hooks on segs. 4 to 9; the first one about one-half or more as long as its segment, the others successively higher to seg. 7, the last three much alike and similar to those of *S. walshii*; lateral spines on 8 and 9, also like those of *S. walshii*, but those of 9 wider at base, being about one-third or more the lateral length of that segment (not including the spine); anal appendages a little longer than mid-dorsal length of segs. 9 + 10; epiproct as in *S. walshii*, about as long as cerci in the male; paraprocts one-fifth longer.

Length 21.0–22.5; abd. 12–13; w.abd. 8–9; h.w. 6.0–6.5; h.f. 6.7–7.0; w.hd. 6.

Habitat and range. Clear, gently flowing, forest streams. N.S. to James Bay; w. to Vancouver I., B.C., s. to Me., N.H., N.Y., s. Ont., Mich., Man., to Wyo., Wash. and the Caribou and Kamloops-Revelstoke districts of B.C. to Vancouver I.

Distribution in Canada. Labr.—Cartwright. *N.S.*—Annapolis, Halifax and Hants (Elmsdale) counties; Cape Breton I. *N.B.*—Queen's county. *P.E.I.*—(Whitehouse, 1948). *Que.*—Covey Hill; Wakefield; Nominingue; Mont Tremblant Park; Saguenay River (Cap Jaseux); Godbout; Eastmain (James Bay); Bradore Bay. *Ont.*—Simcoe and Carleton counties; Muskoka, Parry Sound, Nipissing, Algoma, Thunder Bay, Cochrane (n. to Ft. Albany, James Bay) and Kenora districts. *Man.*—Treesbank; Aweme; Onah;

PLATE 10

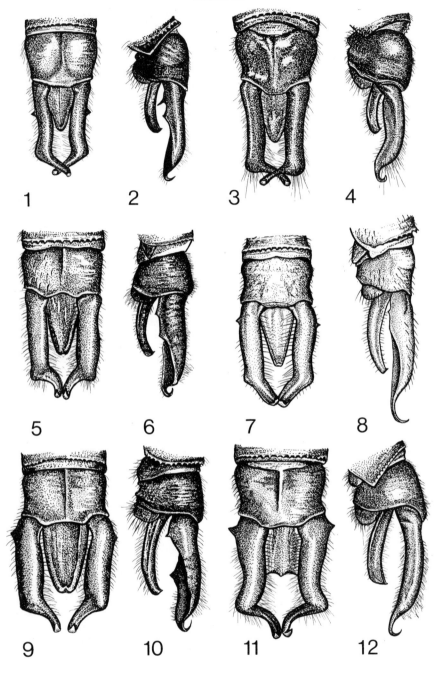

1   2   3   4

5   6   7   8

9   10   11   12

The Pas; mile 256, Hudson Bay Rlwy.; Gillam. *Sask.*—Prince Albert. *Alta.*—Nordegg (Coliseum Mt.). *B.C.*—Sushwap Lake; Squilax; Jesmond; Vancouver I. (Forbes Landing).

The northernmost recorded localities are East Main, Quebec; Fort Albany and Favourable Lake (52°N.), Ontario; Gillam (55°50'N.), Manitoba; and Nordegg (52°N.), Alberta. Nordegg and localities further west where this species has been recorded are presumably far south of its northern limit.

Field notes. From coast to coast, throughout most of the forested regions of Canada, *S. minor* flies from about the second week in June to the end of August or (in British Columbia) to the third week in September. It is everywhere most commonly seen in July and early August, when the males are patrolling small, clear, gently flowing streams, often in association with *Cordulegaster.* At this time they fly usually within a foot of the water, sometimes remaining balanced over a pool for half a minute or more, and then, after making intermittent brief excursions, returning to the same spot.

Emergence may begin as early as the first week in June, our earliest date in southern Ontario (Lake Simcoe) (44°30'N.) being June 6; an individual was found emerging on the Gull River, Lake Nipigon (49°30'N.) on June 10. Tenerals occasionally appear very far away from the breeding places. We have occasionally taken them at De Grassi Point nearly a mile from the creek where they are known to breed, and Whitehouse (1918b) found young adults in the open woods at the top of Coliseum Mt., in the Alberta Rockies. At such times they usually fly back and forth in sheltered glades or along the partly shaded borders of woods. Their flight is generally at a height of four or five feet. We do not remember seeing one beyond the reach of an ordinary net but it is quite possible that they may sometimes fly high, as do other corduliids in early imaginal life.

We have observed the females ovipositing several times in various parts of Ontario, chiefly during the second half of July. "The female flies close to the bank of the stream, often concealed by overhanging foliage, keeping near the surface of the water. Choosing a spot close to a mossy log or hummock she makes irregular flights about this point, tapping the surface of the water with the end of the abdomen and also the wet mossy bank an inch or two higher. The moss is struck once with the ovipositor, the abdomen being turned forward from beneath, and then the water [is struck] by two or three successive taps, the process being repeated more or less rhythmically but not perfectly so. The irregular flights are made after the water is tapped and

---

PLATE 10

Somatochlora—anal appendages of males (odd nos., dorsal view; even nos., lateral view): (1, 2) *S. septentrionalis*; (3, 4) *S. sahlbergi*; (5, 6) *S. albicincta*; (7, 8) *S. brevicincta*; (9, 10) *S. hudsonica*; (11, 12) *S. cingulata.*
(1–6, 9–12 after Walker, 1925.)

82 THE ODONATA OF CANADA AND ALASKA

from time to time a new position is taken up, generally only a few feet away, and the operation is repeated here. Sometimes only the water is tapped and sometimes egg-laying may take place in shallow running water among partly exposed stones as I have observed at Lake Nipigon. Oviposition was observed at Lake Nipigon on a number of occasions from July 16 to 24, but it undoubtedly continues well into August" (Walker, 1925).

Somatochlora elongata (Scudder). (Pl. 8: 5, 6, p. 68; pl. 11: 3, p. 86; pl. 12: 3, p. 92; pl. 15: 2, p. 108; pl. 18: 1, p. 124)

Cordulia elongata Scudder, Proc. Bost. Soc. Nat. Hist., 10: 218, 1866.
Epitheca elongata: Selys, Bull. Acad. Belg., (2) 31: 292, 1871.
Somatochlora elongata: Selys, Bull. Acad. Belg., (2) 45: (reprint p. 27), 1878.

A rather large species, similar to *S. minor* but differing in many details, of which the most significant are given below.

Male. (1) *S. elongata* is larger (see measurements of male below). (2) The abdomen of *elongata* is relatively longer, being as long, without the anal apps. as the hind wing (shorter than hind wing in *minor*).

Average dimensions (in mm.)

	Total length	Length of abd. (excl. apps.)	Length of hind wing
S. elongata	53	38	37
S. minor	45	24	32

(3) In *elongata* the mesepimeral yellow spot is a stripe about three times as long as the metepimeral spot, which is long ovate. In *minor* the mesepimeral spot is long ovate and is much less than twice as long as the metepimeral spot. (4) The hamuli (pl. 11: 3, p. 86) in *elongata* are more obliquely truncated than in *minor*. (5) The sup. apps. (pl. 8: 5, 6, p. 68) in *elongata* are nearer together at base and their distal halves are subparallel, whereas in *minor* they approximate each other from near the base to the apices, usually meeting there.

Female. The females differ in the form of the vulvar lamina. In both species it projects from the abdomen at almost a right angle to the long axis of the abdomen, but in *elongata* it is relatively shorter, its midventral (anterior) length being nearly equal to the ventro-lateral tergal margin of seg. 9. The vulvar lamina (pl. 12: 3, p. 92) is also less compressed in *elongata* and its outline, as viewed laterally, is much blunter and not at all attenuate as in *minor*. The thoracic pattern in both species is similar to that of the males of the same species.

Venation. Anx ♂ ♀ 7–8(9)/5(6); pnx ♂ (9)/7–10, ♀ 6–8/8–10; two rows of post-trigonal cells in fore wings, followed usually by 3 rows with 5 or 6 cells at margin; cells between forks of R_2 ♂ 11–15, ♀ 13–17; pt. 4 or 5 times as long as wide.

Measurements. Total length ♂ 52.0–55.5, ♀ 58.5–62.0; abd. ♂ 37.0–39.5, ♀ 40.0–43.5; h.w. ♂ 34–37, ♀ 35.5–38.0; h.f. ♂ 8.2–8.4, ♀ 8.2–8.5; pt. ♂ 2.5–2.9, ♀ 2.7–3.0.

Nymph (pl. 15: 2, p. 108). Uniform pale brown or with a series of darker brown blotches midway between mid-dorsal line and lateral edges, and also with dark spots alternating with paler areas in the intersegmental membranes. Head as in *S. minor*, occiput with a slight prominence, but no distinct angle, sides of head and occipital angles moderately hairy. Folded labium reaching posteriorly base of mesosternum; premental setae 11 to 12, the 4th to 6th from lateral margin longest, the outer 8 longer and closer together than the remainder; palpal setae 6 or 7; crenations of distal margin or palpus 9, similar to those of *S. minor* and with similar spiniform setae. Hind wing-sheaths extending to middle of seg. 6; hind femora about one-third longer than width of head across the eyes, their length a little less than width of abdomen, hairs sparse. Abdomen (pl. 18: 1, p. 124) widest at seg. 6 with a rather dense marginal fringe of fine hairs, some of which on 9 and 10 are as long as the lateral margin of 9 or longer; dorsal hooks on segs. 4 to 9, that on 4 about one-third as long as that segment, that of 5 one-half the corresponding length, the others about as long as the segments bearing them; each slightly curved and elevated, apices acute; lateral spines present on segs. 8 and 9, not divergent; those on 8 scarcely one-seventh as long as the lateral margin of that segment (not including the spine), slender, the basal width scarcely half the length; those on 9 about one-sixth as long as the lateral length of that segment; cerci one-half as long as mid-dorsal length of segs. 9 + 10, about as long as epiproct, slender, tapering, acute, with straight sides; epiproct four-fifths as wide as long, slightly acuminate; paraprocts about one-fifth longer than epiproct, acute.

Length 23.0–24.4; abd. 14.5–15.0; w.abd. 8.0–9.2; h.w. 7.5; h.f. 7.5–8; w.hd. 6–7.

Habitat and range. Forest streams with intermittent rapids; lakes and ponds along their outlets (Robert, 1953). N.S. to the Laurentian Hills, Que., Ont. n. of Lake Superior and n. Minn., s. to N.C., Pa., Mich. and Wis.

Robert (1953) states that *S. elongata* is the dominant *Somatochlora* in those parts of Mont Tremblant Park which have been visited. It is probably so throughout much of the Laurentian highlands and the northern New England States. Its southernmost distribution suggests that it may be primarily an Appalachian species.

Distribution in Canada. *N.S.*—Annapolis, Colchester, Digby and Halifax counties; "Nova Scotia" (Hagen, 1861). *N.B.*—Charlotte, Northumberland and York counties. *P.E.I.*—Rollo Bay. *Que.*—Outlet of Lake Blanche, Mont Tremblant Park; Saguenay River (Cap Jaseux); Cachée River, on the road to Lake St. John (Provancher, 1877a) (presumably Cachée Creek, 48°39′N., 71°38′W.); near Dolbeau (Lake St. John). *Ont.*—Parry Sound, Nipissing (incl. Algonquin Park), Algoma and Thunder Bay districts.

Field notes. In Ontario and Quebec *S. elongata* has not been recorded before the second half of June, although emergence probably begins a little earlier than any of our records suggest. The climax of the flight period is probably attained in July, although most of our records are for August. Robert's records from Mont Tremblant extend from about the second week of June to the last week of August. We have no June records for the Maritimes because little or no collecting has been done before July. But for the remainder of the season the majority of records are for August and September, specimens having been taken in Nova Scotia from September 1 to 6.

According to Robert (1953), *S. elongata* frequents the same habitats as *S. minor*, the outlets of lakes and the shores of lakes and ponds along the course of a stream. With this statement we would agree, but would add that we have found *S. elongata* in the vicinity of streams larger than those usually frequented by *S. minor* (such as the Oxtongue River, Algonquin Park) but since

S. elongata flies along small streams too, this observation may have little significance.

The only stream on which we have observed *S. elongata* in considerable numbers is one that flows into Lake Timagami, Ontario (Sharp Rock). "In the last half mile of its course this stream falls over two precipitous masses of rock, each about 20 feet high, the second one being only a few yards from its mouth. Between these falls is a long stretch of smooth water, broken here and there by shallow rapids flowing over moss-covered boulders. This part of the stream is exposed to direct sunlight, the low banks being several to many yards distant from the edge of the forest and the intervening area somewhat boggy and covered with shrubs such as *Myrica gale* and *Spiraea salicifolia*. The Somatochloras flew over both the smooth water and the rapids, the males coursing up and down along either bank, in beats of varying length, at an average height of about 18 in. to 2 ft. above the water. They were not difficult to capture when struck with the net from behind."

"Two females were observed ovipositing, in both cases over the rapids among the moss-covered boulders. One of these was seized by a passing male, the pair being then captured, but the other female was closely observed for several minutes at a distance of only a few inches. The method was quite similar to that of *S. minor*. The insect flew within three inches of the water, twisting about in her flight in a most irregular way, now facing one direction, now another. The abdomen was held horizontally with the end slightly raised. Bending the abdomen forward during flight she struck the wet moss on a rock a little above the water line and then, flying back a few inches, she struck the water itself forcibly two or three times, after which there was a pause of two or three seconds during which she made irregular twisting flights, as though seeking another place for her operation, which was then resumed, usually at nearly the same spot" (Walker, 1925).

Although we formerly believed that the eggs were liberated in the water, later observations of various workers support the contrary view that in this method of oviposition the eggs are thrust into the wet bank and the vulvar lamina is then washed clear of debris by being dipped into the water (Münchberg, 1932).

Somatochlora williamsoni Walker. (Pl. 8: 7, 8, p. 68; pl. 11: 4, p. 86; pl. 12: 4, p. 92; pl. 15: 3, p. 108; pl. 18: 2, p. 124; pl. 19: 6, p. 128)

Somatochlora elongata: Needham, Bull. N.Y. State Mus., 47: 499 (nymph), 1901.

Somatochlora williamsoni Walker, Can. Ent., 39: 70, 1907.

A large dark species, similar to *elongata* but, when fully mature, with inconspicuous yellow markings, and the superior anal appendages of the male without an angular ventral prominence near the proximal third.

Male. Mouth-parts and face with the usual pattern (see *S. minor*) but pale markings dull yellow to brownish, labrum and middle of post-clypeus black with a greenish lustre; postfrons and vertex metallic green-black; occiput dark reddish brown, pile of face dark brownish; rear of head with a pale fringe of hair. Pterothorax dark metallic green, varied with brown, with blue or violet reflections, covered with pale brownish hair. Lateral spots similar to those of *elongata* but much darker and inconspicuous brownish, except in tenerals, in which they are orange and distinct; underside of thorax, coxae, fore trochanters and posterolateral surfaces of fore femora brown, legs otherwise black, wings hyaline, venation black, pt. smoky brown to nearly black; membranule smoky brown, segs. 1 to 2 of abdomen and base of 3 polished dark brown, dull on genital lobes, with following parts of 2 yellowish; paired dorsoventral spots, lateral spots in front and behind, the front ones larger; 3 to 10 dull yellowish black; 3 with paired dorsolateral and ventrolateral yellowish spots; 5 to 7 or 8 also with inconspicuous laterobasal yellowish spots.

Abdomen (excluding apps.) as long as hind wing, barely stouter than in *elongata*, widest at distal end of 5, narrowing distally to base of 8 and widening very slightly again on 9, hamuli (pl. 11: 4, p. 86) bluntly rounded and subtruncate.

Sup. apps. (pl. 8: 7, 8, p. 68) a little stouter than in *elongata* and *minor*, with apical recurved parts somewhat convergent, without a distinct ventral angular prominence at proximal third, such as is present in *elongata* and *minor*. Hair on sup. apps. rather long and dense on distal half of dorsal surface, but not forming a conspicuous tuft as in *walshii*. Inf. app. (pl. 8: 7, 8, p. 68) of usual triangular form, about one-half as long as sup. apps.

Female. Yellow abdominal spots larger than in male, lateral spot on 2 unbroken, well defined above but not below; terga of proximal segments margined with brownish yellow, which is wide on 3 but tapers behind with a narrow streak; anterolateral yellowish spots on 3 to 8. Wings as in male but more or less flavescent, the yellow variously distributed, in old females sometimes covering the entire wing. Abdomen in young females widest at seg. 2, tapering continuously to middle of 9, sides of which diverge distally. In old females the abdomen is subcylindrical.

Vulvar lamina (pl. 12: 4, p. 92) perpendicular to axis of body, longer than depth of abdomen between 8 and 9, strongly compressed, triangular in lateral view, slightly acuminate, apex very slender, width at base about three-fifths of length. Anal apps. about a third longer than 9 + 10, two-fifths or more longer than vulvar lamina.

Venation. Anx ♂ ♀ 7–9/5–6; pnx ♂ ♀ 6–9/7–9; two rows of post-trigonal cells, followed in fore wings by 3 rows, beginning about the origin of R_{4+5}; cells between fork of R_2 ♂ 11–23/12–22, ♀ 15–22/16–24; cells between IR_3 and Rspl 6 to 9; pt. 4 to 5 times as long as wide.

Measurements. Total length ♂ 53.5–59.0, ♀ 55.0–59.5; abd. ♂ 36–40, ♀ 38.5–42; h.w. ♂ 35–40, ♀ 38–39; h.f. ♂ 9, ♀ 8; pt. ♂ 2.5–3, ♀ 2.8–3.

Nymph (pl. 15: 3, p. 108). Uniform dark to light brown. Head of usual form with hind margin nearly straight, hair of lateral and dorsolateral areas moderate, elsewhere scanty. Hinge of folded labium reaching posteriorly to mesocoxae; premental setae 11 or 12, the 4th or 5th from the outside the longest; palpal setae 8; crenations of distal margin of palpus broadly and regularly arcuate, the larger ones with a group of about 7 spiniform setae, the longest about as long as the crenation. Wing-sheaths reaching a little beyond base of seg. 6, nearly hairless except along lateral edges. Length of hind femora slightly exceeding width of head and equal to about three-fourths of the abdominal width. Abdomen (pl. 18: 2, p. 124) widest at seg. 6, nearly hairless except along lateral edges where the fringe is of about average length and density, and is much longer on the sides of 9 and 10 than elsewhere; lateral spines on 8 and 9 rather slender, those on 8 one-fifth or one-fourth as long as lateral margin of that segment (not including the spine), and less than one-half as wide at base as long, not divergent; those on 9 one-fourth to one-third as long as the corresponding margin, more than twice as long as their basal width, subparallel; margin of 9 convexly arcuate at base, somewhat concave before base of spine; dorsal hooks (pl. 19: 6, p. 128) present as low knobs on 5 to 9 or 6 to 9, little elevated and not projecting beyond the hind margin of their segment except slightly in the case of those on 8 and 9; cerci and epiproct almost equal in length and about as long as mid-dorsal length of 9 + 10; paraprocts projecting little further back, all evenly taper-pointed.

Length 23–25; abd. 12–13; w.abd. 8–9; h.w. 7.2–7.5; h.f. 7.0–7.3; w.hd. 6.5.

PLATE 11

1

2

3

4

5

6

7

8

9

10

11

12

13

14

15

16

Habitat and range. Quiet, shady forest streams. Me. to Man. and Minn., s. to Conn., Pa. and Tenn.

Distribution in Canada. Que.—Sherbrooke; near Lake Témiscouata; Masham; Kazabazua; Lake Ouareau; Mont Tremblant Park; Nominingue; Laurentides; Abitibi county; Saguenay River (Cap Jaseux). *Ont.*—Essex, Brant, Wellington, York, Dufferin, Bruce, Simcoe and Peterborough counties; Muskoka, Parry Sound, Manitoulin, Nipissing (incl. Algonquin Park), Algoma, Thunder Bay and Cochrane districts. *Man.*—Winnipeg Beach.

Field notes. Since we have observed *S. williamsoni* at the type locality—De Grassi Point, Lake Simcoe, Ontario—far more than at any other locality, the great majority of our notes are from one or the other of two small streams within two miles of this place.

The flight period of this species extends from about the third week in June to the end of the month and continues until some time in September, usually about the first or second week, our latest date being September 15. Although emergence generally begins in late June, it continues for a week or so in July. The only individual that I have found with its exuvia was taken on July 7, but on July 9, 1929, at Franks Bay, Lake Nipissing, I found 11 exuviae floating among stems of emergent vegetation (*Equisetum fluviatile*) and this was suggestive of a very recent emergence, the exuviae having been blown by the wind from the emergent plants.

During the last days of June and in early July, *S. williamsoni* may be seen about the borders of woods or in small clearings, hawking at a height of 30 to 50 feet or more, sometimes keeping within an area of only a few square yards. From the middle of July to the end of their season the males are more often seen along the forest streams where the species breeds. They frequent sluggish streams preferably shady or partly so and seldom appear in the open sunlight except very late in the season when the weather is growing colder.

The males patrol the banks of the stream, usually in the shade, flying in rapid forward movements, interrupted by periods of almost motionless suspension in the air. In sunlight they commonly fly two to four feet above the water, whereas in the dense shade they generally keep within a few inches of the water's surface. Sometimes they may be seen hovering in a dark recess under an overhanging bank and at such times they may be scarcely

---

PLATE 11

Somatochlora—hamulus, right lateral view: (1) *S. walshii*; (2) *S. minor*; (3) *S. elongata*; (4) *S. williamsoni*; (5) *S. ensigera*; (6) *S. tenebrosa*; (7) *S. franklini*; (8) *S. kennedyi*; (9) *S. forcipata*; (10) *S. semicircularis*; (11) *S. whitehousei*; (12) *S. septentrionalis*; (13) *S. albicincta*; (14) *S. sahlbergi*; (15) *S. hudsonica*; (16) *S. brevicincta*.
(1–13, 15 after Walker, 1925.)

visible except for the brilliant green of their eyes. They may remain there suspended for several minutes at a time, occasionally darting away a few yards and then returning to the same spot.

Their preference for still but not stagnant water leads them to frequent not only streams but also lakes of certain types, especially those of the Canadian Shield in Quebec and Ontario. Robert (1953) reports *S. williamsoni* from several lakes in Mont Tremblant Park, Quebec, and notes that it frequents clear waters rather than bog lakes. This is also our usual experience although on one occasion we found *S. williamsoni* in fair numbers over the margins of a typical bog lake about six miles from Lac Témiscouata, Quebec.

Females of *S. williamsoni* are rarely seen except during the teneral stage and the period of early wandering and feeding. In spite of the many visits we have made to the creeks near De Grassi Point we have seen ovipositing females only two or three times there and have only once been able to observe the process carefully. This was in the afternoon of July 21, 1933. Rain had been falling and the sky was still dark although the rain had ceased. A female of *S. williamsoni* was ovipositing in a wet hummock beneath an alder bush. She flew with the abdomen tilted upward, striking the wet muddy bank seven or eight times in succession about three inches above the water line. Then she struck the water once and repeated the movement a number of times. I could not quite reach her safely with the net, but without having been disturbed by me she darted suddenly straight into the woods and did not return. I suspect that the females spend much time in dark woods. Dr. F.P. Ide found females of this species flying during the evening on the beach of Point Pelee, Ontario.

Oviposition of this species was also observed by Mr. N.K. Bigelow near Port Sydney, Ontario, on August 12, 1912. He states that a female was seen flying at 4.30 p.m. on a muddy part of the shore of Diamond Lake. At the water's edge was a strip of wet mud covered with moss and algae left by the receding water line. The insect was flying over this wet surface, two or three inches from the water's edge and striking it with the projecting vulvar lamina. It has also been observed "ovipositing in stems of *Scirpus* and flying back to water" (Walker, 1925).

Somatochlora ensigera Martin. (Pl. 8: 9, 10, p. 68; pl. 11: 5, p. 86; pl. 12: 5, p. 92)

Somatochlora ensigera Martin, Coll. Zool. Selys, 17: 29, 1907 (♀).
Somatochlora charadraea Williamson, Ent. News, 18: 5, 1907a (♂).

A conspicuously marked species of the middle west, easily known by its almost entirely yellow face, the peculiar shape of the anal appendages of the male (pl. 8: 9, 10, p. 68), and the very short anal appendages of the female together with a long projecting vulvar lamina.

Male. Face including labium and labrum orange to brown, the labium with a dark brown margin widening laterally; frons with the blue or green-black middle area extending halfway down the antefrons, vertex brownish red with slight green reflections; occiput polished light reddish brown. Pile of head not very dense, brownish, dark above, elsewhere pale. Pterothorax dark reddish brown with violet-blue and green reflections; dorsal thoracic carina light yellow; lateral thoracic spots clear yellow with an orange tinge, the anterior straight, 4×1 mm., the posterior 2.5×1 mm. with front edge straight; hind edge rounded or subangular; interalar space with large yellow spots; thoracic hair somewhat scanty, pale brown; part of fore femora black beyond the trochanters. Wings hyaline, costal veins dull yellow; pt. brownish black; membranule smoky brown, its basal two-fifths white. Abdomen with segs. 1, 2 and base of 3 brownish black with blue reflections on 2; pale markings bright yellow and distinct, distributed as follows: seg. 1 with a dorsal spot, 2 with three pairs of spots, a small postero-dorsal and two larger pairs of lateral spots; 3 with paired basal dorsolateral spots and much larger and paler basal ventrolateral spots narrowly prolonged caudad. Segs. 3 to 10 greenish black with little lustre and with no pale spots behind 3; intersegmental membrane following 8 and 9 pale yellow. Abdominal hair scanty.

Abdomen (excluding apps.) about as long as hind wing, rather slender, widest at end of 5, but deepening as far as 8. Genital lobes widest at base, subrhomboid. Hamuli (pl. 11: 5, p. 86) hooked with blunt apices. Sup. apps. (pl. 8: 9, 10, p. 68) shorter than 9 + 10, slender, converging as far as middle, then subparallel with apices somewhat convergent and bearing a small tooth, width in dorsal view nearly uniform, greatest about the second third where there is a short, stout spine, narrow in lateral view, apically decurved and showing two additional inferior prominences, a smaller one about midlength and a larger one at the distal end.

Female. Colour pattern similar to that of male, although the pale areas of the head are duller, being yellowish instead of orange brown. Vertex with a reddish brown spot. Wings flavescent at nodus of both pairs and along costal edge of hind pair. Apices, especially of hind wings, dusky, membranule whitish on basal half; otherwise as in male.

Abdomen scarcely longer than hind wing, widest at 3, narrowing in basal half of 3, then very gradually to basal half of 5, widening slightly again to 7, the sides of which are parallel, and narrowing again rapidly from base of 8 to apex. Seg. 8 about two-thirds as long as 7; 9 very short, its lateral tergal edge oblique, about two-thirds as long as 8 and more than twice as long as mid-dorsal length of same segment. Vulvar lamina (pl. 12: 5, p. 92) longer than dorsal lengths of 9 + 10, compressed, strongly projecting; in profile triangular with sides nearly straight. Anal apps. only half as long as vulvar lamina.

Venation. Anx ♂ 6–8/5–6, ♀ 7/5; pnx ♂ 6–8/7–9, ♀ 6–7/7–8; two rows of post-trigonal cells in fore wings, followed by 3 rows, which begin about the level of 1 cell beyond the origin of R_{4+5}, increasing to 4 rows; cells between fork of R_2 ♂ 13–22/14–25, ♀ 22–24/24–25; cells between IR_3 and Rspl usually 6 to 8; pt. about 4 times as long as wide.

Measurements. Total length ♂ 48.5–51.0, ♂ 50.0; abd. ♂ 33–35, ♀ 35; h.w. ♂ 33.0–35.0, ♀ 34.5; h.f. ♂ 7.5–8.0, ♀ 8.0; pt. ♂ 2.3–2.7, ♀ 2.6.

Nymph. Unknown (1972).

Habitat and range. Small streams in woods. Ohio, Ind. and Iowa, s.w. to Okla., Colo. and Mont., n. to s. Sask., Man. and Ont.

Distribution in Canada. Ont.—Simcoe county (Stroud, Aug. 8, 1959, 1 ♂, E.M. Walker). *Man.*—Onah; Westbourne. *Sask.*—Maple Creek.

Field notes. Very little has been added to our knowledge of the habits of *S. ensigera* since Williamson's paper (1922b) was published. This is a species of small streams, Williamson having taken the type specimen of *S. charadraea* flying over a mountain torrent in Bear Canyon, Colorado (Williamson, 1907b). His other specimens were taken along small forest streams in Wells county, Indiana. One of these streams known as Davis Creek has been described by Williamson (1922b) as a habitat of both *S. charadraea* (= *ensigera*) and *S. linearis.*

Williamson refers to it as a "fine little stream three to eight feet wide, flowing mostly over gravel, with many gentle ripples and frequent pools, some of the latter almost waist deep." "The single female of *charadraea* seen ovipositing was flying back and forth tapping her abdomen on a damp clay surface at the edge of the creek and about a foot above the water. Occasionally a *Somatochlora* will fly along, striking the water with its abdomen and rarely throwing itself into the water, but in every case where positive observation was possible these individuals were males." "The males of *linearis* were never observed fighting, but on two occasions two males of *charadraea* were observed to fly at each other and fall to the ground in a rough and tumble scrimmage. In flying the creek males of *charadraea* habitually fly at a lower level than males of *linearis*" (Williamson, 1922b).

Tenerals apparently of both these species were observed hawking at an elevation of 20 to 30 feet, "at sunny openings among the trees over or near the creek," but even at this stage they appear not to wander from their breeding places.

The few Canadian records we have are unaccompanied by observations except the following in the case of the specimens from Maple Creek, Saskatchewan. The data for these specimens are as follows: June 14, 3 ♂ ; July 2 and 3, 3 ♀ ; and July 24, 1 ♂. Specimens taken on the first and last dates are labelled "creek," whereas those taken in early July are "from tumbling mustard on farm." The females are all somewhat juvenile, the males mature.

Since the latest date for the Canadian specimens is July 24, the flight period in Saskatchewan as indicated by this material is June 14 to July 24. On the flight period in Indiana, Williamson (1922b), writes as follows: "The imaginal life of the two species, *linearis* and *charadraea*, in northern Indiana, is thus about thirty days or a little more, including the last few days of June and practically the entire month of July. Their period of oviposition coincides with the time of rapidly falling water level in the creek, thus exposing successive clay banks and fine gravel bars on which the eggs are placed while the surface is moist, thus insuring the distribution of eggs over practically the entire creek bed. Oviposition was observed only where the forest, a heavy second growth mostly of white elm, lay on both creek banks."

Somatochlora tenebrosa (Say). (Pl. 7: 3, p. 58; pl. 8: 11, 12, p. 68; pl. 11: 6, p. 86; pl. 12: 6, p. 92; pl. 18: 3, p. 124; pl. 19: 3, 7, p. 128)

   Libellula tenebrosa Say, J. Acad. Phila., 8: 19, 20, 1839.
   Cordulia tenebrosa: Hagen, Syn. Neur. N. Amer., p. 137, 1861.

Cordulia tenebrica: Hagen, Syn. Neur. N. Amer., p. 138, 1861 (no desc.).
Epitheca tenebrosa: Selys, Bull. Acad. Belg., (2) 31: 289 (reprint p. 55), 1871.
Somatochlora tenebrosa: Selys, Bull. Acad. Belg., (2) 45: 206, 218, 1878.
Somatochlora linearis: Wadsworth, Ent. News, 9: 111, 1898.
Somatochlora provocans: Muttkowski, Bull. Wis. Nat. Hist. Soc., 8: 176–179,
    1910a.

A very dark and rather slender species, the male easily known by the high
and angular arches of the sup. apps. and the narrow strongly curved form of
the inferior appendage. The female is not so easily recognized, the vulvar
lamina being much like that of *minor* and *williamsoni*, but less tapering, and
the labrum being mostly orange instead of black.

Male. Face dark yellow to brown, labium and labrum often orange, the latter with a black
margin and median basal spot; frons, except lower margin and sides, black with metallic
reflections above and on upper part of antefrons, vertex brownish purple, occiput dark yellow,
brown or black with a brown centre, hair of face dark brown. Pterothorax coppery bronze with
metallic reflections, lateral spots pale yellow, distinct but usually not sharply defined,
mesepimeral spot 4.0–4.5 mm. long and 1 mm. wide, metepimeral spot a little shorter but wider
than the mesepimeral, interalar spots pale yellow, underside of thorax and bases of legs and a
streak on outer surface of fore femora light brown. Hair of thorax pale brownish, thin and
short. Wings hyaline or nearly so; costal margin dark, venation blackish; pt. blackish, or smoky
brown; membranule dark smoky brown with a well defined whitish area at base. Interalar space
with a large pale area between bases of front and hind wings. Seg. 1 of abdomen dull brown; 2
and base of 3 dark brown, polished on the sides, 2 with pale yellow areas as follows: a large
anterolateral blotch and a smaller posterolateral spot extending upon the genital lobes; between
2 and 3 a broad pale intersegmental area, 3 with small basal dorsal yellow spots and a larger
ventral basal yellowish area, the remaining segments black without yellow spots; hairs dark
brownish, sparse except on segs. 7 to 9.
    Abdomen (excluding apps.) a little shorter than hind wing, slender, widest at end of 5 and on
6, narrowing very gradually distad.
    Genital lobes well rounded, about as wide as long, hamuli (pl. 11: 6, p. 86) hooked, apices
bluntly pointed. Sup. apps. (pl. 8: 11, 12, p. 68) divergent, strongly and angularly arched, distal
limbs directed downward, convergent except at apices; inf. app. (pl. 8: 11, 12, p. 68) about
three-fourths as long as sup. apps., strongly curved upward, distal third almost vertical, sides
subparallel near the apex, which is blunt and rounded.

Female. Coloration almost as in male. Wings hyaline or lightly washed with amber. Abdomen
longer than in male, being distinctly longer than hind wing.
    Anal apps. and vulvar lamina black. Seg. 9 in mature females considerably compressed and
somewhat constricted; vulvar lamina (pl. 12: 6, p. 92) strongly projecting but not quite
perpendicular to the long axis of the body, triangular in profile with straight sides, which are
almost as long as 9 + 10, the base equal to about one-half the length of the front margin; apex
bluntly pointed; anal apps. longer than 9 + 10, equal to about one-half the length of the front
margin.

Venation. Anx ♂ 9–11/5–6, ♀ 7–10/5–6; pnx ♂ 5–8/7–9, ♀ 5–7/6–9; two rows of post-trigonal
cells in fore wings, generally followed by 3 rows, which begin at a distance of 1 to 3 cells beyond
the origin of MA, 2–4 cells at margin of wing; cells in fork of R_2 ♂ 8–14/10–17, ♀ 12–19/12–20;
cells between IR_3 and Rspl usually 6 to 9; pt. 4 times as long as wide.

Measurements. Total length ♂ 48–55, ♀ 51–64; abd. ♂ 32.5–38, ♀ 38–45.5; h.w. ♂ 34–38, ♀
35.5–40.5; h.f. ♂ 7.5–8.0, ♀ 8.5; pt. ♂ 2.4–2.5, ♀ 2.3–2.8.

PLATE 12

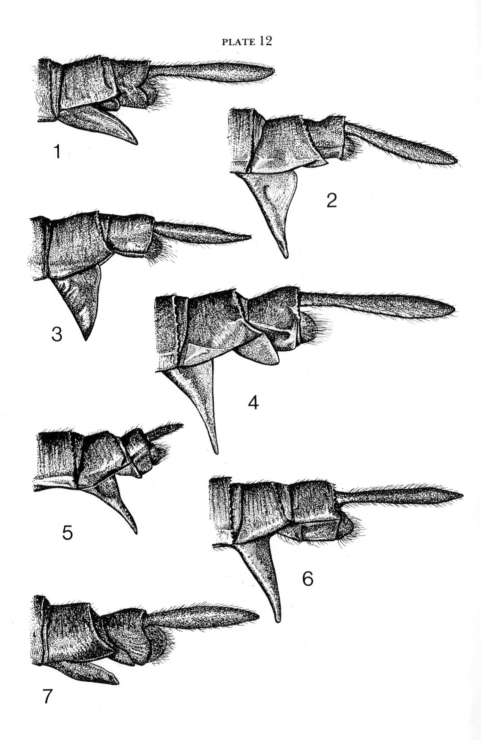

Nymph. Uniform dark brown, not very hairy. Lateral margins of head arcuate, hind margin moderately concave; fringe of hairs between bases of antennae, another between eyes in line with their hind edges; posterolateral margins of head with thick tufts. Hinge of folded labium reaching base of mesosternum; prementum nearly as wide as long; premental setae 11 to 12, the 3rd to 6th from the outside longest; palpal setae 8; distal margin of palpus with 9 or 10 crenations, middle ones about three times as long as deep, and bearing 6 or 7 spiniform setae in a graded series, the lowest of which are generally the longest and are about as long as their corresponding crenations. Hind wing-sheaths reaching nearly to middle of seg. 6. Hind femora about as long as width of head across the eyes or three-fourths the width of abdomen; hind tibiae one-third longer. Abdomen (pl. 18: 3, p. 124) widest at seg. 6 with numerous fine hairs, longest on 9; lateral spines on segs. 8 and 9, those on 8 about one-sixth the length of the lateral margin of that segment (not including the spine), slightly divergent; those on 9 not divergent, about one-fifth as long as the corresponding segment margin and about 3 times as long as their basal width; dorsal hooks (pl. 19: 7, p. 128) on segs. 4 to 9 or 5 to 9, vestigial on 4 when present, very small on 5, well developed on remaining segments, being about as long as the segments that bear them; similar to those of *S. minor* and *S. elongata*, though slightly less elevated and less arcuate; the hook on 9 projects well over the hind margin of seg. 10; epiproct (pl. 19: 3, p. 128) about as long as mid-dorsal length of 9 + 10, triangular, one-fifth longer than wide, in profile somewhat concave above in male with a slightly elevated ante-apical prominence; cerci a little longer, the apices rather abruptly pointed and somewhat bent inward; paraprocts barely longer than the other appendages, straight and acute.

Length 20; abd. 13; w.abd. 8; h.w. 6.5; h.f. 6; w.hd. 6.

Habitat and range. Small forest streams. N.S. and Me. to central Ont., Mich., Iowa and Mo., and s. to S.C., Ala. and Miss.

Distribution in Canada. N.S.—Annapolis county; "Nova Scotia" (Hagen, 1861). *Que.*—Masham; Lanoraie. *Ont.*—Norfolk, Halton, Wellington, Ontario and Simcoe counties; Nipissing (Algonquin Park) and Timiskaming districts.

Field notes. This is another inhabitant of small forest streams with intermittent rapids and pools. It seems to be more partial than other species to shade, as we noted on the few occasions when we have seen this insect in flight. It was particularly noticeable on August 29, 1939, when we saw two males flying over the stream nearest De Grassi Point, the same stream where *S. minor* flies usually earlier in the season. The two males of *tenebrosa* were apparently keeping entirely in dense shade and flew higher than *minor*, at least two feet instead of two or three inches. Another occasion was at Highland Creek near Toronto, on July 6, 1941, when we captured a male in a dark wood, not far from the creek. *S. tenebrosa* was also noted by Williamson (1923a) in the Cumberland Highlands, Tennessee, flying until 7.30 p.m., half an hour after sundown. Wilson (1912) also observed ten or a dozen

---

PLATE 12

Somatochlora—vulvar lamina and anal appendages of females, left lateral view: (1) *S. walshii*; (2) *S. minor*; (3) *S. elongata*; (4) *S. williamsoni*; (5) *S. ensigera*; (6) *S. tenebrosa*; (7) *S. franklini*. (1–7 after Walker, 1925.)

"patrolling back and forth just after sunset in one corner of an old pasture near a small brook at the foot of the mountains. They were strong and rapid fliers and extremely difficult to capture."

The flight period is probably the latest of all our species of *Somatochlora*. Our records, which are not many, range from July 1 to September 9, the majority being in August. The only September record is for Nova Scotia, but September records are common in other parts of the range of *tenebrosa*.

The only record of oviposition we have found is a note by Williamson (1923a), who observed the operation at a small stream near Tracy City, Tennessee. Referring to several small streams which are "slow flowing in sandy beds and are overgrown with brush," he states that at one of these streams "a female was ovipositing in shallow water by dipping the abdomen."

Variations. A male and two females of *S. tenebrosa* from Shannon county, Missouri, received from the late E.B. Williamson, are somewhat larger than any of our specimens from Canada. The male is the same in total length (55.5 mm.) as our largest male, which comes from the Timagami Provincial Forest, Nipissing district, Ontario, but the hind wings are slightly longer (Timagami, h.w. 38 mm., Shannon, 40 mm.). The females from Shannon, however, are distinctly larger than any from Canada, as indicated by the following measurements (in mm.), which are compared with those of the two largest females from Canada that we have seen.

	Total length	abd.	h.w.
Shannon, Missouri	65.0	46.0	43.0
Shannon, Missouri	64.0	46.0	42.5
De Grassi Point, Ontario	59.0	41.0	38.0
Masham Mills, Quebec	56.0	41.0	39.0

Somatochlora franklini (Selys). (Pl. 9: 1, 2, p. 76; pl. 11: 7, p. 86; pl. 12: 7, p. 92; pl. 16: 1, p. 112; pl. 18: 4, p. 124)

Cordulia franklini Hagen, Syn. Neur. N. Amer., p. 138, 1861 (no desc.).
Epitheca septentrionalis: Selys, Bull. Acad. Belg., (2) 31: 298 (reprint p. 65), 1871.
Somatochlora franklini Selys, Bull. Acad. Belg., (2) 45: 195, 205, 217, 1878.
Somatochlora septentrionalis: Harvey, Ent. News, 12: 275, 1901.
Somatochlora macrotona Williamson, Ent. News, 20: 77–79, 1909a.

A small, short-winged species with a slender abdomen, which is very elongate in the male.

Male. Labium and anteclypeus yellow; labrum, entire frons, except a dark yellow spot on each side vertex and occiput, black with greenish reflections, hair of dorsal parts of head blackish and rather dense; rear of head black with a pale fringe of hair. Pterothorax brassy, covered with long pale brownish hairs; mesepisternum with a large orange spot at lower end and a smaller one above in front of the ante-alar sinus; mesepimeron with an oval yellow spot, conspicuous in tenerals; metepimeron without a definite spot, very hairy, legs black, wings hyaline or with a slight wash of yellow, the hind pair with a dark brown spot on the anal triangle; venation blackish, membranule with the basal third pale gray. Segs. 1 and 2 of abdomen dark brown; 2 with a large postero-dorsal and a smaller anteroventral yellowish spot. Segs. 3 to 10 greenish black with a dull lustre; 3 with pale ventral margins. Hair fairly thick on sides of 1 and 2, scanty elsewhere.

General form slender, thorax small, abdomen more than 4 times as long as thorax and one-fourth longer than hind wing, very slender at 3, but expanding to end of 9.

Genital lobes subangular below, strongly curved mesad. Hamuli (pl. 11: 7, p. 86) moderately bent at almost a right angle, the apices somewhat abruptly and bluntly pointed. Sup. apps. (pl. 9: 1, 2, p. 76) as long as 9 + 10, slender sinuate with apices convergent, not reflexed. Inf. app. (pl. 9: 1, 2, p. 76) slightly more than half as long as sup. apps., of the usual triangular form with blunt apex bearing a reflexed spine.

Female. Head and thorax similar to those of male; abdomen of variable length, usually much shorter than that of male (heteromorphic form) but often more or less elongated as in male (homomorphic), the most male-like forms even having small genital lobes. In coloration the female differs from the male as follows: lateral spot of 2 smaller, dorsal spot more obscure; a similar obscure anterodorsal spot and a large pale ventrolateral basal spot on 3. Tergal margins more or less narrowly edged with yellowish, this colour becoming more general on 8 and 9, including the vulvar lamina. Wings with basal brown spot sometimes more diffuse; the yellowish or brownish pigment sometimes extending along the costal borders, and sometimes covering the greater part of the wings. Tenerals frequently have the wings entirely smoky.

Abdomen generally much shorter, being scarcely one-sixth longer than hind wings, less slender at seg. 3 and showing the usual ontogenetic changes. Vulvar lamina (pl. 12: 7, p. 92) horizontal, reaching nearly to end of sternum of 9, less than twice as long as wide, scoop-shaped, distal edge broadly rounded. Anal apps. about one-fourth, or less, longer than vulvar lamina and about as long as 9 + 10.

Venation. Anx ♂ 7-8/5, ♀ 6-8/4-5; pnx ♂ ♀ 5-7/6-8; only two rows of post-trigonal cells in fore wings, often narrowing to 1 row at the wing margin; cells between fork of R2 ♂ 6-10/6-10, ♀ 6-9/7-9; cells between IR3 and Rspl usually only 3 to 5, sometimes 6 or 7; pt. 4 or 4.5 times as long as wide.

Measurements. Total length ♂ 47-54, ♀ 44-55; abd. ♂ 33.0-39.5, ♀ 32-38; h.w. ♂ 25-28, ♀ 26.6-29.5; h.f. ♂ 6.6-7.0, ♀ 7.0-7.5; pt. ♂ 1.9-2.2, ♀ 2.0-2.5.

Nymph (pl. 16: 1, p. 112). Small, uniform light brown and very hairy. Dorsum of head with numerous short hairs besides the usual fringes of coarser hair which are rather long and thick particularly on the clypeus, basal joints of antennae and behind the eyes. Folded labium reaching posteriorly to middle of mesocoxae, not covering any part of the eyes; prementum about as wide as long; premental setae 13, the 4th to 6th from lateral margin longest, the mesal 4 or 5 short; palpal setae 7 or 8; upper edges of palpi very hairy; distal margin of palpi with 10 crenations, of which the larger are almost three times as long as high and bear groups of about 4 spiniform setae, the lowest usually much larger than the others, being about as long as the crenation which bears it. Pronotal and propleural processes of thorax with heavy tufts, and thoracic ridges also heavily fringed; hind wing-sheaths reaching middle of seg. 6, their outer surfaces hairy. Legs heavily beset with coarse hairs, short, the hind tibiae being about as long as width of head, hind femora barely shorter. Abdomen (pl. 18: 4, p. 124) rather long oval, widest at segs. 5 and 6, sides smoothly arcuate; dorsal surface with numerous hairs, generally distributed but varying in length, those along the hind tergal margins forming a coarse fringe, which on seg. 9 is fully as long as the mid-dorsal length of that segment; lateral margins of segments bearing the usual fringe of soft hairs, which on sides and margins of 9 and on 10, are very long and numerous; dorsal hooks absent; lateral spines represented only on seg. 9 by a small tooth-like spine; epiproct a little shorter than mid-dorsal length of 9 + 10, slightly longer than its basal width, apex acute but not acuminate; cerci about one-fifth longer, tapering from basal

PLATE 13

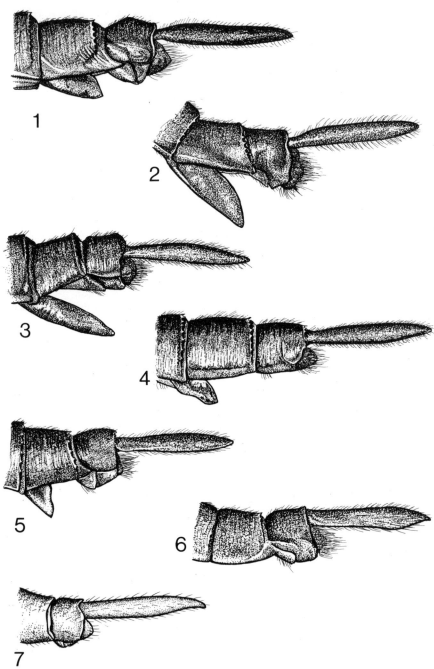

third to apices, greatest width at about one-third of their length; paraprocts barely longer, acute.

Length 17.5; abd. 10.0; w.abd. 6.5; h.w. 5.5; h.f. 5.2; w.hd. 5.6.

Habitat and range. Spring-fed sphagnum bogs. Labr. to Hudson Bay and Mackenzie District, N.W.T., s. to N.B., Me., s. Que., s. Ont., n. Mich., Minn., s. Man., Sask., Rocky Mts. of Alta., B.C. and Yukon T.

Distribution in Canada. Nfld.—(Whitehouse, 1948). *Labr.*—Great Caribou Is. (Battle Harbour); North West River (Hamilton Inlet); Hopedale; Nain. *N.B.*—Charlotte county; *Que.*—Sherbrooke; Lanoraie; Nominingue; Abitibi county; Rupert Bay, Eastmain and Ft. George (James Bay); Ft. Chimo. *Ont.*—Simcoe and Carleton (Mer Bleue) counties; Thunder Bay ("Sucker River" (presumably Suckle Creek), Longlac; Black Sturgeon Lake), Cochrane (Moose Factory; Ft. Albany) and Kenora (Patricia Portion: Favourable Lake; Cape Henrietta Maria; Ft. Severn) districts. *Man.*—Sandilands; Waugh; Assiniboine River, near Aweme; Onah; Winnipeg; Winnipeg Beach; Husavick; Victoria Beach; Berens River; Norway House; The Pas; miles 200, 214, 256 and 332, Hudson Bay Rlwy.; Gillam; Churchill. *Sask.*—Seward; Saskatoon; Waskesiu; Wollaston Lake (D.S. Rawson). *Alta.*—Simpson; Healy Creek; Banff; Boom Creek; Red Deer; Nordegg (6,500'). *B.C.*—Field; Atlin. *N.W.T.*—Great Slave Lake (Ft. Resolution; Caribou Is.); Ft. Simpson.

S. franklini appears to be most abundant in Manitoba, where it has been taken from south of Winnipeg (Sandilands and Treesbank) northward to Churchill at the tree-line. Although mainly a species of the far north, it is found in great numbers at Victoria Beach at the southern end of Lake Winnipeg, as well as much farther north at The Pas. It also occurs eastward, e.g. on Hudson Bay and in Labrador but west of the Rocky Mountains it has been found only at Atlin (60°N.) and probably does not occur in southern British Columbia.

Field notes. The flight period varies greatly in different parts of the country. In general it is earliest at the southern boundary of the species' range but later farther north or at higher altitudes. For instance, at the Mer Bleue, a "boreal island" near Ottawa, Ontario, adults have been taken as early as May 23, whereas those taken by Whitehouse at Nordegg, Alberta (6,500') on July 11, 1917, were still teneral. I took a teneral male at Banff, Alberta, on July 9 and found adults at their breeding place there on August 5, 1912, our

---

PLATE 13

Somatochlora—vulvar lamina and anal appendages of females, left lateral view: (1) *S. kennedyi*; (2) *S. forcipata*; (3) *S. incurvata*; (4) *S. semicircularis*; (5) *S. whitehousei*; (6) *S. septentrionalis*; (7) *S. sahlbergi.*
(1–6 after Walker, 1925.)

earliest record (from Hamilton Inlet, Labrador) being dated July 27, and the others, from Nain and Hopedale, Labrador, having dates from August 21 to 29. The great majority of our specimens, however, were taken during July, although there were some even from Rupert Bay, James Bay, Ontario, taken as early as June 6 to 21.

"At both Banff and Field *franklini* was found flying over mossy bogs at the foot of a mountain. The bog at Field was of small extent and was formed by seepage from a cold spring-fed stream at the foot of a wooded mountain slope. Two deep pockets of spring water added their quota to the stream, although the Somatochloras did not breed here but in the little puddles in the spongy bog. Several males were seen pursuing an irregular course over the bog at a height of two or three feet. One or two female Somatochloras were observed ovipositing but the only one captured proved to be *S. semicircularis*. About the edge of one of the puddles I found [an] exuvia of two species of *Somatochlora*, both of the *arctica* group. There were five of one species, recognizable as *semicircularis*, and only one of the other, but I also found an early full grown nymph of this form in the puddle. From its small size I have no doubt that it is the nymph of *franklini*, in fact all other regional species are excluded. It is noticeably hairier than *semicircularis*" (Walker, 1925).

Whitehouse (1941) found *S. franklini* on July 14, 1939, in the vicinity of Atlin, British Columbia "on a tiny streamlet that drained a large muskeg pool." He netted a male and female, the latter ovipositing, and both "mature but still young." He visited the same place again on July 17 and again took a male and saw a female, "but subsequent visits on July 26th and August 9th were most disappointing, no more being seen." The oviposition of *S. franklini* has not been described but is undoubtedly the same as that of other species of *Somatochlora* in which the vulvar lamina is not compressed or projecting.

Somatochlora kennedyi Walker. (Pl. 9: 3, 4, p. 76; pl. 11: 8, p. 86; pl. 13: 1, p. 96; pl. 16: 2, p. 112)

> *Cordulia (Somatochlora) forcipata*: Harvey, Ent. News, 3: 116, 1892 (in part).
> *Somatochlora forcipata*: Harvey, Univ. Maine Stud., 4: 9, 1902 (in part).
> *Somatochlora semicircularis*: Williamson, Ent. News, 17: 136, 138, pl. V, figs. 2–4, 1906a (in part).
> *Somatochlora kennedyi* Walker, Can. Ent., 50: 366, 371, pl. X, 1918.

A species of medium size and somewhat elongate form, without conspicuous spots, related to *franklini* but larger, and the male less elongate but with longer wings.

Male. Face, including labium, orange to brownish yellow; labrum, middle parts of postclypeus and of antefrons, postfrons and vertex black with a greenish lustre, occiput polished dark brown; rear of head black with submarginal fringe of pale brown hairs, pile of face fairly dense. Pterothorax bronzy with green and bluish reflections, varied with brown chiefly at upper end. Lower ends of pleura, about the metaspiracle and on posterior half of the epimera. Mesepimeral spot ovate, scarcely paler than the other reddish-brown areas, with which it may appear confluent. Metepimeral spot indistinct or absent. Hair of thorax dense, pale brownish, obscuring brightness of thorax. Thorax below and bases of legs reddish brown, legs distal to trochanters black. Wings hyaline usually with a light wash of yellow deepening to amber at bases of hind pair. Venation dark brown, costal veins edged with yellow; pt. yellowish brown; membranule whitish on basal third or fourth. Segs. 1 and 2 of abdomen dark brown, 1 with sides and hind-margin paler, 2 with a large dull-yellow anterolateral spot below the auricles, usually somewhat confluent with a smaller posterolateral spot. Seg. 3 marked as in *franklini*; remaining segments greenish black without yellow spots. Hair dense on dorsum of 1 and 2, moderately developed on the other segments.

Thorax stouter than in *franklini*. Abdomen slightly longer than hind wings, enlarging to end of 6, then narrowing gradually to the end.

Genital lobes subangulate ovate, somewhat bent mesad. Hamuli (pl. 11: 8, p. 86) bent almost at right angles, the lower margin smoothly arcuate, apices bluntly pointed. Sup. apps. (pl. 9: 3, 4, p. 76) slender, smooth, distally tapering to mesally curved apices, which meet at an acute angle with an unbroken lateral carina, in profile horizontal to the apices; inf. app. (pl. 9: 3, 4, p. 76) of the usual triangular form, three-fifths as long as sup. apps.

Female. Similar to male in coloration, with the following differences: yellowish spots of abd. seg. 2 larger and more diffuse, there being a single pair of lateral spots and a pair of posterodorsals or these may be units from a single blotch on each side; the basal dorsal spots of 3 are much larger than in male.

Vulvar lamina (pl. 13: 1, p. 96) about four-fifths as long as the sternum of 9, not elevated, scoop-shaped, widest at base, the free edge broadly rounded. Anal apps. slightly longer than 9 + 10, tapering more abruptly at apices than at base.

Venation. Anx ♂ ♀ 7–9/5(6); pnx ♂ ♀ 6–8/7–9; two rows of post-trigonal cells in fore wings, followed by 3 rows, beginning usually with 1 or 2 cells beyond the origin of MA, often narrowing again to 2 rows before the margin; cells between fork of R2 ♂ 13–19/12–19, ♀ 11–18/18/12–20; cells between IR3 and Rspl usually 5 to 8; pt. 4.5 to 5 times as long as wide.

Measurements. Total length ♂ 51.5–55.0, ♀ 47–55; abd. ♂ 34–38, ♀ 32–38; h.w. ♂ 29.5–32.0, ♀ 29.5–33.5; h.f. ♂ 6.5–7.8, ♀ 7.2–7.4; pt. ♂ 2.3–2.6, ♀ 2.4–3.0.

Nymph (pl. 16: 2, p. 112). Uniform brown, moderately hairy. Head with a hairy fringe between bases of antennae, another short transverse fringe between the eyes, a heavier longitudinal fringe mesad of each eye and a somewhat slighter lateral one, both joining the posterior fringe; the whole dorsal surface of head also covered with shorter hairs. Hinge of labium reaching posteriorly as far as middle of mesocoxae; prementum about as broad as long; premental setae 12 to 13, the 4th to 6th from the outside longest, the innermost 3 or 4 small; a small group of minute setae on outer margins; palp setae 9, rarely 10; distal margin of palpus with 9 or 10 crenations, the largest one about 3 times as long as high and bearing groups of about 4 spiniform setae in a graded series, the longest of which are three-fifths to three-fourths as long as the crenations bearing them; upper margins of palpi hairy. Pronotal and propleural processes of thorax with heavy tufts of hair; thoracic ridges and hind wing-sheaths also fairly thickly fringed. Legs short, moderately hairy, hind femora about equal to width of head across eyes, hind tibiae barely longer. Abdomen long-oval, rather abruptly narrowed on seg. 9; dorsal surface covered with small hairs and with a fringe of hairs along the posterior tergal margins of each segment, increasing in length caudad on the middle part of each segmental margin, particularly on segs. 7, 8 and 9, and a fringe along the lateral margins of all the segments and the posteroventral margin of 9, increasing greatly in length in the last segments; fringe of hair on hind margins of tergites 7 and 8 forming a distinct median tuft; dorsal hooks absent; lateral spines represented on 9 only by a pair of small denticles; epiproct nearly equilateral, barely acuminate and ending in a small sharp spine; cerci projecting farther back, about four times as long as basal width; paraprocts a little longer than the other anal appendages, acute, slender-pointed.

Length 21; abd. 10; w.abd. 7.3; h.w. 6.5; h.f. 6; w.hd. 6.3.

Habitat and range. Flowing water in the open (Robert, 1953); also sometimes cold bog ponds. N.B. to James Bay and N.W.T., s. to Mass., N.Y., Mich., Wis. and s. Man.

Distribution in Canada. N.B.—Charlotte, Queen's, Sunbury and York ("Yaho," presumably Yoho) counties. *Que.*—Knowlton; Iberville; Lanoraie; Hull; Ironside; Mont Tremblant Park; Nominingue; Godbout; La Ferme. *Ont.*—York, Simcoe, Carleton and Haliburton counties; Muskoka, Parry Sound, Nipissing, Algoma and Cochrane districts. *Man.*—Victoria Beach; Dauphin Lake; Berens River; Norway House. *N.W.T.*—Ft. Smith; Great Slave Lake (Ft. Resolution; Caribou Is.); Ft. Simpson; Norman Wells.

Field notes. Although we have taken *S. kennedyi* on many occasions, these have usually been during the period of wandering flight and we have never been satisfied as to the nature of its preferred breeding haunts. We described the habitat (Walker, 1941) as "cold bogs and swamps" and selected for more detailed description (Walker, 1925) a shallow pond in a swampy wood, near St. Andrews, New Brunswick, where both males and ovipositing females were present in fair numbers. Robert (1953), however, states that flowing waters in the open are doubtless the preferred habitat of this species, and this may be true. Or it might be described more specifically as slow streams flowing through open bogs or marshes, such as may be found in the Mer Bleue near Ottawa, and formerly the Holland River marsh near Bradford, Ontario. We have also taken *S. kennedyi* and observed females ovipositing in ponds in the Godbout Valley, Quebec (Walker, 1923).

This species appears early, the flight period in southern Ontario coinciding approximately with the month of June. The earliest and latest dates from the Mer Bleue are June 5 and July 11, but in Quebec it has been taken at Hull or Ironside as early as May 28 and at Lanoraie as late as July 20. Farther north the flight period is later, e.g., at Ko-Ko-Ko Lake, Timagami Provincial Forest, we took a pair on August 5, 1923. Robert's (1953) dates for Mont Tremblant Park, Quebec, are June 27 to July 8, but he remarks that the period is doubtless much longer than this. The records from Manitoba and the Northwest Territories are all from late June and the first half of July.

The males patrol at a height of about three feet above the water (Robert, 1953) or a little above two feet according to our observations at the St. Andrews pond. They hovered over the water chiefly on the colder part of the pond, which was partly constricted off from the main body of the pond and was shadier. They hovered in one spot to a considerable extent, but changed their orientation frequently and did not remain long in one place. They settled quite frequently on low bushes in an oblique position. "One female, observed closely at this pond about 4.30 p.m. on July 9, was flying over the shady end of the pond, keeping about two inches above the water among the thick growth of horse-tails and sedges, and choosing spaces a few square inches in area, in which she had room enough for her operation. She held her abdomen nearly horizontal but slightly drooping, striking the surface of the water at rather irregular intervals, about every three to five seconds on an average, moving after a few strokes to another place, sometimes only a few inches away, sometimes several feet. Often she would fly away a few yards and oviposit there and then return to the original spot" (Walker, 1925).

Somatochlora forcipata (Scudder). (Pl. 9: 5, 6, p. 76; pl. 11: 9, p. 86; pl. 13: 2, p. 96; pl. 16: 3, p. 112)

Cordulia forcipata Scudder, Proc. Bost. Soc. Nat. Hist., 10: 216, 1866.

Epitheca forcipata: Hagen, Proc. Bost. Soc. Nat. Hist., 15: 376, 1873.

Somatochlora forcipata: Selys, Bull. Acad. Belg., (2) 45: 194 (reprint pp. 16, 27, 39), 1878.

A small slender dark species with two pale rounded spots on each side of the thorax, and an abdomen of average length.

Male. Head marked as in *kennedyi*, thorax darker, reddish brown, with metallic reflections, strongest in tenerals, sides darker around the two conspicuous yellow spots, which are somewhat angulate above, rounded below, the first or mesepisternal spot longer than the second or metepisternal spot. Underside of thorax and bases of legs including more or less of fore femora pinkish or brownish gray, legs otherwise black. Wings hyaline, stained with yellow only on the anal triangle; pt. dark yellowish; membranule with basal third whitish. Abdomen sparsely or moderately pubescent; seg. 1 dull grayish brown; 2 polished dark brown with two large ventrolateral yellowish spots, sometimes connected, the second spot encroaching on the genital lobes; other spots as in *kennedyi*; segs. 3 to 10 greenish black, with basolateral yellow spots on 5 to 8, usually largest on 6 and very small on 8.

Thorax slender, abdomen a little longer than hind wing, slender, widest at near end of 5, thence tapering to base of 9.

Sup. apps. (pl. 9: 5, 6, p. 76) about as long as 9 + 10, more arched than in other related species, distal third bent mesad at an obtuse angle, with flattened apices; ventral carina making a rounded prominence near basal fourth and a large blunt ventral "tooth" as seen in lateral profile somewhat beyond midlength. Inf. app. (pl. 9: 5, 6, p. 76) nearly two-thirds as long as sup. apps., of the usual triangular form with the base equal to about one-half the length.

Female. Coloration as in male with the following differences: seg. 2 with paired pale yellow streaks at the transverse carina and a small anterolateral spot with which the streaks are often connected. Basal lateral yellow spots on 3 to 7, decreasing in size caudad. Wings hyaline or slightly yellowish, especially along costal borders; pt. ochraceous.

Abdomen slightly longer than hind wing; in tenerals tapering from base of 3 to base of 9 or 10; seg. 9 with hind margin very oblique. Vulvar lamina (pl. 13: 2, p. 96) about as long as sternum of 9 or a little longer, horizontal or, in dried specimens, often directed slightly downwards, scoop-shaped with apex well rounded and entire; ventral surface slightly convex lengthwise. Anal. apps. about as long as 8 and much shorter than 7.

Venation. Anx ♂ ♀ 7–9(10)/5–6; pnx ♂ ♀ 7–9/5–9; two rows of post-trigonal cells in fore wings

PLATE 14

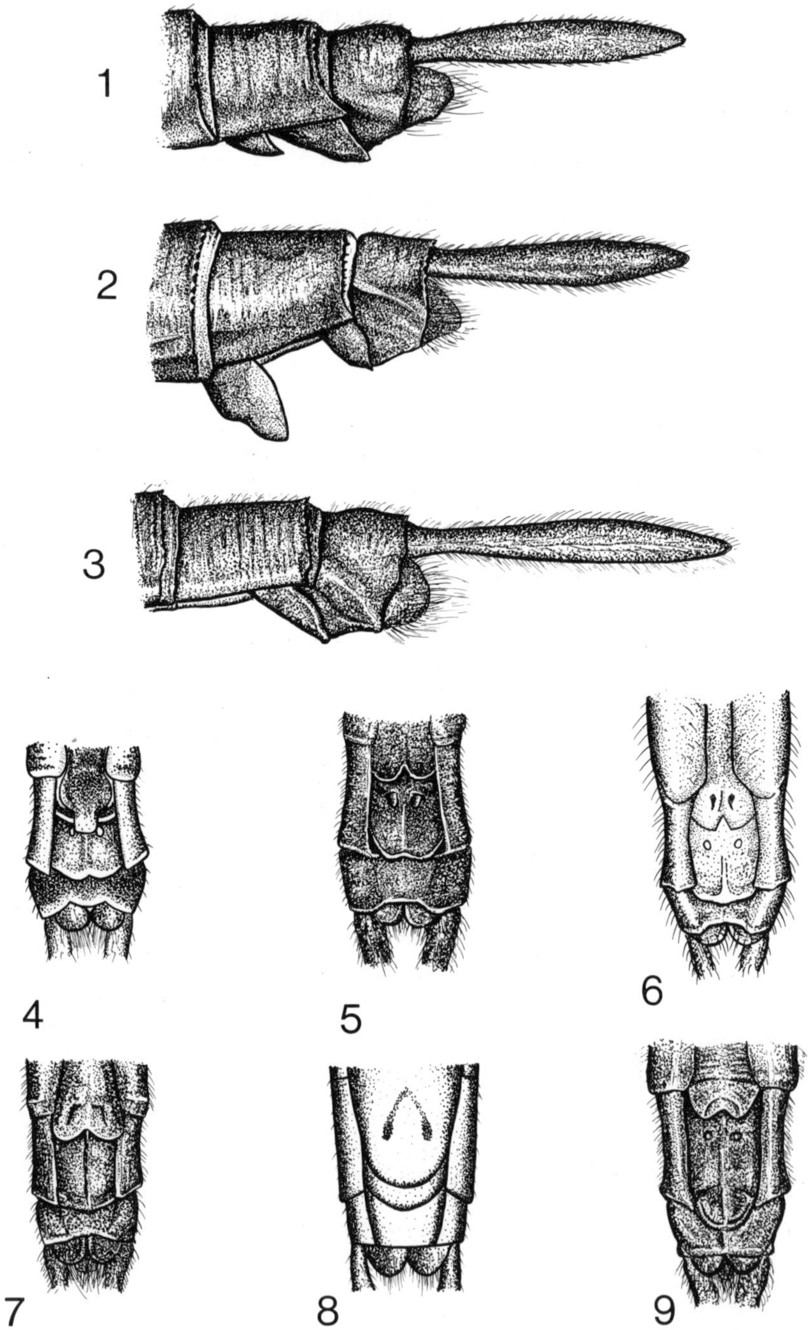

followed in some specimens by 3 rows; cells between fork of R₂ ♂ 11–21/10–19, ♀ 14–21/13–22; number of cells between IR₃ and Rspl 5 to 9.

Measurements. Total length ♂ 43.5–50.8, ♀ 46.5–51.9; abd. ♂ 30.3–35.0, ♀ 33–36; h.w. ♂ 29–32, ♀ 30–33; h.f. ♂ ♀ 6.0–6.8; pt. ♂ 2.2–2.5, ♀ 2.0–2.7.

Nymph (pl. 16: 3, p. 112). A small nymph, uniform dark brown, very hairy. Head of the usual form; fringes of hair across the dorsum as in *S. franklini*, rather coarse, the anteroposterior fringe mesad of each eye very heavy and a somewhat light lateral one, the whole dorsal surface of head also covered with shorter hair. Pronotal and propleural processes of thorax with heavy tufts and the thoracic ridges heavily fringed. Folded labium reaching posteriorly scarcely to level of mesocoxae, the eyes entirely uncovered. Premental setae 12 to 15, the 4th to 6th from the outside longest, the innermost 4 or 5 short; palpal setae 9, rarely 10; crenations on distal margin of palpus 7 to 9, rounded, each with 3 to 5 setae in a graded series, the longest nearly as long as the crenation that bears it, and much longer than the others. Wing-sheaths reaching to base or middle of seg. 6. Abdomen long-oval, widest between segs. 4 and 5, narrowing rather abruptly on 9; dorsal surface with numerous hairs of variable length, those on lateral and posterolateral margins of 9 and on dorsolateral parts of the hind margins of 6 to 9 elongated, forming in the latter situation a series of *somewhat conspicuous tufts on each side*; the dorsolateral surfaces of the more anterior segments tend to be clothed with longer hairs than elsewhere; dorsal hooks absent; lateral spines absent or represented on one or both sides of seg. 9 by a very minute spinule; epiproct of female about as long as mid-dorsal length of segs. 9 + 10 almost equilaterally triangular, apex rather short acuminate; cerci projecting farther than epiproct, by one-fourth or one-fifth the length of the latter, apices acute.
Length 19–20; abd. 11.5–12.5; w.abd. 6.6–7.1 (exuv.); h.w. 6.0–6.6; h.f. 5.0–5.5; w.hd. 5.5–5.8.
Male nymph unknown (1972).

Habitat and range. Small, spring-fed boggy streams. Labr. and Nfld. to Alta., Rocky Mts.; s. to Me., N.H., Vt., Pa., n. Mich. and Wis. Mainly eastern, being common in the eastern Provinces but taken w. of Ont. only twice, in Man. and in w. Alta.

Distribution in Canada. Nfld.—Grand Lake; White Bay. *Labr.*—Cartwright; Hopedale. *N.S.*—Halifax county; "Nova Scotia" (Hagen, 1861). *N.B.*—Charlotte and Northumberland counties. *Que.*—Covey Hill; Hull; Ironside; Lanoraie; Isle of Orleans; Saguenay River (Cap Jaseux); Godbout; Thunder River; Natashquan; Ashwanipi River; near Lake Mistassini. *Ont.*—Simcoe county; Nipissing (incl. Algonquin Park and Temagami Forest Reserve) and Cochrane districts. *Man.*—The Pas. *Alta.*—Simpson's Summit (B.C. boundary); Boom Creek. *N.W.T.*—Great Slave Lake (Ft. Resolution).

Field notes. Where *S. forcipata* and *S. kennedyi* are found in the same locality

---

PLATE 14

Somatochlora—vulvar lamina and anal appendages of females, left lateral view: (1) *S. albicincta*; (2) *S. hudsonica*; (3) *S. cingulata*.
Somatochlora—vulvar lamina, ventral view: (4) *S. whitehousei*; (5) *S. septentrionalis*; (6) *S. sahlbergi*; (7) *S. albicincta*; (8) *S. brevicincta*; (9) *S. cingulata*.
(1, 3, 5, 7, 9 after Walker, 1925; 8 after Robert, 1954b.)

the latter emerges first. Thus when we were visiting the Atlantic Biological
Station at St. Andrews, New Brunswick, in 1923, both species were fairly
common but whereas *S. kennedyi* had already returned to the breeding
haunts by June 24, adults of *S. forcipata* were still scattered through the
woods occupying any sunny opening such as a glade or road, whence they
would dart into the forest when disturbed. *S. forcipata* was absent from the
bog ponds where *kennedyi* appeared "but on July 17 we finally discovered a
breeding place in a small spring run in a little ravine adjoining a sloping
pasture. It was a tiny stream, for the most part only a few inches wide,
following a devious course and disappearing here and there under roots of
trees. Only two or three males (of *forcipata*) were seen here patrolling the
brook and hawking in the more open places, but a female was captured while
ovipositing in little pockets of water close to the stream and formed by the
hoof-prints of cattle. She was flying close to the water in the usual way,
dipping the surface with the abdomen. A large number of eggs were obtained and were kept alive in a spring at Toronto until December but were
finally lost by an accident. Full grown nymphs were found at Godbout in July
1918 and from these two females emerged at Toronto the following year on
May 29. They were found in small pockets or pools in the Godbout river
flats, which are filled at high tide but not completely submerged. The
Godbout river is a typical salmon stream and the pools were thus refilled at
every tide with cold clear water" (Walker, 1925). The water was entirely
fresh. While engaged in hawking at a distance from the breeding grounds, *S.
forcipata* is found in more or less shady roads or glades in the spruce forests,
usually flying at a height of four or five feet but sometimes higher. In early
life adults not infrequently settle on the branches of trees, often flying about
the foliage before coming to rest, apparently examining various spots before
deciding where to alight. These movements are usually upward, the insect
coming to rest at a height of 15 or 20 feet from the ground. Later in the
season their flight is more restless and they may be found patrolling the
more open spaces in the wood, flying back and forth a few feet from the
ground and occasionally darting away out of the depths of the forest returning after a few minutes' absence (Walker, 1925).

Our earliest records, May 27 and June 6, come from Ironside (Hull),
Quebec; the former date, if not both, would be very early for St. Andrews,
New Brunswick, where we made most of our observations. Our latest date
south of Labrador is August 27, 1904, when we took our first specimen of
forcipata – a male. The flight period seems to reach its climax in July in all
parts of its range except perhaps in Labrador, where it may be in August.
Most of our Labrador specimens of this and other Odonata were taken in
August.

Somatochlora incurvata Walker. (Pl. 9: 7, 8, p. 76; pl. 13: 3, p. 96)

  Somatochlora incurvata Walker, Can. Ent., 50: 365, 367, 1918.

Closely related to *S. forcipata* but larger, with the lateral spots narrower and becoming less distinct with maturity. Superior appendages of male less arched and vulvar lamina longer.

Male. Head marked as in *forcipata* with similar clothing of hair; pterothorax metallic blue-green, marked with dull yellow (clear in tenerals) or brown, the metallic reflections obscured by the brownish hair, which is longest on the dorsum; ante-alar sinuses, pleura above and below brownish, the lateral spots in mature specimens sometimes a little paler and more yellowish. The mesepimeral spot narrower and usually more elongate than in *forcipata*. The metepimeral spot shorter and wider than the mesepimeral, angular above when distinct; both of these spots clear-cut in young individuals, becoming obscure with age. Interalar areas dusky, yellowish. Legs black, brownish at base. Wings hyaline with venation dark brown, the costal veins edged with yellowish toward the bases. Pt. reddish brown; membranule pale, in basal third. Abd. segs. 1, 2 and base of 3 brown, mostly polished, somewhat paler on the sides and dorsum of 2 but not forming distinct spots in the few specimens studied. Remaining segments as in *forcipata*, including the laterobasal yellow spots on 3 to 7 or 3 to 8.

Abdomen slender, widest at end of 5, narrowing to end of 8 and again on 9 or 10.

Genital lobes small, bent mesad. Hamuli as in *forcipata*. Sup. apps. (pl. 9: 7, 8, p. 76) similar to those of *forcipata* but less arched, the bent apices being directed more exactly mesad, the ventrolateral prominence not affecting the outline in lateral view. Inf. app. (pl. 9: 7, 8, p. 76) only about half as long as sup. apps.

Female. Elongate as in male, tenerals tapering almost evenly from base to apex, very slender distally. Coloration similar to that of male, seg. 2 entirely brown, except a very small lateral spot and an apical ring which are pale brownish gray, the yellow spots on the abdominal segments usually larger than in male and often diffusely prolonged caudad. Wings hyaline, each with two basal amber streaks and a yellowish cloud, occupying the distal half or less, of the wing. In some individuals the entire wing is flavescent, but even in these the basal streaks appear deeper than the rest of the wing.

Vulvar lamina (pl. 13: 3, p. 96) elongate, extending in all females well beyond the distal margin of the 9th sternum as far as the 10th, horizontal, spoon-shaped, slightly tapering, upper edge before the apical curve slightly arcuate, inferior surface in profile barely convex.

Venation. Anx ♂ 8–9/5–6, ♀ 7–8/5–6; pnx ♂ 6/7–8, ♀ 5–7/6–9(10); two rows of post-trigonal cells in fore wings, followed by 5 rows usually at 1 cell beyond the origin of MA, then 2 rows again, with 2 to 4 cells at the margin; cells between IR3 and Rspl usually 6 to 8 (in both pairs of wings); pt. about 5 times as long as wide.

Measurements. Total length ♂ 55–58, ♀ 49.5–58.5; abd. ♂ 39.0–42.0, ♀ 35.5–43.0; h.w. ♂ 33.0–34.0, ♀ 31.5–36.5; h.f. ♂ 7–8, ♀ 7; pt. ♂ 3.0–3.1, ♀ 3.0–3.5.

Nymph. Unknown (1972).

Habitat and range. Sphagnum pools. N.S. and Me. (White, 1969) to Ont. and n. Mich.

Distribution in Canada. N.S.—Colchester (Truro), Halifax (Moser River), Kings and Queen's (White Point Beach) counties. Ont.—Parry Sound (Bass Creek, Frank Bay; Lake Nipissing) and Nipissing districts.

Field notes. Little has been added to our knowledge of this species since it was first described, apart from its discovery in Nova Scotia (more than a thousand miles east of the type locality), Maine and Pennsylvania.

The type and paratypes were taken in Chippewa county, Michigan, on

July 29 and August 4, 1916 on the shore of Lake Superior "flying over the beach, and in a clearing about a quarter of a mile away from the lake, in both cases flying with swarms of *Aeshna*, in some cases including thousands of individuals of both genera, though the Aeshnas were far more numerous than the Somatochloras. They were found swarming on the beach during the day, when the wind was off-shore, and in the clearing at the close of a warm day from about five o'clock until sundown or later. If the day were cold they would be entirely absent" (Walker, 1925). *Somatochlora franklini* and *S. williamsoni* were flying in company with *S. incurvata*.

We have taken only a single male on Bass Creek, Nipissing district, Ontario; this is a fair-sized forest stream, for the most part quiet, although there are gentle rapids here and there along its course. The specimen taken was a male and was first sighted over the water but flew to shore and alighted on a tree seven or eight feet above the ground. White (1969) watched several males as they hovered over small sphagnum-choked pools in a large bog in Acadia National Park, Maine. "They were quite aggressive, chasing away other males and also males of *Aeshna sitchensis* ... which were also frequent at the same location. The females ... oviposited singly in the sphagnum pools when they were not being pursued by males." In Clinton county, Pennsylvania, at the tamarack bog, a mature female was seen ovipositing alone in a small pool beneath leatherleaf bushes. "She flew furtively, close to the water, turning slowly about in an irregular fashion, and dipped her abdomen to the surface at closely spaced intervals. She visited several of these pools before being netted" (Shiffer, 1969).

The dates, including the specimens from Michigan, Maine and Pennsylvania, range from July 19 (Nova Scotia) to October 10 to 15 (Nova Scotia). The last record is extremely late for any species of *Somatochlora* in Canada; apart from this the latest date is that of White (1969) who took several specimens in Maine on August 30.

Somatochlora semicircularis (Selys). (Pl. 9: 9, 10, p. 76; pl. 11: 10, p. 86; pl. 13: 4, p. 96; pl. 16: 4, p. 112; pl. 18: 5, p. 124)

   Epitheca semicircularis Selys, Bull. Acad. Belg., (2) 31: 295 (reprint p. 61), 1871.

   Somatochlora semicircularis: Selys, Bull. Acad. Belg., (2) 45: 194, 1878.

   Not *Epitheca nasalis* Selys (Bull. Acad. Belg., (2) 37: 32, 1874) referred to *S. semicircularis* with doubt by Walker (1925).

A medium-sized western species with male superior appendages meeting in a curve and the vulvar lamina short and bilobed.

Male. Differs from other related species in having the post-clypeus entirely black; metallic black area of antefrons usually reaching the frontoclypeal suture and thus dividing the yellow of the face otherwise than in *forcipata*. Vertex black with dull metallic reflections; hair of face moderate; occiput dark brown; rear of head black with dull metallic reflections and a white submarginal fringe. Pterothorax metallic green with blue reflections; mesepisternum orange brown above and at lower ends and in ante-alar sinuses. Lateral spots orange yellow, conspicuous in tenerals but becoming duller with age, the mesepimeral spot ovate, 2 mm. long, the metepimeral spot much smaller, often obscure in dried specimens. Hair of thorax pale brownish, longest on dorsum, similar to that of *kennedyi* and other related species. Coxae largely greenish black, legs otherwise black. Wings hyaline, the anal triangle yellow; venation black, costal veins edged with brownish; pt. brown; membranule smoky, paler at base but without a well defined paler area. Abd. segments dull blackish, 2 and base of 3 polished dark brown on sides; 2 with two principal pairs of dull yellowish spots, a smaller anterolateral pair below the auricles and a larger posterolateral pair encroaching on the dorsum, followed by a pale interrupted apical annulus. Seg. 3 usually with a pair of small dorsobasal spots; 4 to 10 greenish black, sometimes with small laterobasal spots on 5 to 8 but often without such spots.

General form more robust than in other related species in our territory; abdomen slightly longer than hind wing, widest at the end of 6, narrowing from 7 to 10.

Genital lobes rather narrowly ovate, bent mesad and meeting at middle line, hamuli (pl. 11: 10, p. 86) stout, hooked, tapering rapidly to acute apices. Sup. apps. (pl. 9: 9, 10, p. 76) slightly shorter than 9 + 10, differing from those of the other related species (*franklini* to *incurvata*) in the appearance, in dorsal view, of the lateroventral prominences and the manner in which the apical ends meet one another in an arc rather than an angle.

Female. Differs from male in the larger size of the dorsal orange spots on 2 and 3, especially those on 3, which usually extend from base to transverse carina. The pale ventrolateral spot on 3 is also larger and prolonged caudad into a narrow streak. The lateral yellow spots on 5 to 8, when present, are larger and sometimes prolonged caudad. Wings generally clear but sometimes partly or entirely yellowish.

Vulvar lamina (pl. 13: 4, p. 96) half as long as 9; nearly flat and scale-like, the hind edge truncate with a narrow median notch, the two lobes with rounded corners. Anal apps. about as long as 9 + 10 or a little longer, apices rather bluntly pointed.

Venation. Anx ♂ ♀ 7–9/5; pnx ♂ ♀ 5–8/6–9; two rows of post-trigonal cells followed by 3 rows with 4–5 cells at the wing margin; cells between forks of R_2, ♂ ♀ 4–26/15–27; cells between IR_3 and Rspl usually 6 to 9, occasionally 10; pt. 4.5 to 5 times as long as wide.

Measurements. Total length ♂ ♀ 47–52; abd. ♂ ♀ 31–35; h.w. ♂ 27.5–31.5, ♀ 29–32; h.f. ♂ ♀ 6.5–7.0; pt. ♂ 2.0–2.7, ♀ 2.1–3.0.

Nymph (pl. 16: 4, p. 112). Pale to dark brown, moderately hairy, the abdomen less so than in any of the other species of the *S. arctica* group. Folded labium reaching posteriorly barely or not quite to middle of mesocoxae; prementum not quite reaching laterally the inner edge of the eyes; premental setae 10 to 13, the 4th or 6th from the lateral margin longest, the innermost 4 or 5 smaller than the others; marginal spiniform setae in a group of about 20, the basal group small and with minute setae; 2 to 4 spinules at the distal joint; palpal setae 7; distal margin of palpus with 9 crenations or sometimes only 8, each with the long edge slightly concave, with 3 to 5 setae in a graded series, the hindmost somewhat longer than the height of the crenation, the others much shorter. Pronotum of thorax with a heavy tuft of hair. Length of hind femur about equal to width of head across the eyes. Abdomen (pl. 18: 5, p. 124) long ovate, widest at seg. 5, shaped as in other related species; hind margins of terga fringed with minute hairs among which are scattered long, slender hairs among shorter and coarser ones, especially on segs. 8 and 9; dorsal hooks absent; lateral spines present only on 9, of very variable size, usually larger than in the other species of the *S. arctica* group; anal appendages as long as mid-dorsal length of segs. 9 + 10; epiproct a little longer than its basal width, sides slightly concave, subapical tubercle of male not elevated; cerci about one-eighth longer, their outer edges in the male broadly and evenly arcuate; paraprocts as long as cerci or a little longer.

Length 21–22; abd. 13.0–13.5 (14)*; w.abd. (6.2)* 7.5–7.7; h.w. 6.3–6.5 (6.9)*; h.f. 5.8–6.0 (6.2)*; w.hd. 5.9–6.5.

PLATE 15

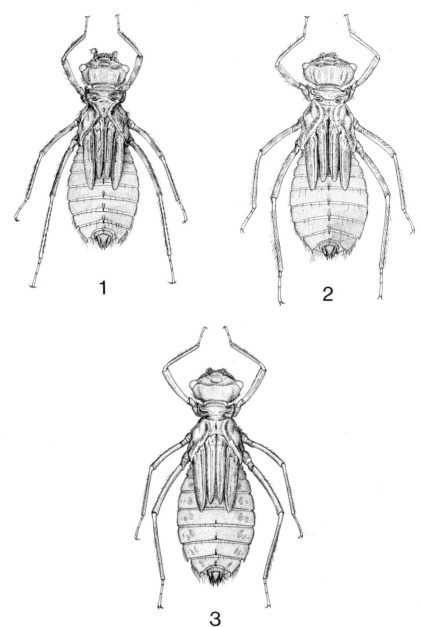

1

2

3

*Measurements in parentheses are those given by Musser (1962) from a long series of nymphs obtained in Utah.

Habitat and range. Whitehouse (1941) summarizes his observations on the habitat of this species as follows: *"semicircularis* seems to prefer reedy swamps, small ponds and muskeg pot-holes; but I have taken it on quite large lakes where these have shallow reedy shores." This statement agrees very well with our own observations on the haunts of this *Somatochlora*. Its general range is s. Alaska, s. through B.C., the Rocky Mts. of Alta. and continuing s. through the western United States to Colo., Utah and Calif.

Distribution in Canada and Alaska. Alaska—Admiralty I.; Juneau; Chitina. *B.C.*—Cultus Lake; Christina Lake; Moyie; Peachland; Ainsworth; Fernie; Powell River; Garibaldi district; Jesmond; Revelstoke; Glacier; Field; Quesnel; Stanley; Chilcotin (near Riske Creek); Bear Lake; Vancouver I. (Ucluelet; Pt. Renfrew; Langford Lake; Nanaimo district; Forbes Landing; Florence Lake district). *Alta.*—Healey Creek; Banff; Nordegg.

Field notes. Within its range this is a very generally distributed dragonfly. At Banff, Alberta, we found *S. semicircularis* common about a number of shallow ponds in the Bow Valley, chiefly along the railway embankment and on the course of a quiet tributary of the Bow River. The mature males kept just above the marsh vegetation or within a few feet of the water, the movements being quite like those of *S. kennedyi* or *forcipata*. Kennedy (1913), who found *semicircularis* fairly abundant in a marsh bordering Bumping Lake, Washington, states that of the two sexes the males "were the more active and the higher flyers. They usually flew about 2 feet above the sedges but occasionally they would take high flights among the black pines growing about the swamp. The males were never at rest except when copulating but the females while usually flying low, frequently rested in the sedges and other plants. In mating the males captured the females while these rested. After capture followed a long nuptial flight, which usually lasted several minutes following which the pair, while yet in copulation, settled on some trees or shrub, where they remained for a long time. The eggs were laid in masses on the surface of the water on the more open pools whereupon the egg masses would immediately disintegrate and fall to the bottom."

We have found this species ovipositing in a spring bog and have found exuviae of the same species about the puddles where the female was ovipositing.

Our dates from British Columbia and Alberta range from June 6, 1909

PLATE 15

Somatochlora—nymphs: (1) *S. minor*; (2) *S. elongata*; (3) *S. williamsoni*.
(1–3 from Walker, 1925.)

(Ucluelet, British Columbia) to September 4, 1924 (Banff, Alberta). White-house (1941), however, found the same species on June 7, 1937, which indicates that emergence must have taken place not later than the end of May.

Somatochlora whitehousei Walker. (Pl. 9: 11, 12, p. 76; pl. 11: 11, p. 86; pl. 13: 5, p. 96; pl. 14: 4, p. 102; pl. 17: 1, p. 118; pl. 18: 6, p. 124)

Somatochlora septentrionalis: Kennedy, Can. Ent., 49: 234–235, 1917b.

Somatochlora whitehousei Walker, Univ. Toronto Stud., Biol. Ser. 26: 154, 1925.

A rather small species of the far north with a dark spot on the anal triangle, and the male with a spatulate abdomen.

Male. Labium and anteclypeus pale yellowish, sides of frons darker yellow, remainder of frons, vertex, postclypeus and labrum shining black to metallic green. Hair of face brownish, that of dorsum of head most dense and dark on vertex and occiput. Rear of head black with an orange-brown fringe of hair. Pterothorax olivaceous with metallic green and blue reflections. Ill defined orange-brown areas on upper fourth of the mesepisterna and a more definite elongate spot on the upper half of the mesepimera; metepimera somewhat coppery but without a definite spot. Ante-alar sinuses and interalar area dark brownish. Hair of thorax pale brownish, rather long and dense, somewhat obscuring the brilliance of the sclerites. Underside of thorax dull metallic green, brownish on the more membranous areas. Bases of legs and a streak along outer face of fore femora brown, legs otherwise black. Wings hyaline; a heavy dark brown spot on hind wing limited to the anal triangle; pt. dark yellowish brown; membranule whitish on basal half or less; seg. 1 of abdomen grayish brown, 2 and base of 3 polished dark brown, 2 with a large anterolateral dark yellow spot beneath the auricles and another large spot of similar colour behind the transverse carina; apical annulus whitish. Remaining segments blackish green; 3 whitish along ventral margins; anal appendages black.

Abdomen slightly longer than hind wing, somewhat spatulate, widest at end of 7, the width of which is almost equal to its length, narrowing again slightly on 8 and rapidly on 9 and 10.

Genital lobes rather small and very hairy. Hamuli (pl. 11: 11, p. 86) slender, consisting of a long arcuate proximal part and a short straight distal part with a distinct angle where the two parts meet. Sup. apps. (pl. 9: 11, 12, p. 76) with wide bases, narrowly separated, convergent on proximal half, then divergent to about the distal third, where they are bent mesad at an obtuse angle. The apical third tapering to the slender recurved apices, which meet one another at a somewhat acute angle. A tooth-like prominence appears in dorsal view just before the distal bend. In lateral view a slender inferior spine appears about the proximal fourth of the lateral carina which is here bent mesad. Inf. app. (pl. 9: 11, 12, p. 76) slightly more than one-half as long as sup. apps., of the usual triangular shape.

Female. Similar to male in colour and general proportions, abdomen somewhat longer, less constricted at 3 and less expanded behind, broadest at end of 5 or 6, distinctly constricted; 2 with a pair of large yellowish dorsolateral spots, prolonged ventrad behind the transverse groove into a narrow streak; 3 with a large, ill defined, dorsolateral yellowish blotch, ventral whitish area larger than in male. Wings hyaline or somewhat suffused with amber; pt. slightly longer than in male.

Vulvar lamina (pl. 13: 5, p. 96; pl. 14: 4, p. 102) entire or with a small narrow notch, more or less compressed and projecting partly as a result of drying, one-half to two-thirds as long as the lower edge of 9; sup. anal apps. about as long as 9 + 10 or slightly longer.

Venation. Anx ♂ 6–7/4–5, ♀ 7–8/4–5; pnx ♂ 5–7/6–8, ♀ 6–8/7–9; two rows of post-trigonal cells followed by 3 rows and sometimes 2 rows again with 3 to 4 cells at margin; cells in fork of R₂

♂ 15–19/14–22, ♀ 15–20/15–21; cells between IR₃ and Rspl 5 to 9; pt. 6 to 7 times as long as wide.

Measurements. Total length ♂ 46.0–47.5, ♀ 45–48; abd. ♂ 29.5–34.5, ♀ 31.0–33.0; h.w. ♂ 26.5–28.0, ♀ 29.0–29.5; h.f. ♂ 6.6–6.8, ♀ 6.0–6.5; pt. ♂ 2.5–2.9, ♀ 2.8–3.0.

Nymph (pl. 17: 1, p. 118). Light to medium brown, nearly uniform or somewhat variegated, with paler blotches on basal half of abdomen and with two pale annuli on each femur, a basal and a median. Head of usual form, setae sparse, especially on the antennae which bear small scattered bristles. Labium large, prementum wider than long, reaching laterally to inner borders of the eyes; premental setae usually 9 or 10, of which the 4th or 5th from the outside is generally the longest; the 3 or 4 innermost being very small; there is also an irregular series of minute marginal spiniform setae on the prementum in two groups, a smaller proximal one of 4 to 10 and a larger distal one of about 8 to 13, some of them vestigial; and a small group of spiniform setae at the distal articulation, 3 or 4 in a row and 3 or 4 scattered small ones; palpal setae 6 or 7, rarely 8; distal margin of palpus with 7 or 8 deep crenations, each with 2 or 3 spiniform setae in a graded series, the lowermost much the largest but not longer than the depth of a crenation, the smallest usually vestigial. Pronotum of thorax little produced laterally with a group of about a dozen setae. Legs moderately hairy, rather short, the hind femora reaching the base or middle of seg. 6, the length slightly less than width of head. Wing-sheaths reaching nearly or quite to end of seg. 5. Abdomen (pl. 18: 6, p. 124) moderately broad ovate, widest at 6, narrowing somewhat abruptly on 9, the mid-dorsal length of which is scarcely one-half the midventral length. Hair rather sparse, the terga except on their margins almost glabrous, the hairs of the general surface being scattered and extremely minute; those of hind tergal margins very short on first 3 or 4 segments, thence increasing in length caudad, 6 to 9 having many hairs as long as the segments which bear them, although these marginal hairs do not form a very dense fringe. Dorsal hooks and lateral spines entirely absent; epiproct about as long as mid-dorsal length of segs. 9 + 10; cerci of about the same length, rather stout but with acute apices; paraprocts a little longer, stout at base but decidedly acuminate.
  Length 20.0–21.5; abd. 11.3–12.5; w.abd. 7.0–7.4; h.w. 6–6.5; h.f. 5.5; w.hd. 5.6–6.1.

Habitat and range. Sphagnum bog pools at the edge of spring runs. Labr. to Hudson Bay, Sask., Alta. and B.C.

Distribution in Canada. Labr.—Hopedale. *Que.*—Rupert Bay (James Bay); Poste de la Baleine. *Ont.*—Kenora district (Patricia Portion: Cape Henrietta Maria; Ft. Severn). *Man.*—Mile 332, Hudson Bay Rlwy.; Churchill. *Sask.*—Prince Albert. *Alta.*—Banff; Nordegg. *B.C.*—Jesmond; Revelstoke Mt. (6,000'); Cariboo.

Field notes. In British Columbia *S. whitehousei* has been taken from June 12 until the first week in September. Early and late recorded dates elsewhere include June 28 (Manitoba, Churchill) and August 28 (Labrador, Hopedale).

  Although this northern *Somatochlora* has been recorded nearly across the continent it has been found in considerable numbers only at Fort Severn, Hudson Bay, Ontario, chiefly by the late Clifford Hope and Dr. W.B. Scott during an expedition of the Royal Ontario Museum in July, 1940.

  The terrain here is chiefly muskeg or sphagnum bogs, wooded with stunted black spruce and tamarack except within a distance of about 10 or 15 miles of the Hudson Bay Coast where the bog is treeless. The specimens

PLATE 16

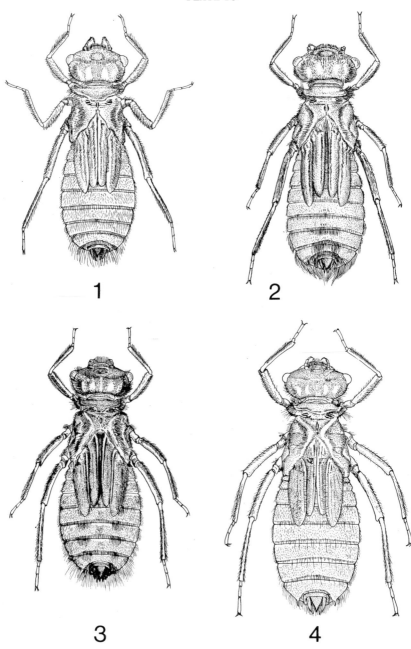

1

2

3

4

taken here, 36 males and 56 females, were mostly tenerals; there were many that were mature but no old ones. Although the adults flew over the open bog there appeared to be no nymphs or exuviae there.

We have met with *S. whitehousei* only at Banff, Alberta, where we found it at one of its breeding places during the second week of August, 1921. This spot was an open mossy bog at the foot of Mt. Rundle and was formed by the seepage from a cold mountain brook, which flowed through it. The greater part of the bog, a few hundred square yards in area, lay on one side of the brook and contained only a few small puddles, the largest being not more than two or three feet in diameter. The water in these puddles was nearly stagnant and was warmer than the water nearer the stream and free from emergent plants. Several species of Odonata flew about these puddles but none visited the stream or the cold pools near it. *Aeshna sitchensis* was the commonest species, followed by *Somatochlora whitehousei*, *S. franklini* and *Sympetrum danae*; single individuals were captured of *Somatochlora semicircularis* and *Aeshna palmata*. Usually not more than one or two Somatochloras were seen at a time. Most of them were males and were hawking in the usual manner, following no definite route and, as a rule, keeping within two or three feet of the bog's surface. They evidently made frequent excursions from the bog into the surrounding woods of spruce and lodgepole pine, since frequently none were seen for several minutes at a time. The few females taken were all engaged in ovipositing in the small puddles or in the wet moss at their margins. The method of oviposition was similar to that of *S. kennedyi* or *S. semicircularis*.

"About the edges of the puddles a number of slime-covered exuviae were found, most of them just above the surface of the water while others were floating. A single nymph was found and was brought back to Toronto, where it lived until the following January, but was unfortunately frozen" (Walker, 1925).

Somatochlora septentrionalis (Hagen). (Pl. 10: 1, 2, p. 80; pl. 11: 12, p. 86; pl. 13: 6, p. 96; pl. 14: 5, p. 102)

Cordulia septentrionalis Hagen, Syn. Neur. N. Amer., p. 139, 1861.

Epitheca septentrionalis: Selys, Bull. Acad. Belg., (2) 31: 298 (reprint p. 71), 1871.

Somatochlora septentrionalis: Selys, Bull. Acad. Belg., (2) 45: 205, 217, 1878.

Somatochlora hudsonica: Williamson, Ent. News, 17: 136–138, pl. VI, figs. 14–17, 1906a.

PLATE 16

Somatochlora—nymphs: (1) *S. franklini*; (2) *S. kennedyi*; (3) *S. forcipata*; (4) *S. semicircularis*. (1–4 from Walker, 1925.)

A small hairy northern species, similar to *S. whitehousei* but with the abdomen of male more slender, and the superior appendages wider apart and more abruptly bent mesad; the vulvar lamina is bilobed.

Male. Colour pattern and distribution of hair on the head similar to that of *whitehousei*. Pterothorax also similar to the latter species, being coppery with green reflections (also blue in tenerals) varied with reddish brown at upper ends of episterna and epimera, and a better defined clear-orange or reddish-brown spot on the mesepimera extending from a little above the metaspiracle to the upper margin. The spot is sharply defined in tenerals, less so in mature insects and all these markings are somewhat obscured by the thick pale-brownish pile, which covers most of the thorax. Ante-alar sinuses dark brown, interalar area grayish yellow. Coloration of legs and wings as in *whitehousei*, including the dark brown spot at base of hind wing, covering the anal triangle. Seg. 1 of abdomen with posterolateral margin pale yellowish, 2 with an obscure dorsolateral yellowish patch; apical annulus whitish; 3 sometimes with a pair of obscure yellowish basal dorsal spots, ventral surface drab. Remaining segments dull metallic blackish green, hairy on posterior half.

Abdomen slightly longer than hind wing, of medium build, more slender than that of *whitehousei* and widest farther back, i.e., generally at end of 7 narrowing on 9 and 10.

Genital lobes rather small, hairy; hamuli (pl. 11: 12, p. 86) small, rounded below, hooked, with small decurved apices. Sup. apps. (pl. 10: 1, 2, p. 80) viewed from above wider apart than in *whitehousei*, subparallel, up to the mesal bend, which is more abrupt than in *whitehousei*, the apices generally crossing one another near the tips; lateral carina with the first ventral spine nearer the base than in *whitehousei* and with a second much smaller spine or angle at about the proximal third. Inf. app. (pl. 10: 1, 2, p. 80) more exposed in dorsal view than in *whitehousei*, of the usual form, about one-half as long as sup. apps.

Female. Similar to male in colour and general proportions (but see *Variations*). Abdomen a little longer, widest at distal end of 5, thence tapering caudad and constricted on 9. Females of the two species *S. septentrionalis* and *whitehousei* are hardly distinguishable except by the form of the vulvar lamina. This structure is much shorter in *septentrionalis* (pl. 13: 6, p. 96; pl. 14: 5, p. 102), being only one-fourth to one-third as long as the sternum of 9, not at all constricted nor projecting, and deeply bilobed. Anal apps. a little longer than 9 + 10, tapering more abruptly on the distal than the proximal half, apices acute.

Venation. Anx ♂ ♀ 7–9/5; pnx ♂ ♀ 6–9/5–8; two rows of post-trigonal cells in fore wing, usually followed by 3 rows which are generally reduced to 2 again before reaching the margin, where there are 2 to 5 cells, generally 3 or 4; cells between fork of R2 ♂ 11–15/10–17, ♀ 13–18/13–18; cells between IR3 and Rspl usually 5 to 7, occasionally 6 or 9; pt. about 6 times as long as wide.

Measurements. Total length ♂ 39.5–46.5, ♀ 41.0–47.5; abd. ♂ 26.5–32.0, ♀ 29–33; h.w. ♂ 25.8–28.0, ♀ 27.5–29.5; h.f. ♂ 7.0–7.3, ♀ 7.0–7.1; pt. ♂ 2.5–2.8, ♀ 2.8–3.0.

Nymph (described from an exuvia). Pale brownish or almost colourless, the only distinct markings being those of the abdominal segments most of which are marked with two pairs of darker brown lateral spots (muscle scars) and two pairs of dark dorsal puncta. Head as in *S. whitehousei*. Eyes less prominent than in the *S. arctica* group, a dorsal pattern of rather long hairs next to the eyes and along the lateral margin. Antennae nearly bare. Labium large, the prementum wider than long, premental setae 11 to 13, the 3rd to 8th from the outside being longest, the 4th or 5th innermost much smaller than the others; a few spiniform setae, including 3 or 4 at the palpal articulation; palpal setae typically 8, the first smaller than the others, more rarely 7; distal margin of palpus with about 8 deeply-cut crenations, each typically with 3 spiniform setae in a graded series, the longest about equal to the height of the crenation. Pronotum of thorax with anterolateral angles somewhat acute and elevated, bearing a tuft of hairs. Legs short, as in *whitehousei*, very hairy, hind femora reaching to about the middle of seg. 6; wing-sheaths extending a little beyond the base of seg. 6. Abdomen elliptical, greatest width at segs. 5 or 6, at about three-fifths of its length, narrowing rather abruptly on 9, without trace of lateral spines on dorsal hooks; hairs of dorsal surface small and scattered except along hind margin of segments, those on 5 to 9 being increasingly longer, some of them being as long as the segments that follow them; lateral margins also with an irregular fringe of hairs, longest on the last 3 segments; anal appendages about as long as the mid-dorsal length of segs. 9 + 10; epiproct about as long as wide, slightly acuminate, cerci of about the same length as epiproct, outer edges in male slightly bent; paraprocts slightly longer than the other appendages.

Length 19–20; abd. 11–12; w.abd. 6.5–7.0; h.w. 6.0–6.5; h.f. 5.7–6.0; w.hd. 6.

Habitat and range. Sphagnum ponds. Labr. to Hudson Bay and Mackenzie District, N.W.T., s. to Nfld., Anticosti I., Que., James Bay and n. B.C.

Distribution in Canada. Nfld.—Grand Bruit; Bay of Islands; Grand Lake; St. Anthony. *Labr.*—Cape St. Charles; North West River (Hamilton Inlet); Cartwright; Hopedale. *Que.*—Seven Islands; Anticosti I.; Pt. Comfort (James Bay); Poste de la Baleine. *Ont.*—Cochrane (Ft. Albany; Big Piskwanish) and Kenora (Patricia Portion: Cape Henrietta Maria; Ft. Severn) districts. *Man.*—Churchill. *B.C.*—Atlin. *N.W.T.*—Ft. Simpson.

Field notes. This is a species from the far north. Most of our specimens came from Labrador and were taken during August, although one is dated July 15, and is mature. Nearly all our records from other localities are dated July, our earliest being July 6, 1939, when Whitehouse took his first specimen, a male at Atlin, British Columbia. His total series of 14 males and seven females was taken during July up to the 22nd, except one female, captured on August 22, apparently the only individual left at the pool, following cold and windy weather. As none of Whitehouse's specimens was teneral, he estimated that emergence probably begins in June, possibly about the middle of the month. In Labrador it would probably not begin before July.

"The habitat of this species is small muskeg pools of 25 to 50 feet in length, with level wet edges, as favored by *Aeshna c. septentrionalis* [= *A. septentrionalis* Burmeister], not large muskeg pools, with firm peaty banks, where *Somatochlora albicincta* and *hudsonica* fly" (Whitehouse, 1941). Whitehouse remarks that on the only occasion when he saw a female of *septentrionalis* attempting to oviposit on one of these large pools, she was driven off by males of these two larger species, and on returning was driven off again.

Males of *septentrionalis* "hawk over the pools awaiting mates and they are slightly less cautious in approaching the edge. They will hover momentarily on occasion but restless flight is more usual. More than one on a pool at the same time is rare. The female is seized over the water and they fly back over the muskeg and scrub willow: presumably to come to rest ... In about fifteen minutes she is back ovipositing; and her male (should he return) while taking no interest in the proceedings, will not interfere with her" (Whitehouse, 1941).

"Ovipositing is performed in the centre of the pool, with a deliberate dip,

dip, dip, of the tip of the abdomen, whilst the insect moves around in alert flight. She is indifferent as to whether the eggs go into clear water or into the creamy-pink slime of decayed vegetation arising in places to the surface. For her task, she appears (apparently from nowhere) well in the middle of the pool, and once this is completed she leaves it, unobtrusively as she came ... I followed one female 100 feet back from the water, as she flew low selecting a place to alight with nice discrimination: her choice being a small pot-hole, surrounded with tall grass. Unable to see her, but positive that she was there, I placed my large kite-net on top, and she flew up into it. On another occasion I followed a male, and netted him in precisely the same way. I feel sure that such small grass-ringed pot-holes are the natural resting place of this species; for to fly back to the trees in a large muskeg flat would be too far from their water, and the scrub willow of the muskeg would offer no protection from the ever-prevailing wind of north latitudes" (Whitehouse, 1941). On this trip the two cloudy days of July 10 and 11 were spent in hunting for exuviae, all that could be collected around the four pools where this species had been flying on the previous sunny days. This material, which was kindly submitted to us for examination contained at least two species, one of which was undoubtedly *septentrionalis*, the other probably *albicincta*.

Variations. Whitehouse (1941) noted that a belated female taken at Atlin, British Columbia on August 21 had small basal lateral dull yellow spots on segments 5, 6 and 7.

Somatochlora sahlbergi Trybom. (Pl. 10: 3, 4, p. 80; pl. 11: 14, p. 86; pl. 13: 7, p. 96; pl. 14: 6, p. 102)

Somatochlora sahlbergi Trybom, Kongl. Svenska Vetensk.-Akad. Handlingar, 15 (4): 7, 16, 20, 1889.

Somatochlora hudsonica: Martin, Coll. Zool. Selys, 17: 27, pl. 27, fig. 28, 1907.

Somatochlora walkeri Kennedy, Can. Ent., 49: 229–236, 1917b.

A medium-sized, robust, hairy species of the extreme northwest, the males with abruptly bent superior appendages, resembling those of *septentrionalis* and *albicincta* but with abdomen lacking the whitish abdominal half-rings, and the female having a vulvar lamina with a small terminal notch.

Male. Labium, anteclypeus and a lateral spot on each side of the antefrons yellow; labrum, postclypeus, remainder of frons and vertex shining black with greenish reflections, occiput dark shining brown; hair of face gray, that of dorsum darker and somewhat denser. Rear of head black with a submarginal fringe of pale gray hairs. Pterothorax coppery, upper ends of mesepisternum and a spot in the ante-alar sinus light reddish brown, lateral thoracic spots absent; interalar areas dull grayish. All the metallic areas somewhat obscured by the thick vestiture of grayish brown hairs. Legs, including coxae, black; wings hyaline, venation black; pt. dark brown; membranule pale gray, darkened toward anal triangle or behind. Abdomen greenish black; seg. 1 brown, dorsum and venter of 2 dark brown, hairy, sides very dark and polished, intersegmental annulus behind 2 pale orange brown; 3 basally smooth and polished, except below where it is hairy. Distal part of 3 and 4 to 10 greenish black, hairy especially on 5 to 10; anal apps. black.

Occiput very convex; abdomen (excluding apps.) about as long as hind wing, robust, widest at end of 6, scarcely narrower on 7, thence tapering slightly to the end.

Genital lobes small, rounded or subangular, hairy. Hamuli (pl. 11: 14, p. 86) broad, hooked, similar to those of *S. albicincta* and *hudsonica*. Sup. apps. (pl. 10: 3, 4, p. 80) somewhat shorter than 9 + 10, of the same type as in *albicincta* but differing as follows: viewed from above less enlarged at base, with no basolateral prominences, the stout proximal parts cylindrical, the slender distal parts more abruptly bent mesad, slightly longer but similarly curved laterad. Carinae as viewed laterally precurrent, without tooth-like processes in the specimens seen. Inf. app. (pl. 10: 3, 4, p. 80) one-half, or slightly less than one-half, as long as sup. apps., triangular with convex margins.

Female. Coloration similar to that of male, but with a single small orange mesepisternal spot on the single specimen examined in detail (probably not a constant character).

Abdomen depressed, slightly constricted at 3, widest at end of 4 or middle of 5, tapering beyond 6 and constricted at base of 9. Anal apps. about as long as 9 + 10, slender, tapering to apices. Vulvar lamina (pl. 13: 7, p. 96; pl. 14: 6, p. 102) less than half as long as sternum of 9, nearly twice as wide as long, the posterolateral edges curved with a V-shaped median notch nearly half as long as the vulvar lamina.

Venation. Anx ♂ ♀ 8–9/5–6; pnx ♂ 6–7/8, ♀ 7–8/8–9; three rows of post-trigonal cells almost to the wing margin; cells between fork of R2 ♂ 20/22, ♀ 24/24; cells between IR3 and Rspl 11 or 12; pt. about 6 times as long as wide.

Measurements. Total length ♂ 48–50, ♀ 48; abd. ♂ 31.6–33.0, ♀ 35.0; h.w. ♂ 32, ♀ 35; h.f. ♂ ♀ 8.5; pt. ♂ 3.

Nymph. Not yet taken in North America but described by Valle (1931) from material collected in Parkkina, Finnish (now Russian) Lapland from pools where the adults were also taken. The nymphs, although not identified beyond doubt, clearly belong to a species of the *S. alpestris* group of which *S. sahlbergi* is the only other known species in subarctic Europe and Asia. It is similar in form to the nymphs of *S. albicincta* and *S. cingulata* and is more like the former in the lack of much indication of a body pattern, except the transverse stripes of the legs, which are usually variable, but are shown in Valle's figure of the young nymph as present on both femora and tibiae. The lateral spines on abd. segs. 8 and 9 are somewhat larger than in *S. cingulata*, the species in which they are most prominent in American forms, those on seg. 8 being shown (in Valle's figure (1931) of full-grown nymphs) as more than one-fourth of that segment's lateral length, the spine included, whereas the spine on seg. 9 is about one-third of the length of seg. 9 (or about one-half if the spine is not included). The cerci are longer than the epiproct but shorter than the paraprocts. The abdominal hairs are rather sparse, but regularly distributed. As in other species of the *S. alpestris* group, there are no dorsal hooks. On the labium the crenations on the distal margin of the palpus are rather lower than is usual in the *S. alpestris* group and the spiniform setae are reduced to 2 or 1. This character, if constant, is unusual. Valle (1931) records 14 premental setae and 9 palpal setae.

Habitat and range. Bog ponds at the edge of the arctic tundra (Valle, 1931, 1952). A palaearctic species of subarctic range, being known from the Yenesei and Lena Rivers in Siberia, Finnish and Russian Lapland, the Mackenzie District, N.W.T. and Alaska. Surprisingly, *S. sahlbergi* has recently been found also in the Irkut River valley at the western end of Lake Baikal in southern Siberia (approximately 52°N.), a locality far to the south of its previously known distribution (Belyshev, 1973).

PLATE 17

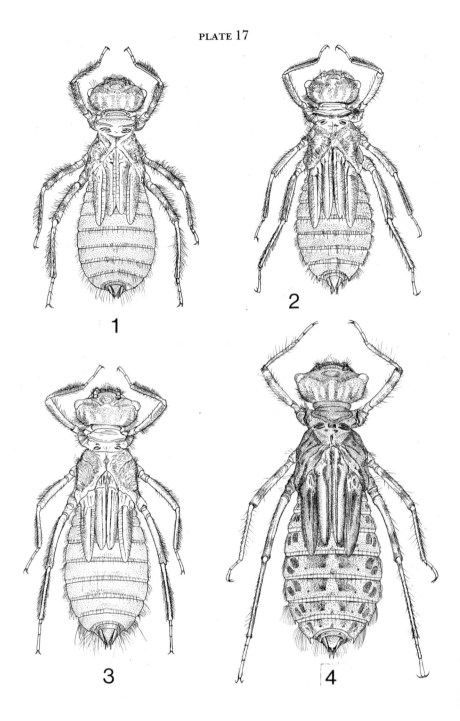

1

2

3

4

Distribution in Canada and Alaska. Alaska—Kuskokwim River; Sagwon, Sagavanirtok River (approximately 69°21′N., 148°16′W.) (Gorham, 1972). *N.W.T.*—Reindeer Depot (68°42′N., 134°06′W.) (July 6 and 8, 1948, W. J. Brown).

Material studied. This comprised the North American material listed above and one male adult from Parkkina, Finnish Lapland (69°33′N.) kindly donated by Dr. K. J. Valle. This specimen is indistinguishable from the males from Reindeer Depot, in colour pattern, size, and form of the anal appendages.

Field notes. The locality in Finnish (now Russian) Lapland where Dr. Valle caught 28 males and one female in July and August, 1928 to 1930, was near the village of Parkkina (approximately 69°33′ N., 31°15′ E.). There, in the high fells that surround the fjord and rise 100 to 150 metres above it, are exposed bog pools with sedges growing around the marshy edges. Beyond the dwarf birches that grow on the gentle slopes around the pools lies the bare Arctic tundra.

It was at a fairly large bog pool in this austere highland region that Dr. Valle came upon his first males of *S. sahlbergi* on July 25, 1928—a clear day with a cold northerly wind, when all the insects were staying in the shelter of the dwarf birches. On a subsequent visit he saw *S. sahlbergi* flying in company with *S. alpestris* (Selys). Males of *S. sahlbergi* patrolled the pool margin apparently on the look-out for females (Valle, 1931). The single male caught by Hämäläinen (1967) in the lake area of Tsoakketenjärvi (69°21′ N., 26°20′ E.), Finnish Lapland was likewise flying along the peaty bank of a marshy pond. In the Irkut River valley, Belyshev (1973) found this species in a fir grove next to a pool through part of which cold, clear water was flowing.

There is good reason to suppose that the habitat of *S. sahlbergi* in Arctic North America closely resembles that described by Valle in Lapland.

Near Parkkina adults were encountered between July 8 and August 12; at Tsoakketenjärvi a male was taken on July 12, 1966; and at Reindeer Depot, Northwest Territories, adults were captured in 1948 by Drs. W. J. Brown and J. R. Vockeroth between July 6 and 19.

Somatochlora albicincta (Burmeister). (Pl. 10: 5, 6, p. 80; pl. 11: 13, p. 86; pl. 14: 1, 7, p. 102; pl. 17: 2, p. 118; pl. 19: 4, p. 128)

Epophthalmia albicincta Burmeister, Handb. Ent., 2: 847, 1839.

PLATE 17

Somatochlora—nymphs: (1) *S. whitehousei*; (2) *S. albicincta*; (3) *S. hudsonica*; (4) *S. cingulata*. (1–4 from Walker, 1925.)

Cordulia albicincta: Hagen, Syn. Neur. N. Amer., p. 138, 1861.
Cordulia eremita Scudder, Proc. Bost. Soc. Nat. Hist., 10: 215, 1866.
Epitheca albicincta: Selys, Bull. Acad. Belg., (2) 31: 303 (reprint p. 69), 1871.
Somatochlora albicincta: Selys, Bull. Acad. Belg., (2) 45: 205, 218, 1878.

A northern species of typically moderate size and compact build with a brassy thorax and white-ringed abdomen. Although not showing marked variations in most of its vast range, it becomes distinctly larger and darker at successively lower altitudes on the Pacific slope, so that its appearance in flight suggests quite a different species. The increase in size reaches a maximum on the Queen Charlotte Islands (subspecies *massettensis* Whitehouse) where it is comparable in dimensions with *S. cingulata* (Whitehouse, 1941). Possibly two or three subspecies are involved in this complex, but the material we have available is insufficient to solve this problem. Only the female of *massettensis* is known, but we have not seen it. Without the male we prefer not to recognize this form as a subspecies.

Male. Labium and anteclypeus pale yellowish brown; labrum and postclypeus black, the lateral lobes of the latter sometimes brownish, frons ochraceous on the sides, in front and above black with green reflections; occiput red-brown; pile chiefly dark brown on occiput and vertex, whitish on face. Pterothorax coppery with brassy green reflections, somewhat obscured by the thick coat of pale brownish hair. A pale reddish brown spot at upper end of mesepisternum and a small narrow pale yellow spot on the upper half of same area; ante-alar sinus pale brown, ante-alar carina very dark brown. Legs black, bases of front legs, including outer surface of tibiae, brownish. Wings hyaline without brown basal spots; pt. dark yellowish brown above, light yellowish below; membranule with basal half or less, white, this portion rather sharply defined. Abdomen with seg. 1 dull coppery, 2 and base of 3 polished dark brown, 2 and dorsum of 3 with large ill-defined paler blotches. Segs. 3 to 10 bronze black with little metallic lustre, all but the last segment with whitish apical annuli, most of which are mesally slightly interrupted; 10 with a pair of dorsal yellowish spots; 4 to 8 with small basal lateral dull yellow spots; anal apps. brown or black, undersides of terga pale drab.

Abdomen as long as hind wing, widest at end of 5, width of 4 equal to two- or three-fifths of its length, scarcely narrowed on 6, but noticeably so on 7 to 10.

Genital lobes rounded, small; hamuli (pl. 11: 13, p. 86) small, hooked, tapering rapidly to acute apices. Sup. apps. (pl. 10: 5, 6, p. 80) as long as 9 + 10, the stout proximal parts with subparallel sides, widened only at the extreme base and at the distal angle, slender distal parts curving mesad and then caudad, also dorsad as seen in lateral view, tapering to small apices. Outer edges of appendages without projecting teeth or angles except a small prominence at base and another where the proximal and distal parts meet, lateral carina bent ventrad but not forming a projecting angle although bearing two very minute teeth, the second tooth near the middle; inf. app. (pl. 10: 5, 6, p. 80) three-fifths as long as sup. apps., triangular with sides nearly straight, basal width about three-fifths of the length.

Female. Like the male in size and colour. Abdomen scarcely longer but showing the usual differences in form, being generally widest at base of 3, gradually tapering to 9, which may be more or less constricted in old females, somewhat constricted at 3 and widest about 6. Dorsolateral spots at base of 3 considerably larger than in male and basolateral spots somewhat larger, especially those on 4. Wings varying from hyaline to deep yellowish brown. Pt. slightly longer and paler than in male.

Vulvar lamina (p. 14: 1, 7, p. 102) about 1 mm. long and less than half as long as sternum of 9, not projecting in live insects, the free margin distinctly bilobed, lobes rounded and separated by a wide angle in specimens not shrunken by drying. Anal apps. about as long as 9 + 10, moderately stout, with acute apices.

Venation. Anx ♂ 7–8/7–10, ♀ 6–9/6–10; pnx ♂ 6–9/7–10, ♀ 6–9/6–10; two rows of post-trigonal cells on fore wings followed by 3 rows as a rule, although these are often imperfect as there may be only 2 rows to the margin, where there are usually 3 or 4 cells; cells between IR3 and Rspl usually 6 to 9, rarely 5 or 10; pt. about 6 times as long as wide.

Measurements. Total length ♂ 45.3–50.0, ♀ 46.0–52.2; abd. ♂ 30.0–32.5, ♀ 31.0–36.7; h.w. ♂ 28–31, ♀ 29–33; h.f. ♂ 6.8–7.6, ♀ 6.9–7.4; pt. ♂ 2.6–3.1, ♀ 2.9–3.2. (See also *Variations*.)

Nymph (pl. 17: 2, p. 118). Uniform brown, moderately hairy. Head behind eyes narrowing little for a short distance, then curving rapidly mesad to meet the slightly concave hind margin. Hinge of folded labium extending posteriorly nearly to hind margin of mesocoxae; prementum reaching laterad to edge of eyes; premental setae 11 or 12, the 4th or 5th from the outside longest, the mesal 3 or 4 very short; palpal setae 5 or 6, slender; crenations on distal margin of palpus 7 or 8, somewhat irregularly rounded or subangulate, each crenation bearing typically 3 or 4 spiniform setae in a graded series, the longest of each series being much shorter than the length of the crenation bearing it. Pronotum bearing moderately thick tufts of hairlike setae. Legs fairly hairy, the length of the hind tibial setae being equal to about twice the tibial width. Hind femora about as long as width of head, the tibiae a little longer. Abdomen (pl. 19: 4, p. 128) slightly wider than head, widest at seg. 6 or 7, the general surface with evenly distributed minute hairs and sparsely scattered longer ones and the hind margins of each tergum with a fringe of shorter setae with longer and more slender hairs scattered among their lateral margins with a somewhat denser fringe consisting of both the shorter setae and the longer slender hairs, the latter much more numerous than on the dorsum, particularly on segs. 8 and 9 and on the ventro-posterior edge of 9; dorsal hooks absent; lateral spines on segs. 8 and 9, those on 8 usually rather less than one-fifth as long as the segment's lateral margin and not divergent, those on 9 about one-fifth the length of the margin (excluding the spine); the lateral spines are, however, variable and sometimes much longer than as described (e.g., as in a reared specimen from Grouse Mt., Vancouver, British Columbia); epiproct nearly as wide as long, distinctly longer than the cerci, which are as long as the mid-dorsal length of segs. 9 + 10, tapering regularly to acute apices; paraprocts longer than epiproct, with slender apices.
Length 20.5–23.5; abd. 12.7–14.6; w. abd. 7.2–8.0; h.w. 6–7; h.f. 5.8–6.5; w. hd. 6–6.4.

Habitat and range. Ponds, lakes and slow streams in boggy places, especially where there is slight water movement. Hudsonian and upper parts of Canadian zone, Labr. and Nfld. to Alaska, s. to the White Mts., N.H., the lower St. Lawrence, n. shore of Lake Superior, and n. of Great Plains to Rockies; thence s. at considerable elevations to s. Alta., B.C. and Wash.

Distribution in Canada and Alaska. Typical *S. albicincta. Nfld.*—Spruce Brook; Bay of Islands. *Labr.*—Cartwright; Hopedale; Nain. *Que.*—Lake Témiscouata; Mt. Albert; Mont Tremblant Park; Saguenay River (Cap Jaseux); Godbout; Seven Islands; Moisie River; Thunder River; Bradore Bay; Amos; Eastmain (James Bay); Great Whale River; Schefferville; Ft. Chimo. *Ont.*—Essex, Kent and Carleton counties; Thunder Bay (Nipigon), Cochrane (Smoky Falls; Lake Abitibi) and Kenora (Patricia Portion: Cape Henrietta Maria; Ft. Severn) districts. *Man.*—Churchill. *Sask.*—Pike Lake. *Alta.*—Banff; Nordegg; Jasper. *B.C.*—Lihumitson (Liumchen) Mt.; Salmon Arm; Revelstoke Mt. (6,300'); Field; Atlin; Vancouver I. (Nanaimo Lakes;

Courtenay; Forbes Landing); Queen Charlotte Is. *N.W.T.*—Great Slave Lake (Jones Point; Yellowknife); Great Bear Lake. *Yukon T.*—Whitehorse. *Alaska*—Kodiak; Katmai; Ft. Yukon.

The larger form of *S. albicincta* has been taken in many localities in southern British Columbia. Representative localities on the mainland are Harrison Bay district (Harrison Bay; Echo Lake; Elbow Lake; Squawkum Lake; Little Mountain; Lihumitson (Liumchen) Mountain); Hope district (Kawakawa Lake); Chilliwack district (Cultus Lake); and on Vancouver Island, Nanaimo Reservoir.

The third form of *S. albicincta*, named *massettensis* by Whitehouse, occurs in the Queen Charlotte Islands.

Field notes. S. albicincta frequents a variety of quiet waters usually in somewhat open boggy situations. The males fly along the edges of boggy streams showing a tendency to pass in and out of small recesses along the banks, flying usually at a height of somewhat more than a foot above the water. They are particularly likely to be found about the mouths of small tributary streams. They also develop in certain types of ponds but in our experience not in stagnant ponds. We found a newly emerged female of the larger form with its exuvia beside a pond on the top of Grouse Mt. near Vancouver, British Columbia. This pond was fed by a small stream from nearby melting snow.

Whitehouse (1918b) found the typical form with *S. hudsonica* at a small bog-margined pool near Nordegg, Alberta (altitude *ca* 4,000′) where it had certainly emerged, since exuviae were found there.

Both larger and smaller *albicincta* breed in small lakes. We found males of the former patrolling the Nanaimo reservoir usually within a few feet of the banks, and Whitehouse found the same form in various lakes at Hope, Chilliwack and Harrison Bay districts. We have rarely seen *albicincta* away from the breeding waters and never flying at a height beyond reach of an ordinary net.

Robert (1944) noted that it is very aggressive, sometimes giving chase to other dragonflies larger than itself, even such large species as *Aeshna canadensis*.

Oviposition has been frequently observed but all that has been recorded is that the female strikes the surface of the water rhythmically with the end of the abdomen, as do all the species of *Somatochlora* that have a short vulvar lamina.

The majority of records are dated July and this is undoubtedly the month when *albicincta* is most abundant in the greater part of its range. Emergence begins late in June or early in July according to locality and weather. Specimens taken on June 27 in Abitibi county, Quebec, by Robert (1944) were already mature. The only nymph we have reared successfully was full-grown when found at Godbout, Quebec, in July, 1918. It emerged at Toronto on June 1, 1919, a date that was probably a little earlier than it would have been had it emerged at Godbout. The earliest record we have of emergence in the field appears to be June 26 when specimens were taken by Dr. J. H. McDunnough at Nordegg, Alberta.

The only specimen that we have found with its exuvia was a female of the larger darker form, which we took on the summit of Grouse Mt., Vancouver, British Columbia, on July 3. Although usually reaching its maximum numbers in July it is still common in Labrador and other northern regions in August, and we have taken it in northern Ontario as late as September 1.

Variations. Throughout most of its geographical range *S. albicincta* is fairly uniform in size and coloration but on the Pacific slope it becomes distinctly larger and darker with decrease in altitude. These changes, however, are not those of a simple cline. In southwestern British Columbia at altitudes below about 5,000' the common type of this species is about as large as it usually becomes, even at sea-level. The general colour is darker and the whitish annuli of the abdomen are less distinct than in the typical form, so that in flight they are hardly noticeable and the insect has the appearance of some other species. At the Nanaimo Reservoir we took a fair series of males of this form but no females. A single female, however, of corresponding size, was taken on the top of Grouse Mountain, Vancouver (*ca* 4,000') and among a considerable series of *albicincta* taken by Dr. W. E. Ricker on Liumchen Mountain, Cultus Lake in the Chilliwack district, there are three females. This series was taken at various heights, from 4,500' to 5,000'. These specimens are, for the most part, similar to those from the Nanaimo Reservoir but within their range of variation in size are some that are small enough to be included in typical *albicincta* as far as can be determined from dried specimens, in which differences in colour are not always well preserved. The question of the taxonomic value of these variations is complicated by the occurrence in Alaska at Reflection Lake of specimens that are as large as those from southern British Columbia though somewhat more robust, and of much larger specimens discovered by Whitehouse (1941) on the Queen Charlotte Islands, British Columbia, and named by him *massettensis* from their more exact location: near Massett, Graham Island. Only young females were seen, the males probably not having yet emerged at the time the females were observed. Until males of this form are found it is impossible to be sure of its status, but Whitehouse (1941) states that, in 1937, two years after his discovery of the large females on the Queen Charlotte Islands, he found similar large females at Forbes Landing, northern Vancouver Island. These were ovipositing females and the only males observed were of the smaller size, commonly found in southern British Columbia at low altitudes.

Thus we have in Canada and Alaska three or four subdivisions of *S.*

PLATE 18

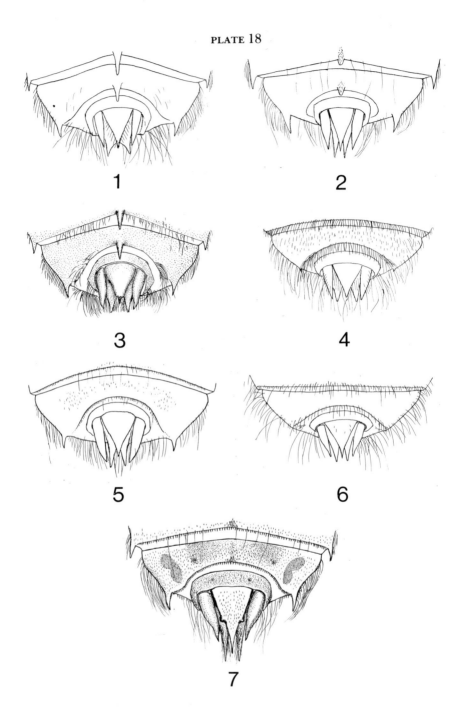

1

2

3

4

5

6

7

albicincta, and possibly two or three subspecies not yet well enough known to be clearly defined. Measurements and distribution of two of these are as shown below.

		Range of measurements (mm.)		
Form of *albicincta*		Total length	Abdomen length	Hind wing length
1 Typical	♂	45.0–49.0	30.0–33.0	29.0–32.0
	♀	45.0–48.0	31.0–32.5	29.0–34.5
2 Southern British Columbia	♂	49.0–53.0	33.0–36.0	31.0–34.0
	♀	49.0–52.0	31.0–34.0	31.0–33.0

The third form of *albicincta* from the Queen Charlotte Islands, discovered by Whitehouse and named by him *massettensis*, is as large as *S. cingulata*, as shown by Whitehouse (1941, p. 540) in his illustration taken from a photograph of a female of each form, with one of a typical *albicincta*. The female of *massettensis* would be easily mistaken for *cingulata* but the pterostigmata are decidedly shorter. The length of *massettensis* as given by Whitehouse is 58 mm. and the wingspread is 79 mm. The length of a female *S. cingulata* is 57–68 mm. and the wingspread 82–85 mm.

Somatochlora brevicincta Robert. (Pl. 10: 7, 8, p. 80; pl. 11: 16, p. 86; pl. 14: 8, p. 102)

Somatochlora brevicincta Robert, Can. Ent., 86: 419–422, 1954b.

A rather small species closely related to *S. albicincta* but with the abdominal annuli widely interrupted dorsally and the abdomen of the male more slender and contracted beyond the middle; vulvar lamina entire. The following description is translated freely from Robert (1954b).

Male. Labium and anteclypeus pale dull yellow, labrum shining black, postclypeus black in the middle dark yellowish on the sides, frons green-black with yellow-green reflections, the sides and middle of the lower margin yellow. Vertex black with green reflections. Occiput black. Vestiture somewhat scantily brownish pilose, more densely on the sides; blackish above. Rear of head black. Thorax metallic green, somewhat obscured by the pale brownish hair, an elongate pale yellow spot on the mesepimeron similar to that of *albicincta*. Legs black, except at base and the posterolateral surfaces of the fore femora. Wings hyaline except the anal triangle and neighbouring cells, which are clear amber; pt. yellow brown, venation black, membranule whitish at base, dark on distal two-thirds. Abdomen hairy from rear half of 3 to 10, the hair longer than in *albicincta*. Abdomen with seg. 1 dark brown, 2 polished brown, auricles black, 3

PLATE 18

Somatochlora nymphs—terminal abdominal segments, dorsal view: (1) *S. elongata*; (2) *S. williamsoni*; (3) *S. tenebrosa*; (4) *S. franklini*; (5) *S. semicircularis*; (6) *S. whitehousei*; (7) *S. cingulata*. (1–7 from Walker, 1925.)

black, except at base where the coloration is like that of the preceding segment, 4 to 8 black, except a small brownish spot at the anterolateral angles, 9 to 10 entirely black above. Intersegmental membrane pale on segs. 1 and 2 and partly on segs. 8 and 9; elsewhere it shows the colour of the segment.

Abdomen scarcely longer than the hind wing, gradually enlarged from constriction on 3 to end of 4 and narrowing again on 6 to 8 but widening very slightly again on 9 and 10.

Sup. apps. (pl. 10: 7, 8, p. 80) similar to those of *albicincta* but differing as follows: the longer proximal parts are less distinctly parallel-sided, being somewhat constricted and the intervening space is somewhat narrowed distad; the sub-basal lateral prominence bears a distinct spine, visible from above, but there is no such spine midway on the margin such as is present in *albicincta*; the meso-distally bent apical parts are somewhat wider and less abruptly angulate. In lateral view the anal. apps. are very like those of *albicincta*. The inf. app. (pl. 10: 7, 8, p. 80) is quite similar both in form and relative length. Hamuli (pl. 11: 16, p. 86) longer and more broadly arcuate than in *albicincta*, the apex less acute; resembling those of *S. whitehousei* in outline.

Female. Similar to male in size and coloration of face and thorax. Hind margin of occiput bearing long reddish hairs. Wings lightly washed with brownish yellow, probably owing to age of the specimen, membranule brown on the apical third only, white at base. Abdomen tapering from base to extremity. Anterior lateral spots of segs. 3 and 4 very large, similar to those of the male on the following segments; segs. 10 yellowish around the points of attachment of the anal appendages, the latter black, a third longer than 9 + 10.

Vulvar lamina (pl. 14: 8, p. 102) horizontal, nearly as long as the 9th sternum, slightly truncate at apex, yellow at the periphery, marked in the middle by two diffusely marginal black spots.

Venation. Anx ♂ 7–9/5–7, ♀ 7–9/6; pnx ♂ 5–6/7–9, ♀ 5/7–9; two rows of post-trigonal cells in fore front wings, followed by 3 rows about opposite the last antenodal vein or well beyond pt., and 4 or 5 cells at margin; cross-veins under pt. ♂ 1/1.

Measurements. Total length ♂ 47–50*, ♀ 44*; abd. ♂ 28–31*, ♀ 29*; h.w. ♂ 29–31*, ♀ 30*; h.f. ♂ 5–7; pt. ♂ 2.4–3.0*.
*These measurements are derived from Robert (1954b).

Nymph. Unknown (1972).

Habitat and range. Known only from peat bogs in two localities in Que.

Distribution in Canada. Que.—Southern end of Lake Mistassini (type locality); La Ferme, Abitibi county.

Field notes. The first specimens were taken by Brother Robert on July 12, 1953, in a moist meadow bordering a small pond in the centre of a peat bog. Another visit to the same spot on August 19 furnished two more specimens. On September 4 a third individual was observed at exactly the same spot.

Of this dragonfly, Brother Robert (1963) writes: "It is without doubt a denizen of cold northern bogs. Females oviposit among waterlogged vegetation; males fly over moist grassland and adjoining lakes."

Somatochlora hudsonica (Selys). (Pl. 10: 9, 10, p. 80; pl. 11: 15, p. 86; pl. 14: 2, p. 102; pl. 17: 3, p. 118; pl. 19: 5, p. 128)

Epitheca hudsonica Selys, Bull. Acad. Belg., (2) 31: 301 (reprint p. 67), 1871.

Somatochlora hudsonica: Selys, Bull. Acad. Belg., (2) 45: 205, 217, 1878 (subgenus of *Epitheca*).
Somatochlora hudsonica: Kirby, Cat. Neur. Od., p. 49, 1890.

Very similar to typical *S. albicincta* but slightly larger and more robust, the male with distinctive anal appendages, the female with the vulvar lamina entire or nearly so.

Male. The colour pattern is so similar to that of *S. albicincta* that it will be unnecessary to do more than indicate the differences from that species. Head as described for *albicincta* with the submarginal fringe of hairs on the rear surface more conspicuously whitish. Pterothorax with metallic green surface somewhat more obscured by the dense pale brownish pubescence, the brownish yellow spots at upper end of mesepisternum more distinct than in *albicincta*, the other spots scarcely different from the corresponding spots in the latter species. Wings hyaline, costal veins dully yellow, pt. dark brown above, paler below. Legs black with rear surfaces of basal segments and outer surface of fore femora brown. Abdomen somewhat hairy, especially on dorsal and ventral surfaces of 1 and 2 and part of 3; 2 and base of 3 with large dull yellowish blotches, distal part of 3 and remaining segments greenish black. All the segments but 10 with whitish apical annuli which are not mesally divided as in *albicincta*, 4 to 7 with small laterobasal yellow spots.

Thorax robust, abdomen about as long as hind wing, expanding from constriction from 3 to end of 5, the width of which is about four-fifths of its length, width of 4 three-fourths of its length (*cf. albicincta*).

Genital lobes and hamuli (pl. 11: 15, p. 86) similar to those of *albicincta*. Sup. apps. wider apart than sides of 10, with prominent lateral angles near base, apices similar to those of *albicincta*, in profile showing a prominent sub-basal angle and the inferior carina with a prominent angle that is wholly lacking in *albicincta*; inf. app. (pl. 10: 9, 10, p. 80) of the usual triangular form with rounded sides and apex, all of which are usually easily seen from above in the space between the sup. apps.

Female. Colour pattern like that of the male with the usual differences. Wings hyaline to strongly flavescent; pt. yellowish. Lateral yellowish areas on 2 and 3 larger than in male, confluent and occupying most of the lateral surface of 2 and most of 3 anterior to the transverse carina and also extending more or less beyond the latter. Basic lateral yellow spots on 4 to 7 larger than in male, decreasing in size caudad.

Vulvar lamina (pl. 14: 2, p. 102) three-fifths to two-thirds as long as sternum of 9, more or less projecting, the free edge entire or with a small median notch, arcuate or slightly truncated. Anal apps. a little longer than 9 + 10, similar to those of *albicincta*.

Venation. Anx ♂ 7–8/5, ♀ 7–9/4–6; pnx ♂ 6–9/7–10, ♀ 5–8/6–9; two rows of post-trigonal cells in fore wing followed by 3 rows; cells between IR₃ and Rspl usually 6 to 9, occasionally 10 or 11; pt. of hind wing about 6 times as long as wide.

Measurements. Total length ♂ ♀ 50.0–53.5; abd. ♂♀ 32.5–35.0; h.w. ♂ 30.5–34.0, ♀ 32.0–34.0; h.f. ♂ 7.5–8.0, ♀ 7.8–9.0; pt. ♂ 2.5–3.3, ♀ 3.3–3.5.

Nymph (pl. 17: 3, p. 118; described from an exuvia). Uniform dark brown to gray. Head as in *S. albicincta*, hair with its usual distribution, rather thick. Labium large, when folded reaching posteriorly to middle of mesocoxae; prementum as wide as long; premental setae 10 to 14, the 5th from the lateral margin longest, the 4 or 5 innermost vestigial; palpal setae 7 or 8, distal margin of palpus with 7 or 8 deeply and subangularly rounded crenations, each with 2 or 3 spiniform setae of which the hindmost is much the longest, though scarcely as long as the crenation (measured from base to apex). Prothoracic processes hairy. Hind wing-sheaths

PLATE 19

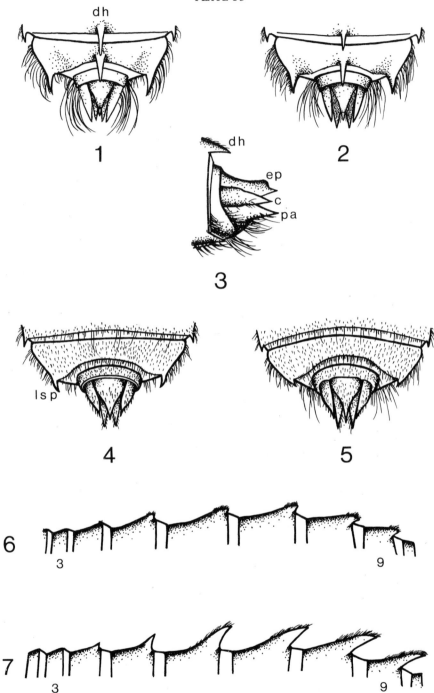

reaching to about the middle of 6, about as long as width of head and hardly longer than hind femora; hind tibiae about one-seventh longer than hind femora. Abdomen (pl. 19: 5, p. 128) broadest at seg. 6, narrowing rapidly on 9. Dorsum with numerous minute scattered hairs, a fringe of setae along hind margin of each segment, most of which are very short, but many are one-half as long as the segments and fairly evenly distributed; the marginal fringe of setae is more dense, those on 9 as long as the lateral margin of that segment; dorsal hooks absent; lateral spines on segs. 8 and 9 extremely small though variable, those on 8 being from one-twelfth to one-ninth of the lateral margin of that segment (not including the spine); those on 9 being from one-tenth to one-sixth of the lateral margin of that segment; in all cases the spines are slender, not at all divergent, those on 9 directed slightly mesad, being in line with the lateral margins; epiproct one-fourth longer than mid-dorsal length of segs. 9 + 10, acuminate and very slender-pointed; in the male abruptly narrowed in the distal third; cerci a little shorter and paraprocts one-third longer than the cerci, acute with apices barely arched in profile, otherwise as in *S. albicincta*.

Length 24.0–24.7; abd. 15–15.3; w.abd. 8.7–9.1; h.w. 6.9–7.6; h.f. 6.7–7.0; w.hd. 6.5–6.7.

Habitat and range. Slow streams and bog-margined ponds. B.C. and w. Alta. to Alaska and Mackenzie District, N.W.T.; e. to n.w. Ont. and s. to the Rocky Mts. of Colo. at high elevations (11,000′).

Distribution in Canada and Alaska. Alaska—Ft. Yukon. *Yukon T.*— Whitehorse; Dawson. *N.W.T.*—Ft. Smith; Great Slave Lake (Ft. Resolution; Gros Cap; Caribou Is.); Reindeer Depot. *B.C.*—Falkland; Salmon Arm; Kamloops; Jesmond; Bridge Lake; Prince George; Burns Lake; Atlin. *Alta.*—Banff; Red Deer; Nordegg; Jasper Park (Caledonia Lake). *Sask.*—(Whitehouse, 1948). *Man.*—Norway House; Churchill. *Ont.*— Thunder Bay, Cochrane (Sucker River; Moose Factory; Ft. Albany) and Kenora (Patricia Portion: Favourable Lake; Attawapiskat Lake; Ft. Severn) districts.

Field notes. This is a northwestern species, apparently the commonest species of *Somatochlora* in the Mackenzie district and the second most abundant of this genus in British Columbia (Whitehouse, 1941). It resembles *S. albicincta* in habit and haunts, and therefore, since the distribution ranges of these two species overlap widely, they are often found together. According to Whitehouse the most characteristic habitat is a "muskeg pool with a firm peaty edge." He refers in particular to ponds of this sort in the vicinity of Atlin in northern British Columbia where both species were flying, and he also took the same two species together at bog-margined ponds at Nordegg,

---

PLATE 19

Somatochlora nymphs—terminal abdominal segments, dorsal view (except 3, left lateral view): (1) *S. walshii*; (2) *S. minor*; (3) *S. tenebrosa*; (4) *S. albicincta*; (5) *S. hudsonica*.
Somatochlora nymphs—dorsal hooks of abdomen, left lateral view: (6) *S. williamsoni*; (7) *S. tenebrosa*.
(1, 2 after Walker, 1941c; 3–7 after Walker, 1925.)
(Abbreviations: c—cercus; dh—dorsal hook; ep—epiproct (superior appendage); lsp—lateral spine; pa—paraproct (inferior appendage); 3, 9—abdominal segments.)

Alberta. But these two species are not always associated even within the area where their ranges overlap. Whitehouse found only *S. hudsonica* at Burns Lake between Prince Rupert and the Alberta border. It was found at two large sloughs and was very common here and variable in size. Whitehouse (1918b) also reports *S. hudsonica* alone from a ditch at Red River, Alberta. We have met *hudsonica* on several occasions in the vicinity of Banff, Alberta, but never associated with *S. albicincta*. Our first specimen was a teneral female taken in a clearing on Sulphur Mountain, Alberta, on June 29, 1913 (Walker, 1925). The other individuals seen were all mature and flying over or near water. "On July 14 I captured 2 males and 2 females at the mouth of a quiet shady stream emptying into the third Vermilion Lake. The stream was three or four yards wide at the mouth and for a distance of 15 or 20 yards but above this point it was too narrow and choked with bushes to be a habitat for dragonflies. The two males of *S. hudsonica* were flying up and down over the low section of the stream, keeping usually near the bank and about a foot above the water. Both females were captured at the mouth while ovipositing, a function which they exercise after the manner of *albicincta*, tapping the surface of the water with the tip of the abdomen. Whitehouse says the females are easy to net since they oviposit around the edge of the pool.

The flight period in British Columbia, as based on Whitehouse's records is about two months. On the Mackenzie district from which we have nine records the dates range from June 12 to July 16, and from all localities together they range from June 12 to September 28. We have, however, only one September record, which comes from Moose Factory, James Bay, Ontario. Most of our dates are from late June and July. Emergence in British Columbia probably begins about June 12 to 15 (Whitehouse, 1941) but evidently occurs sometimes much later, as indicated by the September record and by our capture of a teneral female on June 29.

Somatochlora cingulata (Selys). (Pl. 10: 11, 12, p. 80; pl. 14: 3, 9, p. 102; pl. 17: 4, p. 118; pl. 18: 7, p. 124)

Epitheca cingulata Selys, Bull. Acad. Belg., (2) 31: 302 (reprint p. 68), 1871.
Somatochlora cingulata: Selys, Bull. Acad. Belg., (2) 45: 206, 218, 1878.
Somatochlora cingulata: Kirby, Cat. Neur. Od., p. 49, 1890.

A large robust species, very dark with a white-ringed abdomen; inferior anal appendage of male truncate.

Male. Labium and anteclypeus light yellow (in immature adults) to pale brownish; anteclypeus and sides of frons dull light yellow, labrum and postclypeus black, vertex and frons except sides blue-black with green reflections, rear of head black. Pile of face brownish, rather thin, that of vertex denser and blacker. Thorax brassy with green reflections varied with reddish brown, somewhat obscured by the brownish pile. Ante-alar sinuses brown with black ridges. Interalar area without pale spots. Legs black, and reddish brown at base and in outer surface of fore femora. Wings hyaline, with anal triangle and extreme base of front wings usually clear amber yellow. Venation black, the costal veins alone sometimes brown, pt. reddish brown, membranule smoky brown, the basal fourth or fifth whitish. Abdomen black with a dull greenish lustre and with whitish or pale pinkish intersegmental annuli, which are not interrupted mid-dorsally. Seg. 2 polished dark brown on the sides of 3 with a pale brownish ventro-lateral basal patch, 3 or 4 to 8 with small dark yellow latero-basal spots.

Abdomen slightly longer than hind wing, widest at end of 5, width of which is about three-fourths of its length, beyond 5 narrowing to about two-thirds of this width at base of 8.

Genital lobes small, hamuli hooked, similar to those of *S. hudsonica*. Sup. apps. (pl. 10: 11, 12, p. 80) widest at base, near which each bears a prominent external spine. There is also a prominent lateral angle near the distal third, beyond which the slender apical part is bent mesad, curving backward and upward, the sup. apps. usually meeting each other near the apices. Laterally they appear slender, gently curved downward and without any inferior prominence. Inf. app. (pl. 10: 11, 12, p. 80) three-fifths as long as sup. apps., apex broad, subtruncate and bifurcate, not much narrower than the base.

Female. Similar to male in size and colour. Wings sometimes apically dusky and often somewhat flavescent in tenerals.

Vulvar lamina (pl. 14: 3, 9, p. 102) about one-third as long as sternum of 9, which is prolonged well beyond the tergal margin, a little broader than long, hind margin obtusangularly excavate, surface with a distal depression. Apps. somewhat longer than 9 + 10, straight or slightly incurved, lateral margins carinate.

Venation. Anx ♂ 8–10/5–6, ♀ 8–9/5–6; pnx ♂ 6–9/6–10, ♀ 7–9/7–10; two rows of post-trigonal cells, followed by 3 rows, which are very regular nearly to the margin, where they are usually 4 or 5, more rarely 3 or 6 cells; cells between fork of R_2 ♂ 14–31/12–33, ♀ 12–31/10–32; cells between IR_3 and Rspl 5 or 6 to 11 in fore wings and 8 or 9 to 11 in hind wings; pt. 6 to nearly 7 times as long as wide.

Measurements. Total length ♂ 54–63, ♀ 57–68; abd. ♂ 36.5–41.5, ♀ 37.5–44.0; h.w. ♂ 33.0–38.5, ♀ 35.5–41.0; h.f. ♂ 9–10, ♀ 10.0; pt. ♂ 3.5–4.0, ♀ 3.5–4.3.

Nymph (pl. 17: 4, p. 118). Large and with a blotched colour-pattern. Eyes a little more prominent than usual in this genus; hair on head relatively sparse; labium when folded extending posteriorly nearly to the hind limit of the mesocoxae; prementum about as wide as long, extending laterally over the inner edges of the eyes; premental setae usually 10 to 13, the 4th or 5th from the outside longest, the innermost 3 or 4 (sometimes 5 to 7), and sometimes the first, much shorter than the others; marginal setae usually extending most of the distance along the premental margin, consisting of a variable distal group and an irregular scattered series extending proximad, and 5 or 6 short setae at the distal extremity; palpal setae usually 6 or 7, rarely 5; distal margin of palpus with 7 or 8 deep, subangularly rounded crenations, the first often bifid, most of them bearing 1 to 3 stoutish spiniform setae in a graded series, the hindmost much the longest. Propleural processes on thorax prominent, with a noticeable tuft of setae; thorax elsewhere sparsely hairy; vaguely blotched with paler and darker brown. Legs pale, the femora with dark, often obscure annuli, an apical, an ante-apical and a subapical; hind femora, when extended, reaching from the middle to the end of seg. 6. Abdomen (pl. 18: 7, p. 124) ovate, broadest between segs. 5 and 6, pale brownish with darker brown blotches, marking the attachment of the tergo-sternal muscles on both dorsal and ventral surfaces; hind tergal margins with regular rows of short setae and with fine long hairs of irregular length at intervals; lateral fringe comparatively sparse, except on 8 and 9, on which the hairs are long; a similar fringe also on ventrocaudal margin of 9; dorsal hooks absent but represented by low elevations bearing short setae; slender lateral spines on segs. 8 and 9, those on 8 one-sixth to one-fifth as long as the segment's lateral margin (not including the spine), those on 9 one-fifth (or slightly more) of the corresponding margin; anal appendages a little longer than mid-dorsal length of 9 + 10; epiproct of male with a pair of prominent lateral tubercles just beyond the middle; cerci a little shorter than epiproct, stout in proximal half, rather rapidly contracted beyond the slender, pointed apices, slightly sinuate; anal appendages of female similar except that the epiproct is simply triangular with a long acuminate apex and the cerci have the outer margins straight, the inner margins concave.

Length 26–28; abd. 15.0; w.abd. 8.7–9.1; h.w. 8.0–8.5; h.f. 7.4–7.7; w.hd. 6.5–7.4.

Habitat and range. Lakes of various kinds, especially expansions of rivers, in both calcareous and peaty waters, and large rivers. Labr. and Nfld. to Hudson Bay and B.C., s. to N.S. and N.H., Mich. and n. shore of Lake Superior, w. through the prairie Provinces to s. B.C.

Distribution in Canada. Nfld.—Bonavista Bay; Spruce Brook; Glenwood; Bay of Islands; St. Anthony (July 29, 1957, B. Moore). *Labr.*—Great Caribou Is. (Battle Harbour); Hopedale. *N.S.*—Digby county. *N.B.*—Charlotte county. *Que.*—Devlin; Saguenay River (Cap Jaseux); Godbout; Pointe des Monts; Trinity Bay; headwaters of Moisie River; Cascapédia; Gaspé; Ste. Béatrix; Mont Tremblant Park; Abitibi county; headwaters of St. Margaret's River. *Ont.*—Nipissing (Lake Timagami), Thunder Bay ("Longuelac, headwaters of Sucker River" (presumably Longlac, Suckle Creek); Silver Islet and St. Ignace I., Lake Superior), Cochrane (Smoky Falls; Sucker Lake near Lake Abitibi; Hurricanaw; Hannah Bay; Big Piskwanish; Ft. Albany) and Kenora (Patricia Portion: Favourable Lake; Attawapiskat Lake) districts. *Man.*—Miles 250, 256 and 332, Hudson Bay Rlwy.; Gillam. *Sask.*—Wollaston Lake. *Alta.*—Nordegg (Coliseum Mt.). *B.C.*—Okanagan Valley.

Field notes. This large robust species is strictly boreal, inhabiting the northern coniferous forests from the Atlantic coast to British Columbia, but it is not quite transcontinental and is commoner east of the Mackenzie Valley. It appears to be mainly an inhabitant of lakes of various types. Robert (1953), for instance, reports it from fluvial lakes and from both clear and bog lakes. We have only once met with it in considerable numbers. This was at Godbout, Quebec, on the north shore of the lower St. Lawrence River. They were abundant at the headwaters of a rapid stream, the outlet of a small lake, and doubtless one of a series, for the current was powerful at the immediate discharge, although the lake shore was largely boggy. At the discharge the shore and stream-bed were rocky and it was only here that we found the exuviae. Adults, chiefly males, appeared frequently over the lake and usually beyond reach, pursuing a rather irregular course, now following the shoreline, now wandering out over the water in a manner suggestive of an *Epitheca.* They were not observed to hover over one spot in the usual manner of the Corduliidae. Several times the males were seen to attempt the capture of a female during flight. One or two of these attempts resulted in copulation, the pair immediately flying off into the trees. A pair in copula was also taken on July 24 in the pine barrens a mile or so from any of the lakes. Several times on July 21 females were observed ovipositing after the manner of *S. albicincta.* This always took place near the outlet of the lake where there was some current, but not actually in the stream. Careful search among the bushes along shore rewarded us with the discovery of the exuviae, easily recognized by their large size and the lateral prominence on the median appendage of the male, indicating the bifurcate inferior appendage of the adult. They were found, together with a much larger number of *Cordulia shurtleffi*, a few feet from the water line, but this had undoubtedly receded considerably since the insects had emerged. In a few cases the exuviae were hung up on a bush but the majority were on the ground (Walker, 1925).

Newly emerged individuals were found at Wheaton's Lake, New Brunswick, on July 19, 1923, and in Digby county, Nova Scotia, on July 14, 1940. A large number of tenerals of both sexes were taken at Attawapiskat Lake, northern Ontario, June 24 to 26. Most of our records are dated July and August, but the flight period, based on all reports, extends from June 24 to September 4, the majority of dates being in July and August.

Genus **Cordulia** Leach

Edinb. Encycl., 9: 137, 1815. Type species *Libellula aenea* Linné.

A genus of two closely related species, *C. aenea* (Linnaeus) of the palaearctic region and *C. shurtleffi* Scudder of the nearctic. These are moderate-sized corduliids of rather compact form and are very similar to species of *Somatochlora*, particularly those of the *alpestris* group, both in adult and nymphal stages. The head is entirely like that of *Somatochlora*, the reduction of pale markings of the thorax and abdomen being even more marked than in most species of Somatochlora, the body coloration being mainly metallic green on the thorax, its brilliance reduced by a thick vestiture of brownish hairs, and the abdomen being greenish black and much less hairy than the thorax. The only known character found in both sexes which distinguishes *Cordulia* from *Somatochlora* is the absence in the hind wing of the second cubito-anal cross-vein, which, when present, forms the inner side of the subtriangle (pl. 1: 1, p. 6; pl. 20: 1, p. 134).

The males of *Cordulia* differ from those of *Somatochlora* also in the presence of a vestigial tibial keel on the middle legs and in the form of the anal appendages, especially in their possession of a doubly bifurcated inferior appendage.

The nymph of *Cordulia*, although separable from those of *Somatochlora* in the thoracic markings, appears to have no good generic characters of its own. In the fully grown male nymph, however, the truncate and sharply defined 'male projection,' which is the precursor of the quadrate and doubly bifurcate inferior appendage of the adult, is fairly well differentiated from the homologous structure in Somatochlora. Nymphs of *Cordulia* (unlike those of *Dorocordulia*) lack dorsal hooks. It is open to question, however, whether *Cordulia*, *Somatochlora* and *Dorocordulia* should be separated generically.

PLATE 20

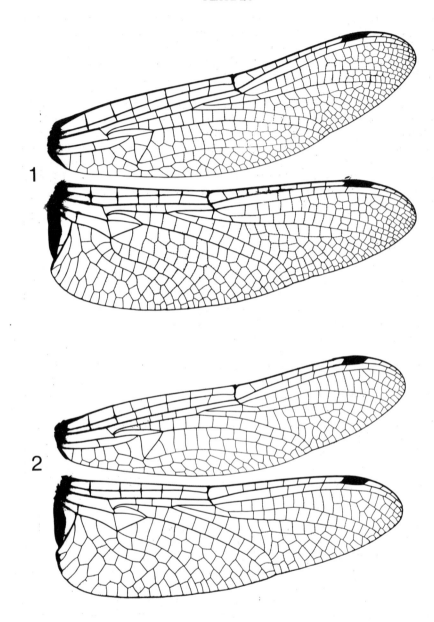

1

2

Cordulia shurtleffi Scudder. (Pl. 1: 1, p. 6; pl. 20: 1, p. 134; pl. 21: 1–4, p. 138)

Cordulia shurtleffi Scudder, Proc. Bost. Soc. Nat. Hist., 10: 217, 1866.

A medium-sized, rather compactly built corduliid with brilliant green eyes and a dark body without pale spots. The male is easily recognized by the doubly bifurcate inferior anal appendage.

Male. Labium light brownish yellow; labrum black with blue and green reflections; anteclypeus with a pale yellow median spot, postclypeus brownish yellow, darkened in the centre and with some bluish reflections. Frons with lower margin and sides ochraceous, above and in front metallic blue with green reflections. Coarsely punctate vertex, metallic green with smaller pits; occiput greenish. Hair of face pale, rather sparse, on dorsum blackish; rear of head black with a subvertical strip of pale hairs. Thorax metallic green above, bronzy on the sides, the brilliance subdued by the rather thick pale-brownish vestiture; dull grayish brown below. Coxae and mesal surfaces of trochanters brown, legs elsewhere black. Wings hyaline, amber yellow only on the anal triangle, membranule with whitish base of variable size, costal veins dark brown, venation otherwise black; pt. dark yellow-brown. Abd. seg. 1 dull brown, 2 above with a tawny geminate spot; a black lateral stripe covering the auricle (pl. 21: 1, p. 138) and a tawny spot below on each side; intersegmental membrane following 2 whitish; 3 with tawny ventrolateral basal spot, 4 to 10 greenish black with dull lustre and little hair except on the last few segments. Anal apps. black.

Abdomen (excluding apps.) hardly longer than hind wing, moderately robust, increasing in width from the constriction of seg. 3 to 7 and 8, slightly narrowed on 9 and 10.

Sup. apps. (pl. 21: 2, 3, p. 138) slightly longer than seg. 9, subcylindrical, little tapering, slightly but increasingly divergent, an obtuse lateral prominence near the base of each app.; blunt; in lateral view horizontal. Inf. app. (pl. 21: 2, 3, p. 138) broad, deeply and angularly bilobed, each lobe bifid in an oblique plane.

Female. Similar to male in general form and coloration except in the usual distinguishing features. Dorsum of 3 mostly tawny, sides of 2 and 3 polished black, except ventrolateral basal spot of 3, which is tawny.

Abdomen less constricted at seg. 3, behind which it is subcylindrical but with a mid-dorsal ridge, sides beyond 3 subparallel, converging slightly on 9 and 10. Vulvar lamina about one-fourth as long as 9, bilobed half way to base, the lobes rounded and separated by a wide angle. Wings hyaline or with more or less flavescence, which is seldom very strong except in teneral individuals.

Venation. Anx ♂ 7–9/5–6, ♀ 7–9/5; pnx ♂ 8–10/10–13, ♀ 7–9/8–10; two rows of post-trigonal cells in hind wing (♂ ♀), becoming 3 after 2 or 3 cells, and then 4; 1 cross-vein under pt.

Measurements. Total length ♂ 42–47, ♀ 44–48; abd. ♂ 27–32, ♀ 29–33; h.w. ♂ 27–32, ♀ 29–32; h.f. ♂ ♀ 7.0–7.5; pt. ♂ 2.2–2.8, ♀ 2.3–2.7.

Nymph. An average-sized, somewhat robust corduliid nymph, moderately hairy, the thorax with a broad, dark brown longitudinal lateral stripe. Head twice as wide as long; anterolateral outline broadly arcuate, including the eyes, which are more prominent posterolaterally; post-ocular margins of head straight, scarcely angulate as they pass into the hind margin which is

---

PLATE 20

Wings of Corduliidae: (1) *Cordulia shurtleffii*; (2) *Dorocordulia libera*.
(1, 2 from Needham and Westfall, 1955.)

barely concave. Antennae about as long as head, 7-segmented, the relative lengths of the segments being 1.9:1.8:2.7:2.0:2.3:2.3:2. Folded labium extending to middle of mesocoxae; sides concave, base of prementum scarcely one-third as wide as width at origin of palpi; premental setae 11 or 12 (the 4 mesial setae being smaller); palpal setae 7; crenations on distal margin of palpus one-third to one-half as long as wide, each with 3 or 4 spiniform setae in a graded series, the first being much the longest. Wing-sheaths reaching to base or middle of seg. 6. Legs moderately long, the hind femora reaching to the middle of seg. 8, moderately hairy; tibiae very hairy. Abdomen rather hairy, broadly oval, being four-fifths or two-thirds as wide as long (exuv.), widest at seg. 6 or between 6 and 7; narrowed on 9; dorsal hooks nearly obsolete, represented by very low knobs; lateral spines on seg. 8 and 9 only, small and slender and nearly equal; those of 9 slightly bent inward; cerci longer than mid-dorsal length of 9, straight, acute; epiproct straight, longer than cerci; paraprocts a little longer than epiproct, male projection occupying the proximal three-fourths of the epiproct, defined posteriorly by a sharp arcuate margin, the sides contracted in the middle.

Colour greenish brown, marked with dark brown as follows: sides of mouth-parts, sides of thorax each with a longitudinal stripe from propleura to base of hind wing, and another shorter stripe beneath, ending on abd. seg. 1. Fore femur pale with three dark annuli, a proximal, a post-median and an apical; middle and hind femora with similar but less distinct annuli; tibiae also triannulate but less distinctly so than the femora. Abdomen with a mid-dorsal pale longitudinal stripe, widening behind, flanked by three dark stripes on each side, the widest ones nearest the pale stripes.

Length 20–23 (exuv.); abd. 12.5; w.abd. 7–8; h.w. 5; h.f. 7; w. hd. 5.5–6.0 (exuv.).

Habitat and range. Quiet marshy or boggy waters, particularly sphagnum bog ponds or small lakes, from Hamilton Inlet, Labr., Fort Albany, Hudson Bay, Ont. w. to Great Slave Lake, N.W.T., Watson Lake, Yukon T. and Alaska, s. to Nfld., N.S., N.B., Conn., Pa., Ohio, Mich., Wis., Man., Sask., Alta., Colo., Utah., Nev. and Calif.

Distribution in Canada and Alaska. Nfld.—St. John's; Spruce Brook; Gander; Bay of Islands; White Bay. *Labr.*—Hamilton Inlet. *N.S.*—Digby, Halifax and Pictou counties; Cape Breton I. (Victoria county). *N.B.*—Charlotte and Gloucester counties. *Que.*—Covey Hill; St. Anne's; Hull; Ironside; Lake Témiscouata; Godbout; Thunder River; Mont Tremblant Park; Nominingue; Kipawa Lake; Laniel; La Ferme; Abitibi county; Saguenay River (Cap Jaseux); Lake St. John; Lake Mistassini; Poste de la Baleine. *Ont.*—Peel, York, Dufferin, Simcoe, Victoria, Carleton, Haliburton and Renfrew counties; Muskoka, Parry Sound, Manitoulin, Nipissing (incl. Algonquin Park), Algoma, Sudbury, Thunder Bay, Cochrane and Kenora (incl. Patricia Portion) districts. *Man.*—Treesbank; Aweme; Onah; Douglas Lake; The Pas; mile 17 Hudson Bay Rlwy.; Gillam; Churchill. *Sask.*—Lac Vert; Prince Albert; Prince Albert National Park (Waskesiu Lake, Crean Lake); Churchill River; Nameless Lake; Wollaston Lake. *Alta.*—Banff; Red Deer; Nordegg; Jasper Park (Marjorie Lake); Fort Chipewyan; Athabasca Lake; Fort Smith. *B.C.*—Vancouver; Lower Fraser Valley; Sumas Canal slough; Hope; Christina Lake; Princeton; Nelson; Ainsworth; Cranbrook; Kaslo; Kootenay district; Salmon Arm; Kamloops; Revelstoke; Prince George; Prince Rupert; Boundary district; Ft. Nelson; Atlin; Vancouver I. (Sooke; Thetis Lake; Nanaimo district; Newcastle I.; Courtenay; Forbes Landing); Queen Charlotte Is. (Massett). *N.W.T.*—Great Slave Lake (Gros Cap; Wilson I.; Caribou Is.; Pearson Pt.; Christie Bay; Yellowknife; Rae); Norman Wells. *Yukon T.*—Watson Lake; Whitehorse. *Alaska*—Fox Pt.; Admiralty I.; Juneau; Kukak Bay; Anchorage; Palmer; Gulkana; Big Delta; Ft. Yukon.

Field notes. By and large we would consider *Cordulia shurtleffi* to be the commonest and most widely distributed species of its family in Canada. In the more southern parts of the eastern Provinces it is outnumbered by *Epitheca cynosura* and *E. spinigera* but its distribution in Canada is far wider than that of any species of *Epitheca*.

C. shurtleffi shows a preference for sphagnum bogs and ponds. At Go Home Bay, Georgian Bay, near its southern limit we found a few nymphs in the bottom debris of bog ponds where there was little or no drainage and the oxygen content of the water was very low.

It is an early species to emerge, the first tenerals appearing about the beginning of June or sometimes early as May 24, depending on the season and the locality. Robert (1953) gives the flight period of this species as June 7 to July 25 and this appears to be about the usual time in eastern Canada. This period begins earlier, as a rule, in British Columbia, Whitehouse's dates (1941) being from May 2 to August 26, the peak being from about the fourth week in May to the fourth week in June. *C. shurtleffi* flies chiefly in sunshine but sometimes in shade. The males fly over the edge of a floating bog within two or three feet of the water, this being typically the mating flight. Adults may also be seen flying in the woods, sometimes passing from sunshine into deep shade. They may also be seen sometimes patrolling a glade, at the edge of a wood, keeping a little above the lower vegetation. The females oviposit by striking the surface of the water with the tip of the abdomen after the manner of most Corduliidae and Libellulidae. They may be concealed from view, while thus engaged, by sedges or other emergent vegetation.

A study of the prey of this species, conducted near Flatbush, Alberta, revealed that nymphs fed heavily on gammarid Crustacea and on chironomid larvae (in a lake particularly rich in these creatures) and adults on a wide variety of smaller insects, including Trichoptera, Chrysopidae, Chironomidae and teneral Zygoptera and Anisoptera (Pritchard, 1964a, b).

Genus **Dorocordulia** Needham

Bull. N.Y. State Mus., 47: 485, 504, 1901. Type species *Cordulia libera* Selys.

This genus of two species, both from eastern North America, is based upon venational characters of doubtful generic value. It is related to *Somatochlora* and *Cordulia* but the species are both more slender and fragile than those of the other two genera. Needham (1901) separated *Dorocordulia* from *Somatochlora* by the following venational differences.

The triangles of both the fore and hind wing (pl. 20: 2, p. 134) are open in

PLATE 21

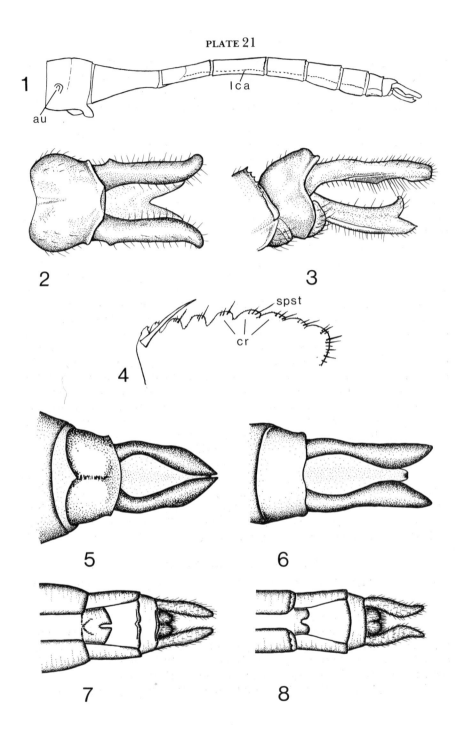

Dorocordulia and crossed in *Somatochlora*; the triangle of the fore wing in *Dorocordulia* is followed by a single row of three to five cells, this becoming a double row at about the level of the middle fork; the triangle of the fore wing in *Somatochlora* is followed by two rows of cells, becoming three or more rows at the wing margin; the second cubito-anal cross-vein of the hind wing is absent in *Dorocordulia* but present in *Somatochlora*; and lastly, in *Dorocordulia*, there is a long space beyond the single cross-vein behind the pterostigma, whereas in *Somatochlora* the spaces are more nearly equal and there are often two cross-veins behind the pterostigma.

The anal appendages of the males of *Dorocordulia* are more like those of the most typical species of *Somatochlora* than those of *Cordulia*, the superior appendages having the apices tapering upward, and the inferior appendage being triangular, like that of the former genus. On the other hand the vulvar lamina of the female is small, flat and bilobed as in *Cordulia*, although there are some species of *Somatochlora* that have a similar type of vulvar lamina. Possibly such species should not be retained in the same genus as typical *Somatochlora*.

The coloration of *Dorocordulia* is like that of the other two genera, having a metallic green background with the eyes in life being brilliant green and the legs black.

Nymphs lack evident generic characters. They resemble those of *Cordulia* rather closely. Like that genus they differ from *Somatochlora* in possessing a broad dark longitudinal thoracic stripe; and they differ from *Cordulia* in being smaller and in their possession of dorsal hooks on the abdomen.

Key to the Species of Dorocordulia
Adults

1 Rear of occiput blackish; abdomen strongly spatulate, its width behind seg. 2 more than twice as great as the narrowest part; distal half of sup. apps. of male convergent, the apices meeting, or nearly so (pl. 21: 5, p. 138); vulvar lamina divided by a narrow cleft (pl. 21: 7, p. 138)
libera (p. 140)

Rear of occiput yellowish; abdomen slender, the width behind seg. 2 less than twice as great as the narrowest part; distal half of sup. apps. of male with lateral margins parallel (pl. 21: 6, p. 138); vulvar lamina divided by a shallow rounded notch (pl. 21: 8, p. 138)   *lepida* (p. 142)

PLATE 21

Cordulia shurtleffi: (1) abdomen of male, left lateral view; (2, 3) anal appendages of male: (2) dorsal view; (3) left lateral view; (4) distal margin of left labial palpus of nymph.
Dorocordulia: (5, 6) anal appendages of males, dorsal view: (5) *D. libera*; (6) *D. lepida*; (7, 8) vulvar lamina of females, ventral view: (7) *D. libera*; (8) *D. lepida*.
(5, 6 after Needham and Westfall, 1955; 7, 8 after Needham, 1901.)
(Abbreviations: au—auricle (oreillet); cr—crenations; lca—lateral (or ventrolateral) carina; spst—spiniform seta.)

Nymphs

1 Low dorsal hook on seg. 9   *lepida* (p. 143)
   Neither a dorsal hook nor a ridge on seg. 9   *libera* (p. 140)

Dorocordulia libera (Selys). (Pl. 20: 2, p. 134); pl. 21: 5, 7, p. 138)

Cordulia libera Selys, Bull. Acad. Belg., (2) 31: 263 (reprint p. 29), 1871.
Somatochlora libera: Williamson, Drag. Indiana, p. 314, 1900.
Dorocordulia libera: Needham, Bull. N.Y. State Mus., 47: 505, 1901.

A small, very slender corduliid with bright metallic coloration and a spatulate abdomen terminating in very small anal appendages; eyes in life brilliant green.

Male. Labium and anteclypeus yellow or orange, face elsewhere and dorsum of head shining black with green and blue reflections, particularly on the frons and vertex; rear of head with a marginal ocular border of metallic green, bearing a fringe of gray hairs, the more hidden part black; occiput hardly yellow at all. Pterothorax metallic green with blue reflections, subdued by a fairly dense covering of reddish brown hair; ante-alar sinus with close-set dark hair; legs beyond the coxae black; wings hyaline, yellow at the extreme base; venation including pt. black, membranule sooty, paler close to the front. Abdomen greenish black, slightly metallic; anal apps. black.

Head with frons rounded above and in front; vertex transversely convex, slightly emarginate in front; eye-seam short, owing to a narrow forward thrust of the occiput. Legs slender, tibial keel on distal two-fifths of fore tibia, vestigial on middle tibia, and on almost the entire length of the hind tibia, on which it is acute at tip. Thorax and base of abd. seg. 3 narrowing to the middle, thence very slender and cylindrical to the end of 5, widening from base of 6 to end of 7, where its least width is nearly trebled, narrowing again on 9 and 10, the general plan of the abdomen being thus quite strongly spatulate.

Anal apps. (pl. 21: 5, p. 138) not much longer than 10 with sup. apps. converging and their apices meeting. In profile the base is slender, the middle part being the deepest, whereas the distal part is curved upward, tapering to acute apices. Inf. app. about as long as sup. apps., triangular, slightly concave above, with the minute tip bent upward. Genital lobes rather prominent, narrow and directed obliquely downward.

Female. Similar to male in form except for the stouter abdomen which begins to widen at the middle or end of seg. 4, reaching maximum width at 7; segs. 5 and 6 much wider than in male; anal apps. about two-thirds as long as seg. 9; vulvar lamina (pl. 21: 7, p. 138) half as long as sternum of 9, its free margin truncate, distolateral edges rounded, with a narrow median notch half way to the tergal margin of 8.

Venation. Anx ♂ 7–8/5–6, ♀ 7–9/5–6; pnx ♂ ♀ 6–7/6–8.

Measurements. Total length ♂ 37–41, ♀ 38–43; abd. ♂ 26–28, ♀ 28–30; h.w. ♂ 26–27, ♀ 28–31; h.f. ♂ 7.2–7.5, ♀ 7.3–8.0; pt. ♂ 1.6–1.8/1.7–2.0, ♀ 1.7–2.0/1.8–2.0.

Nymph. Very like a diminutive *Cordulia*, but a little more slender, with the trunk and legs less hairy and less distinctly marked. Anterolateral margin of head, including the eyes, forming a broad curve; eyes more prominent posterolaterally, posterior outline of head broadly curved and passing into the hind margin with scarcely any suggestion of an angle; hind margin gently concave; dorsal surface of posterior arch of head with 5 or 6 hairs. Antenna about as long as head, 7-segmented, ratio of segment lengths: 1.8:1.8:2.4:1.7:2.0:2.4:1.5. Folded labium extending back to middle of mesocoxae, width at palpal articulation about 70 per cent of length, sides somewhat flaring distally, premental setae 10 to 12, the 4 inner ones forming a discrete group; palpal setae 7; movable hook shorter than the longest palpal seta; crenations of distal margin of palpus immediately next to the movable hook somewhat longer than wide but, following the first 3 or 4, much lower, most being 3 or 4 times as wide as they are long, each bearing 3 or 4 spiniform setae in a very steeply graded series. Legs moderately long and slender; hind femur extending to middle or end of seg. 1; hind tibiae not so hairy as in *Cordulia*; hind wing-sheaths extending to end of seg. 6 or middle of seg. 7. Abdomen depressed, elongate-oval, widest at seg. 7 (exuv.) or between segs. 6 and 7 (alc.); dorsal hooks in the form of minute knobs on segs. 5 to 8; lateral spines on segs. 8 and 9, those of 9 slightly longer being about one-fourth the length of the segmental margin (including the spine); hind margin of 9 with a fringe of long hairs; lateral margin of abdomen with a shorter fringe; epiproct about one-third longer than the cerci, a little shorter than paraprocts; no distinct development of a male projection on the epiproct.

Coloration similar to that of *Cordulia*, especially in the conspicuous dark stripes on sides of thorax, extending from posterolateral corners of head across pronotum and thoracic pleura to base of hind wing; traces of a lower stripe also may be present. Annuli of legs may be indistinct or absent and markings of abdomen are also faint.

Length 16.5–18.0 (Needham, 1901:21((exuv.)); abd. 10–11; w. abd. 6–7; h.w. 5–6; h.f. 6; w. hd. 5.2–5.6.

Habitat and range. Bog ponds, bog or marsh-bordered lakes and sluggish streams. Eastern Provinces of Canada and northeastern United States.

Distribution in Canada. N.S.—Annapolis county. *N.B.*—Charlotte county (St. Andrews; Campobello I.). *Que.*—Eastman; Meach Brook; Kazabazua; Mont Tremblant Park; Nominingue; Saguenay River (Cap Jaseux) (Fernet and Pilon, 1969). *Ont.*—Essex, Waterloo, York, Simcoe, Victoria and Carleton counties; Muskoka, Parry Sound, Nipissing (incl. Algonquin Park), Algoma, Thunder Bay, Cochrane and Kenora (Sioux Lookout) districts.

Probably fairly common east of the Prairies to the Atlantic but unknown from Newfoundland.

Field notes. With its metallic coloration and brilliant golden-green eyes, this beautiful insect looks like a small fragile *Somatochlora*, but its very slender middle abdominal segments and broadly spatulate distal part give it a distinctive aspect which makes it easy to recognize.

During its period of flight, which begins about the first week in June and continues until about the third week in July, *D. libera* may often be seen hawking back and forth over open marshes and sphagnum bogs or patrolling a small stream usually at a height of three to five feet, but not very regularly. It also flies in sunny glades in the woods or along their borders. It generally moves rather slowly but with rapidly vibrating wings, the body tilted with the abdomen slightly curved. When approached it darts away swiftly but, if the collector be stationed on the regular path of its flight and strike with the net from behind, it is not very difficult to capture. In an open marsh it follows no very definite course but still tends to keep at a height averaging about four feet.

D. libera is essentially a midsummer dragonfly, appearing a little later than the smaller species of *Epitheca* and somewhat before most species of Somatochlora.

Our earliest records for Ontario are dated June 1 and 2 (Severn and Bobcaygeon) but adults appear to reach their peak of numbers from about June 20 to the middle of July. At Lake Simcoe they fly until about the beginning of August and at Go Home Bay (Georgian Bay) from June 20 to July 20. Farther north the flight period is a little later, as indicated by Robert (1953) for Mont Tremblant Park, Quebec, namely from June 28 to August 8.

The nymphs "inhabit edges of water commonly under overhanging turf and clamber up projecting roots and stumps to transform" (Needham and Westfall, 1955). The few nymphs we obtained at Go Home Bay were found at the bottom of sphagnum-bordered ponds and marshy bays, such as those frequented by the adults. The exuviae are usually left on stems or stumps less than a foot above the water.

In southern Ontario the young imagos may be seen early in June along sunny paths or small openings in woods, flying back and forth and coming to rest on the branches of trees at various heights, often out of reach of the net. Later in June or July, depending on the locality, the males are more often seen over water, flying back and forth near the banks, usually at a height of four or five feet. They fly higher than *Cordulia* and do not follow the water's edge so closely but tend to wander freely over the water or the marsh.

We have only one note on the ovipositing habits of *Dorocordulia libera*. On July 15, 1933, we observed a female on the broad pond-like outlet of the stream near De Grassi Point, Lake Simcoe. She was flying over the open water, dipping the end of the abdomen in the usual way, and moving about rather freely. Males were often seen in flight over the broad weedy part of the stream on this occasion, but no indications of pairing were noted.

Dorocordulia lepida (Hagen). (Pl. 21: 6, 8, p. 138)

Cordulia lepida Hagen: Selys, Bull. Acad. Belg., (2) 31: 264 (reprint p. 30), 1871.
Dorocordulia lepida: Needham, Bull. N.Y. State Mus., 47: 506, 1901.

Similar to *D. libera* in size and wingspread, but lacking the wide expanse of the posterior abdominal segments 7 to 9, and differing in the superior anal appendages, the apices of which are turned outwards instead of inwards.

Male. Head marked as in *D. libera*, except that the lateral yellow areas of the postclypeus are larger, paler and more clearly defined than in *libera*; occiput yellowish at its middle and in rear above. Thorax dark orange brown, variegated with dark green, with a brassy lustre, partly obscured by the pale brownish hair, which is somewhat thinner than in *D. libera*; yellow spots absent. Legs black, outer surface of fore femora orange brown. Wings clear, costal veins narrowly yellow, venation otherwise dark brown; membranule whitish on basal third or half, distally smoky. Abdomen similar to that of *libera* but orange areas on seg. 2 larger and paler, covering dorsum and lower half of sides, but not extending on to the genital lobes; seg. 3 with a large basolateral area orange-brown, remainder of abdomen including anal apps. greenish black.

Abdomen smooth on first six segments, moderately hairy beyond, segs. 3 to 5 and proximal half of 6 about equally stout, slightly enlarging on 7 and 8, narrowing again only slightly, the segments to the end successively shorter from 6 to 10, widening from base of 6 to caudal end of 8, then narrowing slightly again.

Sup. apps. (pl. 21: 6, p. 138), viewed dorsally, sinuate, nearly meeting in the middle, then swinging outwards, the pointed apices divaricated, hairy in middle, inf. app. two-thirds as long as sup. apps., triangular, blunt at tip.

Female. Similar to male in colour pattern. Abdomen tapering from base or base of seg. 3 to apex, swollen as usual on segs. 1, 2 and base of 3, beyond which it is depressed. Anal apps. a little shorter than seg. 9, longer than 10. Vulvar lamina (pl. 21: 8, p. 138) about one-third as long as sternum of 9, nearly truncate, with a very shallow median notch, the two lobes being broadly rounded.

Venation. Anx ♂ 7(8)/5, ♀ (6)7/(4)5; pnx ♂ 5/(6)7, ♀ 5–6/6–7.

Measurements. Total length ♂ 36–37, ♀ 37–38; abd. ♂ 25.0–25.5, ♀ 25–28; h.w. ♂ 26–27, ♀ 28.5–29.0; h.f. ♂ 6.5–7.0, ♀ 6.0–7.0; pt. ♂ 1.8, ♀ 2.0.

Nymph (description from Garman (1927)). Premental setae 10 or 11; palpal setae 7, the basal one well separated from the rest; crenations of distal margin of palpus shallow, each bearing 2 or 3 spiniform setae. Dorsal hooks present as small knobs on abd. segs. 5 to 9; lateral spines on segs. 8 and 9.

Length 19; abd. 10; h.w. 5; h.f. 6.

Habitat and range. Bogs and marshes. Maritime Provinces, northeastern United States, N.Y. and N.J. to Md.

Distribution in Canada. *N.S.*—Annapolis (South Milford; Lequille River; Baillie Lake; Lamb Lake) and Halifax counties. *N.B.*—Sunbury (French Lake) and York (Nashwaak) counties.

Field notes. In Canada *D. lepida* has been taken from June 23 to July 26. Garman (1927) records its flight period in New England as being from May 22 to August 31.

In Massachusetts (Cape Cod) it was found at boggy ponds (Gibbs and Gibbs, 1954), and in New Jersey (Burlington county) in and near cranberry bogs (Beatty, 1946).

Very little is known of the habits of this species. Three specimens were taken in a large open marsh adjoining French Lake, New Brunswick, on July 10, 1932. The day was very windy and the three specimens were all in the shelter of tall shrubs at the edge of the marsh. They included a pair in copula, resting on a bush, and a single female.

The male from Baillie Lake, Nova Scotia, was found clinging to herbage near the ground, close to a rapid part of the stream. The female from Lamb Lake was ovipositing.

FAMILY LIBELLULIDAE

The last family to be treated in our survey of the dragonflies of Canada and Alaska is the Libellulidae. This is not only the largest family of Odonata now living but is also believed to be the family of most recent origin. It is in the warmer parts of the world, provided the supply of fresh water is adequate, that the Libellulidae are most numerous in genera and species. It is therefore not surprising that in Alaska and most of Canada there are few species of Libellulidae and very few genera, only two of them being well represented in species, namely, *Sympetrum* and *Leucorrhinia*. The family as a whole is best represented in Canada in southern Ontario.

The main characters of the adults are as follows. Eyes large, globose, meeting above in an eye-seam of variable length, the posterior margin not sinuate and without a tubercle; vertex large, produced forward, in dorsal view concealing the median ocellus, which is much larger than the lateral ocelli, the latter situated close to the bases of the antennae; occiput large, triangular, somewhat convex above and behind; rear of head more or less swollen on each side. Prothorax often with hind lobe raised into a bilobed crest, bearing a fringe of long hairs. Mesopleural suture with a short sinuosity near the middle, not always very noticeable but usually characteristic. Legs without tibial keels. Wings with the anal angle rounded in both sexes, there being no anal triangle. Venation in general similar to that of the Corduliidae, but with the differentiation between fore and hind wings with respect to the triangles and subtriangles carried further. Anal loop (with one exception in our fauna) foot-shaped, both heel and toe present, and each containing a branch of the midrib or bisector. Arculus nearly always with the sectors joined at base and usually for some distance beyond the base, so that they appear stalked. Antenodals either *complete* when both costal and subcostal series are present in equal numbers, or *incomplete* when the last cross-vein of the subcostal series is wanting. Radial and median supplements are distinct.

The abdomen is relatively shorter than in the Corduliidae as a general rule, and is usually more depressed, tapering smoothly from segment 2 to the apex, particularly in the female.

The main characters of the nymphs have been described under the Libelluloidea (p. 14). Within this superfamily, nymphs of the Macromiidae stand clearly apart from those of the Corduliidae and Libellulidae on the basis of characters mentioned in the key (p. 15). Nymphs of the last two families cannot be distinguished incisively by any single character. The most useful one is the depth of the crenations on the distal margin of the labial palpus which in Corduliidae (but also *Pantala*) are usually one-fourth to one-half as long as they are broad, or longer, and resemble scoops; in Libellulidae (except *Pantala*) these crenations are much shallower than this.

Nymphs of Libellulidae usually inhabit warmer, shallower, more eutrophic waters than do nymphs of Corduliidae and tend to be more active and to grow faster.

KEY TO THE GENERA OF LIBELLULIDAE

ADULTS

1 Fore-wing triangle with anterior (costal) margin broken, thus making the space it encloses a quadrangle; anal loop of hind wing indistinct (pl. 22: 1, p. 148) *Nannothemis* (p. 147)

   Fore-wing triangle normal, three-sided; anal loop well developed, sac-like or foot-shaped   2

2 Vein R_3 undulating, not smoothly curved (e.g. pl. 22: 3 and 27: 3, pp. 148 and 188)   3

   Vein R_3 smoothly curved (e.g. pl. 22: 2 and 23, pp. 148 and 154)   5

3 Vein Cu P in hind wing arising from hind angle of triangle (e.g. pl. 22: 3, p. 148)   4

   Vein Cu P in hind wing arising from outer side of triangle (e.g. pl. 26: 1, p. 180)   7

4 Wings with several bridge cross-veins (pl. 22: 3, p. 148) or, if only one, hind wing with large nodal and preapical dark spots *Libellula* (p. 162)

   Wings with only one bridge cross-vein (pl. 27: 3, p. 188); dark spots, if any, on hind wing confined to base near anal angle *Pantala* (p. 271)

5 Midrib of anal loop nearly straight or only slightly bent at ankle (e.g. pl. 22: 2, p. 148); postnodal cross-veins of second series, and first cross-vein under pt. strongly aslant   6

   Midrib of anal loop distinctly bent (e.g. pl. 26: 2, p. 180); postnodal cross-veins much less, if at all, aslant   7

6 Anterior (costal) side of fore-wing triangle three-fifths as long as proximal side, or less (pl. 23: 4, p. 154) *Celithemis* (p. 154)

   Anterior side of fore-wing triangle two-thirds as long as proximal side, or more (pl. 22: 2, p. 148) *Perithemis* (p. 151)

7 Wings with only one cross-vein, or none, under pt.   8

   Wings with two or more cross-veins under pt.   9

8 Wings with triple-length vacant space before single cross-vein which is either under distal end of pt. or just beyond it (pl. 26: 2, p. 180) *Pachydiplax* (p. 193)

   Wings with single cross-vein under middle or proximal half of pt. under which neither of the adjacent vacant spaces is of triple length (pl. 27: 1, p. 188) *Sympetrum* (p. 198)

9 Anterior (costal) side of fore-wing triangle more than one-half as long as proximal side (pl. 27: 2, p. 188); face white *Leucorrhinia* (p. 236)

Anterior side of fore-wing triangle less than one-half as long as proximal side; face not white   10

10 Pt. of fore wing distinctly longer than that of hind wing; anterior and posterior sides of pt. of about equal length; dark area at base of hind wing (pl. 37, p. 252)   *Tramea* (p. 261)

Pt. of fore wing about equal in length to that of hind wing; anterior side of pt. longer than posterior side, giving it a trapezoidal appearance; no dark area at base of hind wing   11

11 Spines on basal half or two-thirds of outer angle of hind femur short and of about equal length, with three to four large spines on distal half or third (pl. 25: 3, p. 168); fore wing with six (occasionally five) paranal cells before subtriangle   *Erythemis* (p. 190)

Spines on outer angle of hind femur gradually increasing in length distally (pl. 25: 4, p. 168); fore wing with five paranal cells before subtriangle   *Erythrodiplax* (p. 196)

NYMPHS

1 Lateral spines on seg. 8 as long as mid-dorsal length of seg. 9 or longer; without dorsal hooks on segs. 5 to 7   2

Lateral spines on seg. 8 shorter than mid-dorsal length of seg. 9, or absent; or, if with long lateral spines on seg. 8 as above, then with dorsal hooks on segs. 5 to 7   3

2 Movable hook of labial palpus long and slender; crenations on distal margin of palpus shallow* (pl. 38: 8, p. 254); lateral spines on seg. 8 only slightly shorter than those on seg. 9; lateral spines on seg. 9 extending posteriorly beyond tips of cerci (pl. 45: 4, p. 276)   *Tramea* (p. 261)

Movable hook of palpus short and robust; crenations on distal margin deep (pl. 38: 7, p. 254); lateral spines on seg. 8 only about one-third as long as those on seg. 9; lateral spines on seg. 9 not extending posteriorly as far as tips of cerci (pl. 45: 5, 6, p. 276)   *Pantala* (p. 271)

3 Total length less than 11 mm.; epiproct usually as long as paraprocts; no dorsal hooks on abdomen   *Nannothemis* (p. 149)

Total length more than 11 mm.; epiproct usually shorter than paraprocts; dorsal hooks present or absent   4

4 Dorsal hook on seg. 9   *Perithemis* (p. 152)
No dorsal hook on seg. 9   5

5 Eyes capping anterolateral part of head (pl. 38: 2, 3, p. 254); abdomen relatively long and tapering   *Libellula* (p. 163)

---

*This character serves as a reminder that diagnostic features in this and other keys are not necessarily reliable in early instars (see Bick, 1951a).

Eyes more smoothly rounded and more lateral in position (e.g. pl. 38: 4,
p. 254; pl. 39: 2, 3, p. 258); abdomen relatively short and blunt   6

6 Paraprocts strongly decurved (pl. 45: 3, p. 276); lateral spines absent
Erythemis (p. 190)

Paraprocts straight or nearly so; lateral spines usually present   7

7 Dorsal hooks absent, or represented only by prominences (on segs. 4 to
9) each bearing a conspicuous tuft of setae   8

Dorsal hooks present   11

8 Dorsal prominences on segs. 4 to 9, each bearing a conspicuous tuft of
setae   *Erythrodiplax* (p. 196)

Dorsal hooks absent   9

9 Lateral spines on seg. 9 equal to or greater than mid-dorsal length of
seg. 9 (pl. 44: 1, p. 272); dark ridge running mesad from mesoposterior
part of eye (pl. 38: 4, p. 254)   *Pachydiplax* (p. 193)

Lateral spines on seg. 9 usually less than mid-dorsal length of seg. 9 (e.g.
pl. 44: 2, 3, p. 272); no dark ridge running mesad from mesoposterior
part of eye   10

10 Lateral spines on seg. 8 minute (one-tenth or less the mid-dorsal
length of seg. 8) or absent; lateral spines of seg. 9 varying from about
one-tenth to one-half the mid-dorsal length of seg. 9
Sympetrum (part) (p. 199)

Lateral spines on seg. 8 of normal size (one-third or more the mid-
dorsal length of seg. 8); lateral spines of seg. 9 usually at least one-half
the mid-dorsal length of seg. 9   *Leucorrhinia* (part) (p. 236)

11 Lateral spines on segs. 8 and 9 long and straight, those on seg. 9
extending to or beyond the tips of the paraprocts (pl. 39: 2, p. 258)
Celithemis (p. 155)

Lateral spines on segs. 8 and 9 short, those on seg. 9 not extending to the
tips of the paraprocts (e.g. pl. 42: 1, p. 266   12

12 Dorsal hook on seg. 3; epiproct and paraprocts about equal in length
Leucorrhinia (part) (p. 236)

No dorsal hook on seg. 3; epiproct distinctly shorter than paraprocts
Sympetrum (part) (p. 199)

Genus **Nannothemis** Brauer

Verh. Zool.-Bot. Ges. Wien, 18: 329, 726, 1868. Type species *Nannophya
bella* Uhler.

A monotypic genus, the single species being the smallest of North American Anisoptera, and one of the smallest dragonflies in the world. With the
small size is associated a reduced number of cross-veins and hence of wing
cells and cell rows (pl. 22: 1, p. 148). Thus the triangles and supratriangles

PLATE 22

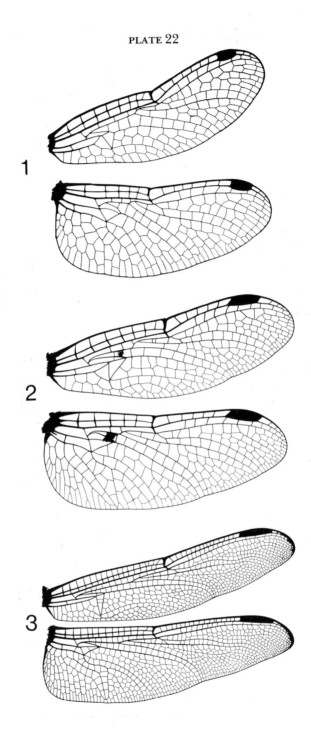

are all clear, the bridge has only one cross-vein, and the antenodals of the fore wing are incomplete. The most distinctive characters are the four-sided triangle of the fore wing and the incomplete anal loop, which opens behind, the midrib running back to the wing margin. These two features, especially the former, are suggestive of primitive characters, but they contrast in this respect with two specialized features: the rather long fusion of the sectors of the arculus and the complete retraction of the arculus of the hind wing to the level of the triangle base.

The nymph is very small and compact and almost uniformly brown, only the femora being inconspicuously annulate. The eyes are rounded and prominent, and the lateral angles of the head well rounded. Crenations on the distal margin of the labial palpus are shallow. The abdomen is truncate, not depressed and about twice as long as the thorax, its greatest width being slightly more than three-fourths its own length; the last three or four segments bear coarse hairs, especially segment 9 on the ventroposterior margins; dorsal hooks are absent; the lateral spines on segments 8 and 9 are small and thorn-like; the wing-sheaths extend to the base of segment 5. Like the adult, the nymph appears to lack consistent indications of being a primitive type. In the form of the eyes and palpal crenations and in the absence of dorsal hooks, it is clearly a libellulid.

Nannothemis bella (Uhler). (Pl. 22: 1, p. 148)

Nannophya bella Uhler, Proc. Acad. Phila., 1857: 87, 1857.
Nannophya bella: Hagen, Syn. Neur. N. Amer., p. 186, 1861.
Nannothemis bella: Packard, Amer. Nat., 1: 311, 1867.

The smallest of native Anisoptera, the total length of either sex being only about 20 mm., the male black, becoming whitish pruinose when mature, the female black and yellow.

Male. Head much wider than thorax, eye-seam much shorter than mid-length of vertex, labium black, labrum dark brown to black, face white or creamy with dark-brown quadrate spots, vertex and occiput dark brown; rear of head black with a white geminate spot below occiput. Thorax and abdomen black, becoming gray pruinose with maturity, legs black. Wings hyaline with blackish venation; pt. brown with pale hind margins.
Abdomen slightly more than three-fourths as long as hind wing, cylindrical in proximal half, widening and flattening distally, narrowest at seg. 5, widest at 8, 10 very short.
Anal apps. slightly longer than seg. 9.

Female. Dark brown, marked with yellow as follows: on thorax, lateral spots on edge of pronotum, dorsolateral stripes on anepisternum of mesothorax and large irregular spots on the sides; on abd., basal spots on segs. 1 to 7, giving an annulate appearance, whole dorsum of seg. 10 and anal apps. Basal half of wings strongly yellow, fading distally to become nearly hyaline.

PLATE 22

Wings of Libellulidae: (1) *Nannothemis bella*; (2) *Perithemis tenera*; (3) *Libellula incesta*. (1–3 from Needham and Westfall, 1955.)

Venation. Anx ♂ 6/4, ♀ 7–8/4–5; pnx ♂ 4–6/5, ♀ 5–7/5–6. Cells in triangle, s. of fore wing and spt. 1.

Measurements. Total length ♂ 20.0–21.5, ♀ 20.0–20.5; abd. ♂ 12.4–13.5, ♀ 12.5–13.5; h.w. ♂ 15.0–15.5, ♀ 15.0–16.5 (NW: ♂♀ 10–15); h.f. ♂♀ 2.5–3.0; pt. ♂ 1.4–1.5, ♀ 1.5–1.7.

Nymph. Very small, compact; without dorsal hooks but with small lateral spines on segs. 8 and 9 only; rather hairy, especially on the distal segments of the abdomen; exuviae showing little or no pattern.

Head with eyes hemispherical, prominent, directed anterolaterad and reaching about as far forward as the frons. Antennae shorter than head, posterolateral surface broadly rounded, hind margin nearly straight. Labium about as long as it is wide; sides of prementum moderately contracted; premental setae 8 to 10; palpal setae 6 or 7, movable hook rather short, distal margin with crenations very low, most of them bearing 2 or 3 setae. Metathorax about as wide as head, legs moderately long, tibiae and tarsi slender; hind wing-sheaths about one-half as wide as long, about three-tenths longer than the hind femur. Abdomen slightly more than three-fourths as broad as long, broadest at about the distal third or fourth; curve of lateral margin smoothly rounded, except where interrupted by the lateral spines, which are present on segs. 8 and 9 only and are each rather more than one-third as long as the lateral margin of the segment (including the spine). Setae particularly long on the margins of the last 3 or 4 abdominal segments, concealing the lateral spines of 8 and 9 to a considerable extent; epiproct with base about as long as the other two sides; cerci scarcely longer, strongly turned outward; paraprocts little longer than epiprocts.

Length 8.5–10; abd. 4.5–5.5; w. abd. 3–4; h.w. 2.3; h.f. 2–3.5; w. hd. 3–3.5 (measurements incorporating those of Weith and Needham (1901)).

Habitat and range. Typically, and in Canada almost exclusively, sphagnum bog ponds. United States e. of the Mississippi River, s. Ont. and Que., probably also N.S. and N.B., Fla. w. to La., n. to Me., Mich. and Wis.

Distribution in Canada. Que.—Sixteen Island Lake; Mont Tremblant Park. *Ont.*—Lambton, Middlesex, Brant, Huron, Peel, Simcoe, Lennox and Addington, and Renfrew counties; Muskoka and Parry Sound districts.

Field notes. This tiny libellulid, as found in Canada, seems to be almost exclusively an inhabitant of floating sphagnum bogs, where it is sometimes quite abundant. It flits about the low vegetation, settling frequently on the cotton grass, leatherleaf, cranberries and other bog plants, the wings when at rest being bent sharply downward on each side of the supporting stem. They fly within a few inches of the bog vegetation and never stray over the open water. Sometimes they are entrapped by sundews (*Drosera rotundifolia* and *D. intermedia*). The two sexes tend to occupy different parts of the habitat. Near Mont Tremblant the males "hunt for small flies over flooded sphagnum bogs whereas the females seek less moist places several hundred feet from the water's edge" (Robert, 1963).

"The females oviposit in the shallow places where the nymphs live, in temporary water of one to two inches depth, and very warm. The female dips the tip of her abdomen to the surface . . . but only about three or four times; then rests; then repeats" (Needham, 1901). Sometimes the ovipositing female is accompanied by the male. Calvert (1929) found such a pair on

July 18, 1925, on a small tributary of the Wading River, New Jersey. The small stream was dammed and the insects were flying over "the tufted and partly overflowed vegetation on the bank." It was here that oviposition was observed, "the male holding the female by her head while she dipped her abdomen repeatedly into the water to discharge her eggs" (Calvert, 1929).

Our recorded flight period was from June 20 to August 9, but we have never determined the first day of emergence for any year. It is probably about the middle of June or somewhat earlier in southern Ontario.

Genus **Perithemis** Hagen

Syn. Neur. N. Amer., p. 185, 1861. Type species *Perithemis domitia* Drury.

This is a Neotropical genus of about twelve species as recognized in the latest revision of the genus (Ris, 1930). Although the genus is very distinct, the species are somewhat homogeneous and difficult to define without large series. The monograph by Ris, however, is an excellent treatise of the group.

Although *Perithemis* is essentially Neotropical, it is distributed throughout the Antilles, Central America and Mexico, as well as South America, and several species have spread through most of the southern and eastern United States. The species *P. tenera* is widely distributed in the United States and barely enters Canada along the Lake Erie counties of Ontario to the vicinity of Hamilton.

The species of *Perithemis* are small, stocky libellulids of a brown and yellow coloration; the wings are deep amber in the males, with or without darker brown spots, and often partly hyaline in the females, in which they have large dark brown spots or stripes.

The head is well rounded, and the face moderately well beset with blackish hairs; the occiput is large. The hind lobe of the prothorax is broadly bilobed, bearing a long fringe of brown hairs. The wings (pl. 22: 2, p. 148) are rather broad, the hind wing at the level of the nodus being more than one-third as wide as long. The abdomen is decidedly shorter than the wings, very stocky and tapering only slightly towards the tip. Most of the segments are only slightly longer than they are deep and segments 2 and 3 are distinctly deeper than long, lacking the usual swelling and constriction. The male genitalia and the anal appendages are very small.

The venation is remarkable in two respects: first, for the peculiar form of the fore-wing triangle in which the inner and anterior sides are nearly equal in length, the outer side being considerably longer than either of the other sides; and second, for the form of the anal loop, the midrib of which is nearly straight, except near the base, where there is a short initial curve. Other venational features are the position of the arculus before the middle of the space between the first and second antenodals; the absence of the last antenodal from the subcostal space; and the presence, as a rule, of two or three bridge cross-veins and three or four cross-veins behind each pterostigma. The venation in general is open, the triangles and sub-triangles being in some species nearly always free, and in others divided into two or three cells. The supplements are distinct but inconspicuous, the radial and median supplements usually subtending a single row of cells.

The nymphs are broad, with a depressed abdomen and slender spreading legs. The eyes are small and lateral. The lateral spines on segments 8 and 9 are about equal in length. A distinctive feature is the presence of a dorsal hook on segment 9.

Perithemis tenera (Say). (Pl. 22: 2, p. 148; pl. 39: 1, p. 258)

Libellula tenera (Say, J. Acad. Phila., 8: 31, 1839.
Perithemis domitia (Drury): Hagen, Syn. Neur. N. Amer., p. 135, 1861.
Perithemis tenera: Needham, Bull. N.Y. State Mus., 47: 512, 1901.

A small stocky golden-brown dragonfly, the male having clear amber wings and the female having wings marked with conspicuous brown stripes.

Male. Face dull light yellow, sometimes greenish; labrum and labium often paler than the ante- and postclypeus; anteclypeus and postclypeus often darker ochraceous; frons with a moderate clothing of blackish hairs, rear of head dull brown. Thoracic dorsum golden brown with pale dull yellowish stripes which are usually not clearly defined, the most distinct parts being on the broad lower parts of the two epimera. Legs light brownish with blackish hairs. Wings deep clear amber, venation and pt. red; other dark wing spots absent in Canadian specimens we have seen. Abdomen dull brown, paler laterally and ventrally; appendages yellow dorsally.

Abdomen short, stout, somewhat compressed at base, depressed toward the middle and tapered toward the end.

Female. Coloration similar to that of male, except that the wings which are partly hyaline, are conspicuously marked with broad dark brown stripes, a proximal pair over the triangles and adjacent wing cells, these being larger on the hind wing, where the blotch swings caudad and then mesad, and a distal pair, often longer and broader than the proximal pair and very nearly crossing the wing between nodus and pt.

Venation. Anx ♂ 6–8/5(6), ♀ 7–8/5(6); pnx ♂ 5–6(7)/5–6, ♀ 5–6/4–5; cells in triangle and st. 1.

Measurements. Total length ♂ 23.0–25.5, ♀ 22.0–24.0; abd. ♂ ♀ 13.0–14.5; h.w. ♂ 17.0–20.0, ♀ 18.5–20.0; h.f. ♂ ♀ 4.0–5.0; pt ♂ ♀ 2.5–2.8.

Nymph. The nymph (pl. 39: 1, p. 258) is peculiar in several respects: its general appearance suggests a miniature corduliid rather than a libellulid, particularly in the cleaner and less sprawling habit, the wide depressed abdomen and slender spreading legs, and the full series of cultriform dorsal hooks ending on seg. 9. Head wider than long, slightly concave along posterior margin. Prementum short, when folded not extending posteriorly beyond the fore coxae; premental setae 9 or 10; palpal setae 5; crenations on distal margin of palpus distinct, each bearing several spiniform setae. Abdomen flattened, triquetral, oblong oval in dorsal view; dorsal hooks on segs. 3 to 9, in lateral view forming a smoothly descending curve resembling "a segment of a circular saw" (Needham, 1901); lateral spines on segs. 8 and 9 short and approximately equal in length; relative lengths of epiproct, paraprocts and cerci 10:10:7.

Length 15; abd. 9; w.abd. 6; h.w. 4; h.f. 5.5; w.hd. 4.5.

Habitat and range. Ponds and quiet streams. Fla. to Coahuila, Mex., n. to Mass., N.H., s. Ont., s. Mich., Wis., Iowa and Nebr.

Distribution in Canada. Ont.—Essex (Pelee I.; Pt. Pelee; Dolson Creek, Oxley), Kent (Thames River near Prairie Siding), Oxford (Chesney Lake, Innerkip), Wentworth (Dundas Marsh, Princess Pt.) and Huron counties.

Field notes. According to Needham (1901) this species appears on the wing in Illinois about the end of May and flies throughout June. Our few records from Ontario range only from June 29 to August 14, most of them being from July. Williamson's records (1917) from Indiana cover many years of collecting and show that *P. tenera* may be on the wing in Indiana within the last ten days of May. It flies throughout June and July, and in smaller numbers in August, the flight period sometimes continuing into early September.

We have not observed these insects during the teneral period but Brimley (1903) states that they "fly in upland fields often resting in the flowers of the ox-eye daisies." At this stage they are a "timid weak species, loving the sunshine" (Muttkowski, 1908). Needham (1901) adds that *P. tenera* "is very sensitive to cloudiness and moisture, being seldom seen in flight except when the sun is shining." It flies "low over the surface of the water, and rests frequently on the tops of low stems and twigs near the shore. It perches horizontally with fore and hind wings often unequally lifted" (Needham and Heywood, 1929). "On Lake Madison (Wisconsin) it was present in great numbers about the beds of yellow water lilies and Potamogeton on the quieter bays, now flitting over and now resting upon the lily pads, and never far above the surface" (Whedon, 1914).

Although the female is sometimes held by the male while ovipositing, she is usually unattended. We have observed ovipositing females several times in eastern Pennsylvania and near Richmond, Virginia, and although never in tandem they appeared sometimes to be closely watched from the air by one or more males.

At a large pond near Doylestown, Pennsylvania, on July 16, 1951, we observed a female of *P. tenera* ovipositing within a foot of the bank in water that was thick with green algae. She kept within a height of about two inches of the water, which she touched repeatedly at about the same spot, moving only a very little when apparently disturbed by our presence.

An account of oviposition in *Perithemis* is given by Montgomery (1937a) who points out that *P. tenera* shows a distinct tendency to oviposit directly over some object in the water to which the gelatinous material enclosing the eggs adheres. On one occasion at a lake near Huntingberg, Indiana, he "saw a female *tenera* ovipositing by striking the tip of her abdomen against a small water-soaked stick which extended a few inches above the surface of the water. She remained near the stick and within a few inches of the surface of the water during the time that she was under observation. Successive 'dips' were made from a height of three or four inches. The completion of each dip took the female to a corresponding altitude on the opposite side of the stick where she turned and began another descent. The tip of the abdomen was struck against the stick at or near the surface of the water. The stick was later removed for examination and found to be covered by a gelatinous mass of about one-fourth inch thickness along that part of its length which had been near the surface of the water, extending from slightly above this point to approximately two inches below."

Another noteworthy feature in the oviposition of *Perithemis* was first recorded, though not published, by the late E. B. Williamson. This is the process whereby the egg mass "explodes" as it sinks in the water after being discharged. This process was observed by Montgomery in the same case as the one described in the preceding paragraph. When the female after being captured was held by the wings and her abdomen struck against the surface of water, a mass of eggs was released. "This mass dropped through the water for a short distance perhaps an inch or two – then 'exploded,' scattering the undivided eggs over a comparatively large area. This was repeated several times; the mass of eggs released each time became smaller and smaller until each appeared to consist of only two or three eggs. However, all of the masses even those consisting of only two eggs 'exploded'" (Montgomery, 1937a).

The reproductive behaviour of this species has been the subject of a detailed experimental study carried out by Jacobs (1955) near Bloomington, Indiana.

Variations. The transverse brown bands that typically occur on the wings of the female vary considerably in size and intensity. When these bands are small, the brown is replaced to a variable extent by yellow-brown or amber. Occasionally females are homoeochromatic, the whole wing being suffused with amber as in the male (Shiffer, 1968). The intensity of wing pigmentation in the female is correlated with the choice of oviposition site: dark individuals tend to select log or stick sites, whereas light ones tend to select patches of floating water plants (Jacobs, 1955).

Genus **Celithemis** Hagen

Syn. Neur. N. Amer., p. 147, 1861. Type species *C. eponina* (Drury).

This is a genus of rather small to medium-sized dragonflies, resembling *Sympetrum* in form but being generally larger and more slender than North American species of that genus, and having a more variegated colour pattern. The nearly straight midrib of the anal loop and the bicoloured spot at

---

PLATE 23

Wings of *Celithemis*: (1) *C. eponina*; (2) *C. elisa*; (3) *C. monomelaena*; (4) *C. martha*.

PLATE 23

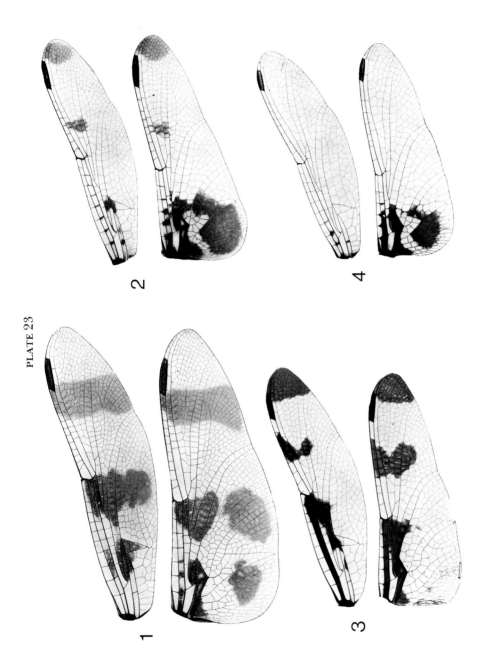

the base of the hind wing make easy the separation of *Celithemis* from *Sympetrum*. The wings of both sexes are usually strikingly spotted with yellow and brown (pl. 23, p. 154).

The head is of medium size with the eye-seam rather long and the frons moderately prominent and well rounded; the vertex is little excavated in front. The hind lobe of the pronotum is large, upright and bilobed, the lobes being fringed with long hairs. The pterothorax is relatively narrow. The legs are long and slender, the femora of males having widely spaced spinules which are gradually elongated distad. The abdomen is moderately long, rather slender, very little constricted and somewhat depressed at segment 3; it is widest between 5 and 6 becoming gradually narrower toward the end. The male genitalia of segment 2 and the genital lobes are very small.

Females resemble the males in colour pattern and form, but the yellow markings tend to be lighter and the abdomen more cylindrical; the vulvar lamina is bilobate and erect, and its lobes are distinctly separate.

The base of the hind wing (pl. 23, p. 154) is rather broad; the anterior side of the fore-wing triangle is no more than three-fifths as long as the proximal side; this triangle is divided into two or more cells; the triangle of the hind wing is usually clear; the anal loop has a nearly straight midrib and mesal margin except at the base; the arculus lies nearer the first antenodal than the second; there is typically only one bridge cross-vein; there are five or six antenodal cross-veins in hind wing; the antenodals are incomplete. The radial supplement has one or two (occasionally three) rows of cells, and the median supplements usually only one row of cells. The pterostigma is brownish red.

The nymphs are greenish, clearly patterned in brown, and of delicate build, being similar in form to those of *Sympetrum* and *Leucorrhinia* except that the head is generally more prominent in front and not much developed behind, and the eyes in some species are conical in shape with the apex of the cone directed laterocaudad. The abdomen is fairly broad; there are long, straight lateral spines on segments 8 and 9, those on 9 being twice or more than twice the length of those on 8, extending to or beyond the tips of the paraprocts. Dorsal hooks are present on segments 3 to 7 or 4 to 7; they are short, spine-like and largest on segment 6 or 7. The legs are rather long, especially the hind pair.

Four of the ten North American species have been recorded from Canada and only from Ontario, Quebec and the Maritime Provinces.

Key to the Species of Celithemis

Adults

1 Wings with postnodal markings   2

   Wings without postnodal markings; spots at base of hind wing only (pl. 23: 4, p. 154)   *martha* (p. 161)

2 Wing membranes stained with yellow or orange; ante-apical brown stripes crossing the wings; apical dark spots absent (pl. 23: 1, p. 154)
eponina (p. 156)

Wing membranes largely hyaline; ante-apical brown stripes not crossing the wings; apical dark spots present   3

3 General coloration blackish, marked with yellow in tenerals. Wing spots very dark; ante-apical spots more than one-half as wide as wing; large nodal spots present (pl. 23: 3, p. 154)   *monomelaena* (p. 160)

General coloration, including wing spots, brown and yellow; ante-apical spots less than one-half as wide as wing; nodal spots absent (pl. 23: 2, p. 154)   *elisa* (p. 158)

Nymphs

Nymph unknown   *martha* (p. 162)

1 Dorsal hooks well developed on segs. 4 to 7, that on seg. 6 being the longest   *eponina* (p. 157)

No dorsal hook on seg. 4 or, if present, evident only as a tubercle   2

2 Dorsal hook on seg. 4 absent   *elisa* (p. 158)

Dorsal hook on seg. 4 represented by a tubercle   *monomelaena* (p. 160)

Celithemis eponina (Drury). (Pl. 23: 1, p. 154)

Libellula eponina Drury, Ill. Nat. Hist., 2: 86, pl. 47, fig. 2, 1773.

Celithemis eponina: Needham, Bull. N.Y. State Mus., 47: 514, 1901 (nymph).

The largest species of *Celithemis*, easily known by the yellow to orange wing membrane and the pattern of the dark brown stripes crossing both pairs of wings proximal to the pterostigmata, and the absence of dark spots beyond the pterostigmata.

Male. Dark yellow, labrum and labium paler yellow; vertex high, yellow as in face; rear of head black. Prothorax dull black, posterior lobe fringed with long pale hairs. Dorsum of pterothorax rather dark, rich, yellow-brown, densely hairy; pleura somewhat paler, with sutures sometimes black, and a pale yellow spot on the mesepimera near the mesocoxae. Legs for the most part black, but coxae, trochanters and proximal parts of femora pale straw-coloured. Wings (pl. 23: 1, p. 154) strongly stained with yellow with a pattern of spots and stripes easily recognized by the distribution of heavy stripes, particularly the wide stripes crossing all the wings proximal to pt. and the absence of apical spots distal to pt.; venation yellow to orange or red. Abdomen black and yellow or orange, black above with a segmental series of dorsal yellow spots on segs. 3 to 7, the spots making a yellow longitudinal stripe interrupted at the intersegmental areas with yellow lateral streaks on 1 to 4, largest on 3. Segs. 8 to 10 entirely black; anal apps. yellow.

Abdomen variable in form, slender and tapering from the intersegmental suture between 4 and 5 to the posterior end. Although widest at this level, it is sometimes only very slightly so.

Anal apps. slightly arched and bowed, meeting near the apices, bluish and finally slightly divaricate.

Female. Usually with larger pale areas on both thorax and abdomen, those of the abdomen being sometimes continued as far as seg. 6, but very small on that segment, yellow spots on dorsum may be continued to the end of 7. Vulvar lamina very short, obtusangularly excised between two obtusangular lobes, which are very short.

Venation. Anx ♂ 8–11/6–7, ♀ 7–10/5–7; pnx ♂ ♀ 7–10/7–11; cells in fore-wing triangle 2 to 5; rows of cells subtended by Rspl 2 or 3.

Measurements. Total length ♂ 36–39, ♀ 33–38; abd. ♂ 23–24, ♀ 22–25; h.w. ♂ 30–33, ♀ 30.5–33.0 (NW: ♂ ♀ 32–34).

Nymph. Smooth, almost uniformly yellowish and nearly glabrous. Head prominent in front; eyes prominent, somewhat horn shaped, being broadly rounded in front with small rounded apices directed posterolaterally; lateral surfaces of head behind eyes broadly curved and passing insensibly into the hind margin. Antennae shorter than head, the first two segments much thicker than the other segments which together form a slender flagellum, the last segment tapering to a fine tip; relative lengths of antennal segments from base to apex about as follows: 9:10:12:8:13:15:13. Premental setae 10 to 14, the innermost 3 or 4 setae smaller than the outer ones but not forming a discrete group; palpal setae 7 to 10; movable hook a little shorter than most of the setae; distal margin of palpus with about 9 low crenations each bearing a graded series of 3 to 5 spiniform setae of which one is much longer than the others. Thorax widening to 4 mm. across bases of hind legs, hind wing-sheaths extending to base or middle of seg. 6. Legs long and slender and with very fine setae that give them a glabrous appearance; femora each with a dark proximal annulus. Dorsal hooks well developed on abd. segs. 4 to 7, the longest being on 6 and sharp; lateral spines on segs. 8 and 9, those on 9 reaching end of anal appendages.

Length 17–21 (Garman, 1927: 16); abd. 11.0–12.5 (Garman, 1927: 9); w. abd. 6.7–7 (Garman, 1927: 5); h.f. 5.0–6; w. hd. 5.7–6.

Habitat and range. Ponds, lakes and slow streams. Fla. to Tex., Kans. and Nebr.; n. to N.H., N.Y., s. Ont., s. Mich., Wis., and s. Minn.; also in Cuba.

Distribution in Canada. Ont.—Essex, Kent, Lambton, Norfolk, Waterloo, Huron, York, Frontenac, Leeds, Carleton and Lanark counties.

Field notes. This is the largest and most striking species of *Celithemis* and has the widest distribution, being the only one that has spread southward beyond the North American continent to Cuba. It is common over most of the eastern United States, but has penetrated only a short distance into Canada, being recorded here only from southern Ontario. In Essex and Kent counties, however, it is abundant and is probably common in the Rideau Lake district in Frontenac and Lanark counties. It was formerly common on Grenadier Pond, Toronto. It is said to frequent the borders of ponds and neighbouring grassy slopes (Needham, 1901).

Needham (1901) remarks that its flight has "a flutter to it suggestive ... of a butterfly" and this is the impression that we have received as a rule, although in bright weather, as noted by Whedon (1914) adults may be "much more agile and quite difficult to capture."

When in copulation, a pair would "ascend fifty or sixty feet and dart off over the lake for a time" (Whedon, 1914). According to Needham and Westfall (1955) the female "in ovipositing is held by the male; both are apt to be seen on windy days when other species are in shelter, dipping to crests of foaming waves, far out from shore. Transformation occurs in early morning, often on stumps about a foot above surface of water."

Not enough records are available to determine the flight period accurately, although our dates, which range from June 15 to August 16, closely resemble those from neighbouring States. In Indiana, it is recorded by Williamson (1917) as flying from June 11 to August 20, and in Ohio by Borror (1937) from June 13 to August 10.

Celithemis elisa (Hagen). (Pl. 23: 2, p. 154)

Diplax elisa Hagen, Syn. Neur. N. Amer., p. 182, 1861.
Celithemis elisa: Needham, Bull. N.Y. State Mus., 47: 515, 1901 (nymph).

A medium-sized brown and yellow species with brown spotted wings.

Male. Face red, in dried specimens orange brown, except a transverse blackish groove in which stands the median ocellus: vertex and occiput brown; rear of head dark brown with a yellow spot below the neck. Prothorax with hind lobe erect, fringed with long brown hair. Pterothorax densely clothed with brown hair, with a mid-dorsal black stripe, which is widest in front, tapering behind, meso- and metapleural sutures with a black line, black spot above middle coxa and over metaspiracle. Legs mostly black, coxae brown, fore femora with ventromesal surface light brown. Wings (pl. 23: 2, p. 154) with cubital-anal areas and costal streaks yellow. Hind wing with a larger basal spot yellow and brown. Front wing with antenodals of second series touched with brown, and a brown spot at the anterior side of the triangle; hind wing with large brown spots on the first four to five antenodals and two dark brown spots extending from base beyond the triangles, the hind spot much broader than the front one. All wings with apical spots and smaller ones between nodus and apices. Abdomen brown with segmentally arranged triangular yellow spots with lateral extensions from the broad basal parts, most of them extending to end of segment, or near it on seg. 7. Anal apps. orange.

Frons very prominent, rounded with the usual transverse dorsal groove with numerous dark hairs; pterothorax not very large; abdomen slender, tapering from base to apex.

Anal apps. about as long as seg. 9, most widely separated at base, convergent but straight, except at base; in profile moderately arched throughout two-thirds of their length; curved upward toward their apices.

Female. Pattern of body and wings similar to that of male but usually of a lighter yellow. Abdomen shorter and less tapering; vulvar lamina about one-fifth as long as the sternum of seg. 9 and scarcely one-third as long as wide, bilobed, the lobes rounded or subangular but separated by an angular notch, and appendages about as long as the lateral margin of seg. 10, slender and acute.

Venation. Anx ♂ ♀ 7–9/5–6; pnx ♂ ♀ 6–8/7–9; cells in fore-wing triangle 1 to 3; rows of cells subtended by Rspl 2 or 3.

Measurements. Total length ♂ 30.0–34.0, ♀ 28.0–32.0; abd. ♂ 18–21, ♀ 18–20; h.w. ♂ ♀ 24–27 (NW: ♂ ♀ 27–28).

Nymph (description from Needham (1901)). Smaller than nymph of *C. eponina*, and with less prominent dorsal hooks on abdomen. Premental setae 11 to 13; palpal setae 7 or 8. Dorsal hooks weakly developed on abd. segs. 5 to 7, short but pointed; lateral spines on seg. 9 only reaching to base of epiproct; epiproct slender and sharp, triangular, about one-third shorter than the paraprocts, its base equal to about three-fourths of its length; cerci extending to about two-thirds of the length of the epiproct.

General coloration dull greenish brown, the fore wing-sheaths and dorsolateral part of abdomen darker than the other parts of the abdomen, with a mid-dorsal pale stripe bordered by an irregular wider and darker stripe, lateral of which the abdomen is paler again; abd. segs. 6 to 8 with a pair of brown spots about midway between the mid-dorsal line and the lateral margin. Femora with generally four dark annuli alternating with pale annuli.

Length 14.5–17; abd. 8–11.0; w. abd. 5; h.w. 5.2–5.4; h.f. 4–5.25; w. hd. 4.

Habitat and range. Marshy lakes and ponds. Fla. to Tex. and Kans., n. to N.B., Me., N.H., s. Que., s. Ont., Mich. and Minn.

Distribution in Canada. N.S.—Halifax county. *N.B.*—Charlotte county. *Que.*—Alcove; Ste-Béatrix. *Ont.*—Essex, Kent, Lambton, Norfolk, Welland, York, Bruce, Simcoe and Frontenac counties; Muskoka, Parry Sound and Nipissing districts.

Field notes. This is the most widely distributed species of *Celithemis* in Canada, but it is ecologically local and, as our experience indicates, it is nowhere abundant in Canada.

Although it breeds in ponds, lakes or slow streams, we have rarely found it in bodies of stagnant water. At Go Home Bay, Georgian Bay, it frequented marshes or shallow bays supporting dense stands of rushes (*Scirpus*) and was usually seen hovering over patches of sweet gale (*Myrica gale*) which are common in the shallower parts of open marshes, especially near the borders of woods. *C. elisa* tends to perch on the tips of the branches of *Myrica* or of any tall plants such as rushes that rise above the general level of the marsh.

We have occasionally found *C. elisa* on various other kinds of ponds but always in small numbers. Examples of some of these in Ontario may be briefly described as follows: (a) water (3–4 inches deep) in an open foundation of an unused and roofless brick building near Niagara-on-the-Lake; (b) gravel-pit ponds in an early stage; these were at De Grassi Point, Lake Simcoe, and had been greatly enlarged and deepened in recent years; *C. elisa* had not been observed here before, although the pond had been visited frequently for many years; and (c) open shallow ponds on Point Pelee, Lake Erie; such ponds are too close to the lake shore to become stagnant.

The flight period of *C. elisa* in southern Ontario, based on somewhat scanty data, is June 18 to August 23.

In Indiana, where many more records are available (Williamson, 1917; Montgomery, 1945, 1950), the flight period is much longer, extending from the first third of April to the first week in October, although there are no records during the first two thirds of May.

At a swampy pond in the Shawangunk Mountains, New York, we saw a few individuals of *C. elisa* flying out over the water about two feet above the surface. They moved along slowly, hovering and then suddenly darting forward. They were apparently males. Females were seen several times late in the afternoon when it was still very warm but there was no sunshine.

Celithemis monomelaena Williamson. (Pl. 23: 3, p. 154; pl. 39: 2, p. 258)

Celithemis fasciata: Hagen, Psyche, 5: 383, 1890c (in part).

Celithemis monomelaena Williamson, Ohio Nat., 10: 153, 155, 1910.

A slender black species, varied with brownish yellow or brown when juvenile but eventually becoming almost wholly black.

Male. Head black, labium yellowish brown, opposed inner edges of labial palpi darker; labrum black, anteclypeus pale brownish; post-clypeus light brown with a darker edge; frons shining black; vertex and occiput greenish black with metallic reflections, rear of head greenish black below. Thorax black or dark brown with sides largely yellowish to brown, especially on the epimera, the mesepimera bearing an irregular angular brown stripe, the metepimera being almost entirely brown. Later the thorax may become entirely black. Legs black. Wings (pl. 23: 3, p. 154) hyaline with heavy black or dark brown stripes. Each wing has a longitudinal proximally doubled stripe extending from wing-base to nodus, widened somewhat at the triangle and more so at the nodus where in the hind wing there is a transverse extension toward the hind margin curving somewhat proximad; a large rounded spot lies in the anal area covering part of the anal loop. There are also two pairs of postnodal spots, ante-apical and apical, the former curved and extending about two-thirds of the distance across the wing. The apical spots extend from about the level of the distal third of pt. to the tip of the wing. Abdomen with proximal four segments laterally more or less yellow or brown, the remaining parts black or dark brown with middorsal reddish streaks of segs. 4 to 7 diminishing in size posteriorly, but often disappearing in older specimens, in which the abdomen becomes entirely black. Abdomen little enlarged on segs. 1 and 2.

Genital lobes and hamuli small and inconspicuous, the inner ramus slightly more slender and curved than the outer ramus.

Female. We have no female available for description, but Williamson (1910) distinguishes the female of *monomelaena* from the male by certain characters of the wing pattern, which, however, may not be quite constant. As a rule the apical spots of the male wings are entirely black, whereas those of the female are black but with clear tips.

Venation. Anx ♂ 7–8(9)/5–6; pnx ♂ 7–9/8–10; cells in fore-wing triangle 2 (rarely 1); rows of cells subtended by Rspl 1 or 2.

Measurements. Total length ♂ 31.0–33.5 (NW: ♂♀ 30–36*); abd. ♂ 20.0–20.5 (NW: ♂♀ 19–26*); h.w. ♂ 20.5–28.0.

*These measurements include the abdominal appendages.

Nymph (description from Leonard (1934)). Head twice as wide as long, front margin broadly convex; eyes prominent, conical, the apices directed caudolaterally far behind the middle; posterolateral corners of head very broadly rounded; posterior margin nearly straight. Antennae about as long as head, 7-segmented. Labial suture reaching level of mesocoxae; distal width of prementum about three times the width at the labial suture; premental setae 13, the outer 7 being much longer than the inner 6, and the fourth from the outside being the longest of all; palpal setae 9; distal margin of palpus with shallow crenations each bearing about 4 spiniform setae. Prothorax with lateral margins obtusangulate; hind margin convex and slightly bilobed; hind wing-sheaths extending to abd. segs. 6 or 7. Abdomen ovate, widest at seg. 6 or 7; dorsal hooks on segs. 4 to 7, the largest on 7, and that on 4 a mere tubercle; lateral spine on 8 about two-thirds as long as seg. 8; that on 9 about as long as segs. 8 and 9 together; lateral s... .1es on both segments almost straight, slender, and very sharp; epiproct triangular, sharply pointed, about three-fourths as long as paraprocts; cerci about one-half as long as epiproct; paraprocts shorter than lateral spines on seg. 9 by about 0.25 mm. Coloration dull greenish brown, the head, wing-sheaths, and mid-lateral part of the abdomen being darker than the rest of the body.

Legs obscurely marked with dark bands. Abdomen with a very narrow, pale, mid-dorsal stripe that includes the weak dorsal hooks on segs. 4 to 7; on either side of this stripe is a slightly darker and broader irregular stripe, merging laterally with a very dark irregular stripe which gives place to a lighter area covering the lateral half of each side of the abdomen; four minute dark spots on segs. 7 and 8; two large brown blotches on segs. 6 and 7.

Length 15.0–17.0; abd. 9.75–11.0; w. abd. 6.25; h.w. 5.25–5.45; h.f. 5.0–5.25; w. hd. 5.25.

Habitat and range. Nymphs have been found by Leonard (1934) in shaded parts of a moderate-sized lake with a boggy margin and thick organic silt on the bottom. Okla. and Mo. e. to Va., n. to Wis., Mich., N.Y. and Mass.

Distribution in Canada. Ont.—The single record (by Hagen (1890b) as *C. fasciata*) from York county needs confirmation.

Field notes. The recorded flight period in Ohio is June 18 to August 23 (Borror, 1937). Leonard (1934) saw emergence taking place by a small lake in Michigan (near Ann Arbor) in the mid-morning of June 12: the nymph was "clinging to a *Carex* stem only three inches above the shallow water, about two feet from shore." The part of the lake preferred by *C. monomelaena* was where the water was protected by a thick growth of trees and under-brush; here the bottom for some distance from the boggy *Carex*-grown shore supported a very thick layer of organic muck, and there was a dense growth of water lilies close to the shore. "In this region dragonfly naiads were abundant, especially those of *Argia violacea*, *Lestes vigilax*, *Gomphus exilis*, and *Leucorrhinia intacta*."

Celithemis martha Williamson. (Pl. 23: 4, p. 154)

Celithemis martha Williamson, Occ. Pap. Mus. Zool. Univ. Mich., 108: 5, 1922a.

A medium-sized species, slightly smaller and more slender than *C. elisa*, with much less maculation of the wings.

Male. Coloration of teneral is as follows. Face light ochraceous; dorsal concavity of frons a little darker around the median ocellus, vertex similarly coloured, occiput somewhat darker brown, anterior and dorsal surface of head with a sparse growth of brown hairs. Rear of head dark brown, paler medially. Prothoracic lobes blackish with pale gray hairs. Pterothorax with moderately dark dorsum, margined with a pair of pale yellow longitudinal stripes; lateral and ventral surfaces an almost unbroken yellow. Legs black, slender. Hind wing (pl. 23: 4, p. 154) with a conspicuous dark area on the basal fourth similar to that of *C. elisa* but a little smaller. Fore wing with only one or two very small brown spots on the first one to three antenodal cross-veins. Abdomen slender, tapering beyond the middle to the slender anal apps.; with a series of broad yellow spots from base to seg. 7, the anterior part of each spot between base and transverse suture wholly yellow, beyond suture narrowing to rounded end. Anal apps. pale and very slender.

In mature individuals the face becomes blackish, and the top of the frons and vertex shining black with metallic reflections. Lateral stripes on thorax are lost, the whole becoming dull black. The abdomen becomes entirely black beyond seg. 7.

Female. Similar to male, differing chiefly in the form of the abdomen, which is only slightly enlarged at base, having nearly parallel sides from 4 to 7; slightly enlarged at the distal end.

Venation. Anx ♂ ♀ 8–9/5; pnx ♂ ♀ (6)7(8)/(6)7–8; cells in forewing triangle 2; rows of cells subtended by Rspl 1 (Williamson, 1922a; Needham and Westfall, 1955).

Measurements. Total length ♂ 28, ♀ 26; abd. ♂ 18, ♀ 17; h.w. ♂ 25, ♀ 24 (NW: ♂ ♀ 23–28).

Nymph. Unknown (1972).

Habitat and range. Nothing appears to have been recorded concerning the habitat of *C. martha*; like other species of *Celithemis* it probably inhabits ponds and marshes (Needham and Heywood, 1929). At the time of its original description (Williamson, 1922a) its range extended over Md., Pa., N.J., N.Y. and Mass. to Me. Needham and Westfall (1955) have added a single Canadian locality, in N.S. We have one of the Nova Scotia specimens, a female, taken at Mount Uniacke, Halifax county on July 20, 1948 by D.C. Ferguson. Robert A. Restifo (*in litt.*) has recently found this species in Ohio.

Distribution in Canada. N.S.—Halifax county (Mt. Uniacke).

Field notes. In Canada *C. martha* has been taken only near mid-July. Its flight period in Massachusetts (Cape Cod) has been recorded as June 25 to September 2, with emergence reaching its peak on July 30, and reproductive activity on about August 6 (Gibbs and Gibbs, 1954).

In Cape Cod a favoured habitat was a small, extremely productive pond adjoining one of the larger lakes. During the height of the breeding season there tandem pairs were very numerous. "Males retained a hold on the female as she dipped her abdomen to lay eggs" (Gibbs and Gibbs, 1954). In New Jersey, Beatty (1946) found adults over a small mill pond near the outlet of a dammed cedar swamp, and amongst the emergent vegetation that extended ten to fifty feet from the shore of a small, shallow artificial lake.

Variations. A detailed study of the variability of many venational characters was published by Williamson (1922a).

Genus **Libellula** Linné

Syst. Nat., 1: 543, 1758. Type species *L. depressa* Linné.

The type genus of the family Libellulidae is *Libellula*. It is the genus which once included all the Anisoptera and it is still the principal Holarctic genus to which belong the large stout-bodied dragonflies with conspicuously marked wings and bodies, which chase one another over ponds and other still, weedy waters throughout the summer. We are using the name in the broader sense preferred by Ris (1910) and Kennedy (1922), mainly because their opinions were based on prolonged and careful study. In this broader sense *Libellula* includes *Plathemis* and *Ladona*, which are given generic rank by most American odonatists including Needham and Westfall (1955). If Plathemis is separated from Libellula, then its Palaearctic relative, *Libellula depressa* Linnaeus, will need to be separated as well.

The species of *Libellula* are large sturdy dragonflies, generally easy to recognize from their colour patterns alone, particularly when they are fully mature. The wings are usually conspicuously striped or spotted with dark brown, sometimes with white, and the abdomen and thoracic dorsum may also become white or bluish gray at maturity, making the insect very conspicuous in flight.

The frons is rounded and not very pronounced, but well excavated above. The vertex is decidedly concave, the eyeseam being shorter than the occiput.

The thorax is stout and hairy, the abdomen being broader than in most of our libellulid genera, depressed and not constricted at segment 3. It is wider in the female than in the male, and in the former there is sometimes a widening of the tergite of segment 8 suggestive of the similar expansion in the Gomphidae. More frequently, however, the abdomen tapers evenly from base to apex, the margins being straight or only slightly curved.

Of venational features (pl. 22: 3, p. 148) we may mention the following. The arculus is nearer the second than the first antenodal cross-vein. The sectors of the arculus are in contact but not solidly fused at the base in the fore wing, and somewhat more so in the hind wing. The triangle of the fore wing is narrow, slightly convex on the outer side, and its anterior side is generally less than one-half as long as the inner side. The triangle of the hind wing is smaller and more nearly equilateral than is that of the fore wing. The fore-wing triangle is most frequently divided into three or four cells, but the full range is two to five. The hind-wing triangle almost always consists of two cells only. There are two to four bridge cross-veins. Vein R_3 is strongly undulate, and IR_3 somewhat less so. The three supplements (or planates) are well marked and generally subtend two rows of cells. The anal loop is long, with two rows of cells and a rather prominent heel.

The abdomen is shorter than the wings, usually being from about three-fourths to four-fifths of their length. It varies greatly in width, species of the subgenus *Plathemis* being so much broader in the abdomen than other species of *Libellula*, e.g. *incesta* or *vibrans*, that they appear to belong to different genera.

Segments 2 and 3, from which the genitalia of the male arise, are very short and the sex organs are correspondingly minute and unsatisfactory as taxonomic characters, as are also the anal appendages.

The nymphs are dark and elongate, and hairy in appearance; they live on the bottom of standing water among silt or debris. The eyes are prominent and typically cap the anterolateral parts of the head; the lateral surfaces of the head narrow behind the eyes. The distal margin of the labial palpus is relatively smooth and not distinctly crenulate. The abdomen is relatively long and tapering. Dorsal hooks and lateral spines are variably developed, often being difficult to discern because of the dense covering of hair-like setae. *Libellula* nymphs (unlike those of *Perithemis*) never have dorsal hooks on segment 9.

Key to the Species of Libellula
Adults

1 Basal dark area of hind wing extending in males from base to nodus, in females two-thirds of the distance from base to nodus and, in both sexes, to the anal angle (pl. 24: 3, p. 164)   *luctuosa* (p. 175)

  Basal dark area of hind wing, if present, extending not more than halfway to nodus and not approaching the anal angle   2

2 Nodal dark spot on wing absent; apical spot usually absent   3

  Nodal spot present or included in a larger dark area; apical spot present or absent   5

3 Total length less than 46 mm.; h.w. less than 38 mm.; dark patch at base of hind wing; front wing with smaller dark basal spot   4

  Total length more than 46 mm.; h.w. more than 35 mm.; without dark areas at wing bases   *incesta* (p. 185)

4 Dorsum of pterothorax pale except for narrow black mid-dorsal carina, becoming wholly pruinose white at maturity   *julia* (p. 169)

  Dorsum of pterothorax with wide stripe of pale brown, bordering black mid-dorsal carina on each side and thus reducing pruinose white area by half   *exusta* (p. 172)

5 Nodal spots small (extending posteriorly for less than one-quarter of wing's breadth), but dark and distinct   6

  Nodal spots large (extending posteriorly for at least one-half of wing's breadth)   7

6 Wings without a yellow front margin and without a large basal black spot   *vibrans* (p. 187)

  Wings with broad anterior yellow margins, the hind wings each with a large basal triangular black spot (pl. 24: 1, p. 164) *quadrimaculata* (p. 165)

7 Wing spots light orange-brown suffused with yellow; nodal spots crossing anterior half of wing's breadth; a similar spot over pt. and distal area, clearer at the extreme tip (pl. 24: 6, p. 164)   *semifasciata* (p. 183)

  Wing spots dark brown, without a yellow suffusion; nodal spots usually crossing more than half of wing's breadth   8

8 Wings with basal, nodal and apical dark areas   9

  Wings with basal and nodal dark areas, but no apical dark areas (although ante-apical clouding may occur)   10

PLATE 24

Wings of *Libellula*: (1) *L. quadrimaculata*; (2) *L. lydia*; (3) *L. luctuosa*; (4) *L. forensis*; (5) *L. pulchella*; (6) *L. semifasciata*.

PLATE 24

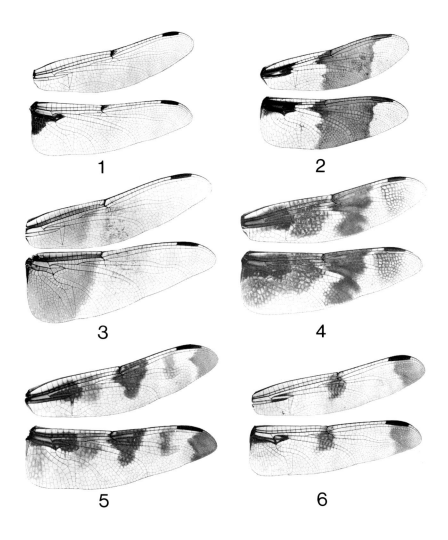

1

2

3

4

5

6

9 H.w. more than 37 mm. (pl. 24: 5, p. 164) *pulchella* (p. 179)

  H.w. less than 37 mm. *lydia* ♀ (p. 173)

10 Nodal spots reaching posterior margin of fore and hind wing, and meeting that margin for at least one-fifth of the length of the hind wing (pl. 24: 2, p. 164) *lydia* ♂ (p. 173)

  Nodal spots not quite reaching posterior margin of fore and hind wing or, if so, meeting that margin for less than one-sixth of the length of the hind wing (pl. 24: 4, p. 164) *forensis* (p. 178)

Nymphs

Nymph unknown *exusta* (p. 173)

1 Distal margin of median lobe crenate (pl. 38: 5, p. 254) 2

  Distal margin of median lobe smooth (pl. 38: 6, p. 254) 3

2 Dorsal hooks on segs. 7 and 8 *julia* (p. 170)

  No dorsal hooks on segs. 7 and 8 *lydia* (p. 174)

3 Dorsal hooks on segs. 7 and (except in *pulchella*) 8 4

  No dorsal hooks on segs. 7 and 8 *forensis* (p. 178)

4 Palpal setae (on each palpus) 5 5

  Palpal setae 7 or more; dorsal hooks on segs. 3 to 8 or 4 to 8 6

5 Epiproct distinctly decurved at tip (pl. 45: 1, p. 276) *incesta* (p. 186)

  Epiproct straight, or only very slightly decurved at extreme tip (pl. 45: 2, p. 276) *vibrans* (p. 189)

6 Palpal setae 8 or 9; dorsal hook on seg. 7 but not on 8 *pulchella* (p. 181)

  Palpal setae 7 or 8; dorsal hooks on segs. 7 and 8 7

7 Dorsal hook on seg. 3 8

  No dorsal hook on seg. 3 *luctuosa* (p. 176)

8 Largest dorsal hook on seg. 5, 6 or 7; premental setae (on each side) 10 to 14 *quadrimaculata* (p. 166)

  Largest dorsal hook on seg. 8; premental setae 16 *semifasciata* (p. 184)

Libellula quadrimaculata Linné. (Pl. 24: 1, p. 164; pl. 38: 2, p. 254; pl. 41: 1, p. 264)

Libellula quadrimaculata Linné, Syst. Nat., 1: 543, 1758.

Libellula ternaria Say, J. Acad. Phila., 8: 21, 1839.

Leptetrum quadrimaculatum: Kirby, Cat. Neur. Od., p. 27, 1890.

A middle-sized species of olive-brown, yellow and black coloration, and with basal, nodal and apical wing spots; widely distributed in the Holarctic region; the commonest large libellulid in Canada.

Male. General colour olive-brown; wings (pl. 24: 1, p. 164) with a black nodal spot of variable size, the hind part with a large black patch at the base of each wing, with pale venation covering the cubital space, triangle, supratriangle and part of the anal area; a broad yellow stripe including the area in front of the black basal spot and a broad stripe along the anterior margin, fading distally. Occasional individuals have a more or less distinct dusky cross-bar across the apices, sometimes including pt. (variety *praenubila* Newman). Face greenish; labrum paler, yellow, mesal edge black and a narrow basal transverse margin black. Prothorax buff above with a black transverse stripe, densely hairy; sides with a clear yellow spot in front of the metaspiracle, bordered below with black; metapleura also with a yellow spot. Abdomen fawn on proximal segments, with short pale pubescence, shading into black on posterior half of seg. 6 and remaining segments, except the orange-yellow lateral margin on segs. 3 to 8. Legs and anal apps. black. Wings in tenerals clear amber along the costal border, the amber area extending about halfway across the wing.

Abdomen fairly broad at base, the sides straight and convergent, the width diminishing from 5 mm. (seg. 2) to 2 mm. (seg. 10); apps. about as long as 9 + 10.

Female. Similar to male in form and colour pattern; abdomen a little stouter. Vulvar lamina consisting of two widely separated rounded prominences, arising from the thickened margins of the sternum of seg. 8, which are bent ventrad.

Venation. Anx ♂ 14–17/11–13, ♀ 14–18/12–14; pnx ♂ 11–13/11–15, ♀ 11–13/11–14; cross-veins under pt. ♂ 2–4/3–4, ♀ 3–4/3–4.

Measurements. Total length ♂ 40.0–43.8, ♀ 39.3–44.1; abd. ♂ 24.0–26.8, ♀ 24.0–28.0; w.abd. ♂ 5.1–5.6, ♀ 5.8–6.2; h.w. ♂ 33.2–36.7, ♀ 31.7–38.1; h.f. ♂ 5.6–6.5, ♀ 5.8–6.9; w.hd. ♂ 7.9–8.1, ♀ 7.2–8.0; pt. ♂ 3.3–3.9/3.4–4.0, ♀ 3.2–4.2/3.8–4.2.

Nymph. Robust, orange-brown, without a distinct pattern, moderately hairy. Head (pl. 38: 2, p. 254) twice as wide as long, frontal margin nearly straight, eyes fairly prominent. Antennae with segments of relative lengths about as follows: 7:10:16:7:10:14:13. Hinge of folded labium extending posteriorly a little beyond front of mesocoxae; prementum nearly as long as wide; angle of distal margin of median lobe about 48°; premental setae 10 to 14 of which about the 8 outermost are the longest; palpal setae 7 to 9 (Musser (1962) records occasionally only 6). Abdomen (pl. 41: 1, p. 264) ovate, fairly convex, greatest width between segs. 6 and 7 at about three-fifths of its length; dorsal hooks on segs. 3 to 8, largest on 5, 6, or 7, but all rather small, straight and closely surrounded by somewhat stiff setae; lateral spines on segs. 8 and 9, of about the same length, each about as long as one-seventh of the lateral length of its respective segment (including the spine); lateral spines diverging only very slightly from the contour of the abdominal outline; anal appendages longer than mid-dorsal length of segs. 9 + 10; epiproct and paraprocts of same length, epiproct slightly more than twice as long as its basal width; cerci about three-fifths as long as epiproct, straight and tapering to sharp apices.

Length 22–26; abd. 13.5–18; w. abd. 7–8; h.w. 5.5; h.f. 5.6–6; w. hd. 6.

Habitat and range. Still waters in marshy or boggy ground, especially peaty waters. Circumboreal: abundant in Canada from Nfld. and Labr. to James Bay and Mackenzie District, N.W.T., w. to Alaska and s. through B.C. to Wash., Idaho, Wyo. and S. Dak., on the east coast from N.S. and N.E. s. to Pa., thence w. and s. to Okla., Tex., N. Mex. and Ariz. Also in Europe and Asia.

Distribution in Canada and Alaska. Nfld.—Holyrood; Grand Bruit; Port aux Basques; Spruce Brook; Bay of Islands. *Labr.*—Goose Bay; Hopedale. *N.S.*—Annapolis, Digby, Halifax and Pictou counties; Cape Breton I. (Cape Breton and Victoria counties). *N.B.*—Charlotte, Gloucester, Northumberland, Queen's, St. John's, Sunbury and York counties. *P.E.I.*—

Charlottetown; Souris (New Harmony Road). *Que*—Covey Hill; Hemmingford; Eastman; Knowlton; Aylmer; Hull; Ironside; Wakefield; Alcove; Vaudreuil; Hudson; Mont Tremblant Park; Nominingue; lower St. Lawrence; Saguenay River (Cap Jaseux); Godbout; Magdalen Is.; Abitibi county; Lake Mistassini. *Ont.*—Essex, Kent, Lambton, Norfolk, Brant, Haldimand, Wentworth, Waterloo, Peel, York, Ontario, Durham, Prince Edward, Simcoe, Hastings, Frontenac, Leeds, Prescott, Carleton, Lanark and Renfrew counties; Muskoka, Parry Sound, Nipissing (incl. Algonquin Park), Algoma, Sudbury, Thunder Bay; Rainy River, Cochrane and Kenora (incl. Patricia Portion) districts. *Man.*—Treesbank; Aweme; Onah; Winnipeg; Stony Mountain; Victoria Beach; The Pas. *Sask.*—Cypress Hills Park; Regina; Hamton; Lac Vert; Lashburn; Duck Lake; Prince Albert; Namekus Lake; Waskesiu Lake. *Alta.*—Banff; Red Deer; Ft. Smith. *B.C.*—Vancouver; Sumas Lake; Sumas Canal; Cultus Lake; Chilliwack; Lihumitson (Liumchen) Creek; Harrison Bay; Hope; Princeton; Osoyoos; Westbank; Christina Lake; Okanagan Falls; Cranbrook; Kaslo; Crawford Bay; Kootenay Lake; Edgewater; Enderby; Salmon Arm; Clinton; Revelstoke; Horsefly; Chilcotin (near Riske Creek); Prince Rupert; Vancouver I. (Langford; Shawnigan Lake; Nanaimo district; Sproat Lake; Victoria Lake). *N.W.T.*—Ft. Smith; Slave River; Great Slave Lake (Ft. Resolution); Ft. Simpson; Norman Wells. *Alaska*—Fox Pt.; Juneau.

Field notes. In southern Ontario and perhaps in all parts of its range, *L. quadrimaculata* is one of the earliest dragonflies to emerge in the spring. Our earliest dates on which the winged insects have been observed are May 7 and 8 but usually they are first noted near the end of May or early in June. Following emergence they stray over the countryside, often through open woods or along their borders, settling on low vegetation, branches or foliage, usually about a foot above ground but not on the ground itself, as *L. julia* and *L. lydia* do. At this stage they frequently appear in large numbers, sometimes in "swarms," and in such aggregations have been observed in migration probably more often than any other species of dragonfly.

During the second half of June in southern Canada they have generally reached their maximum numbers before the end of the month or early in July, depending in which part of the country the observation is made. Many have returned to their breeding places. These are still waters of almost any kind from marsh-bordered lakes to ponds, pools or ditches. Small ponds on flat soils are favourite habitats and these may be open or surrounded by a dense growth of emergent aquatic plants. Frequently they are found about sphagnum pools but are not specially typical inhabitants of such bodies of water. As a contrast to the commoner types of habitat, we found many of this species on a pond occupying a depression among hills near Ottawa. This was a shallow pond with a firm, nearly flat bottom. It was thickly overgrown with

PLATE 25

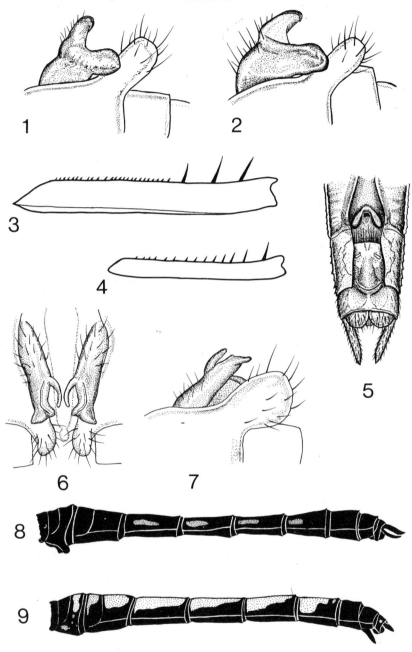

horsetails (*Equisetum fluviatile*) and short rushes, with few other emergent plants and little open water. The greatest depth of the pond was about two and a half feet.

The exuviae of *L. quadrimaculata* are usually left in the grass or herbage near the pond.

This is a vigorous species and a voracious feeder. Adults attack and devour other dragonflies as large as a *Sympetrum* (Whitehouse, 1941). Near Flatbush, Alberta, Pritchard (1964a) found the nymphs fed predominantly on molluscs and chironomid larvae in a lake where these two kinds of organism were well represented.

Pairing takes place during flight and is of variable duration—lasting from a few seconds to more than a minute (Schiemenz, 1953). While ovipositing, the female flies over the water alone, although the male often "stands guard" while his mate is depositing eggs, and drives away other males of his own or any other species, or females, should they attempt to oviposit in the same preserve. According to Schiemenz (1953) the males of *L. quadrimaculata* frequently fight for the possession of a female.

The flight period of *L. quadrimaculata* in Quebec and Ontario, based on our records, is from May 8 to August 28 and in British Columbia (Whitehouse, 1941) May 7 to September 29. Our records from other Provinces are not sufficiently numerous to indicate the entire period.

Libellula julia Uhler. (Pl. 25: 1, p. 168; pl. 41: 2, p. 264)

Libellula julia Uhler, Proc. Acad. Phila., 1857: 88, 1857.
Libellula exusta: Whedon, Rep. Minn. State Ent., 1914: 98, 1914.
Ladona julia: Needham and Heywood, Drag. N. Amer., p. 218, 1929.

This and the next species to be considered, *L. exusta*, were placed by Needham (1897, p. 146) in a separate genus, *Ladona*. A third species of the southern United States also belongs to this group but does not range northwest as far as Canada. The European *L. fulva* Müller also belongs here.

L. julia is a rather small *Libellula* but the largest of the three related species that Needham placed in *Ladona*. Its general colour is brown, the thoracic dorsum and basal segments of the abdomen becoming pruinose in the adult, conspicuously white in the male and gray in the female. The wings are hyaline with small dark basal spots.

Male. Face and mouth-parts fawn in tenerals, labium in adults light brownish orange, labrum

---

PLATE 25

Libellula: (1, 2) hamulus, left lateral view: (1) *L. julia*; (2) *L. exusta*; (5) vulvar lamina, ventral view of *L. pulchella*.
Hind femur: (3) *Erythemis simplicicollis*; (4) *Erythrodiplax berenice*.
Erythrodiplax berenice: (6, 7) hamulus: (6) ventral view; (7) right lateral view; (8, 9) abdomen, left lateral view, of nearly mature adults: (8) male; (9) female.
(3, 4 after Borror, 1945; 8, 9 after Borror, 1942.)

paler, clypeus and frons olivaceous gray, a deep black transverse furrow between frons and vertex, which is darker and browner than frons. Occiput also brown. Post occipital fringe of gray hairs; rear of head variegated with pale yellow and dark brown, polished. Thorax dark brown on the sides deepened to nearly black over the mesopleural sutures, the entire thoracic dorsum, except the extreme internal edges and a median hairline being conspicuously white in adult males. Legs proximally dark brown, the distal ends of the femora and remainder of the legs black. Wings hyaline with black venation and small basal spots black and reddish brown, those of fore wing not extending beyond the first antenodals, those of hind wing extending to second antenodals and including a black triangular spot covering bases of the paranal veins, adjoining the small white membranule. Abdomen with segs. 1 to 4 white, the remaining segments blackish above, grading into orange-brown on the sides with a thin grayish vestiture.

Frons moderately advanced, with a broad median groove above, transversely sulcate in front, the groove more or less medially divided, and surrounded by a peripheral ridge. Pterothorax relatively large, about two-fifths as long as abdomen. Wings about one-third longer than abdomen; their width at nodus about two-sevenths of their length. Abdomen somewhat shorter than usual in Libellula, widest at seg. 8, tapering caudad to base of seg. 9, which widens again slightly to base of 10.

Inner hamulus (pl. 25: 1, p. 168) short, blunt, strongly curved, and directed laterally. Genitalia small. Sup. apps. about as long as seg. 9, convergent from bases, meeting in middle line near angular apices, in lateral view arched at bases but curved upwards apically; inf. app. dorsally concave, slightly more than two-thirds as long as sup. apps.

Female. Coloration similar to the male but, when mature, lacking the intense contrasts, the pruinosity being merely gray instead of white. Similar in proportions to male, the abdomen scarcely stouter, but a little shorter and decidedly blunter, chiefly owing to the replacement of the sup. apps. of the male by the minute anal apps. of the female, which extend only a little beyond the paraprocts. Vulvar lamina consisting of two small, well separated plates, the caudal margins being convex.

Venation. Anx ♂ 13–15/10–12, ♀ 13–16/10–13; pnx ♂♀ 10–12/11–14; cross-veins under pt. 2–4/3–4.

Measurements. Total length ♂ 39–44, ♀ 38–43; abd. ♂ 22–27, ♀ 24–27; w.abd. ♂ 4.0–4.5, ♀ 4.3–4.6; h.w. ♂ 31–36, ♀ 32–37 (NW: ♂♀ 28–34); h.f. ♂♀ 6.5–7.0; w.hd. ♂♀ 7; pt. ♂ 3.5–3.7, ♀ 3.7–4.5.

Nymph. Dark brown with no distinct pattern, sparsely hairy except on legs, which are somewhat longer than those of most species of Libellula. Length of head a little more than one-half its width, the eyes moderately prominent; postocular margins curving inward and not very prominent, bearing many coarse setae. Antennae of usual length, the two proximal segments stout, the remaining segments slender; relative lengths of segments: 1:1:1.5:8:1:1.2:1. Labium with distal margin of prementum distinctly crenate with a prominent apical tooth which is acute to nearly rectangular but rounded at the extreme tip; 3 long premental setae on each side with two irregular median groups of very small setae; palpal setae 6; distal margin of palpus with 9 to 11 broadly rounded crenations. Legs moderately hairy; hind wing-sheaths reaching base of seg. 6. Abdomen (pl. 41: 2, p. 264) widest at seg. 6 or between segs. 6 and 7, the greatest width (exuv.) about 40 per cent of the length; dorsal hooks on segs. 4 to 8 successively less upright, the longest on 6, ridge-like on 7 and 8, with projecting tips; lateral spines on segs. 8 and 9, about equal in length, each about one-sixth as long as the lateral margin of the segment of which it is a part (the spine included), slender, acute and not divergent; epiproct a little longer than mid-dorsal length of segs. 9 + 10, its basal width a little more than one-half its length; cerci about three-sevenths as long as epiproct or paraprocts, which are about equal and very attenuate.

Length 24–25; abd. 15–16; w. abd. 6–7; h.w. 6.7–8.0; h.f. 5.5–6.0; w. hd. 4.5–5.5.

Habitat and range. **Bog ponds and swampy bays.** *L. julia* occurs close to the

Canadian-U.S. boundary from N.B. and Que. to B.C. and from Conn. to Wash. It is transcontinental in distribution although not yet recorded from Mont. It does appear evident, however, that *L. julia* must be rare in (if not wholly absent from) the dry, treeless plains where permanent waters are scarce. *L. julia*, so far as records indicate, is most abundant in the dark waters of the Canadian Shield in Ontario and Quebec. Farther south it appears to be replaced by two closely related species, *L. exusta* and *L. deplanata* Rambur. These are smaller than *L. julia*, particularly *L. deplanata*, which is the smallest North American *Libellula* and absent from Canada. The other species, *L. exusta*, occurs in Nova Scotia but is known from no other part of Canada. (These three species have long been considered by Needham and his followers as constituting a separate genus, *Ladona*, but this name was not recognized by Ris (1910), the leading European authority on the Libellulidae, nor by Kennedy (1922), who made a detailed study of the penis of *Libellula* and its immediate allies. *Ladona* is perhaps worth of retention as a subgenus, but is not accepted by European odonatists.)

Distribution in Canada. N.S.—Annapolis and Halifax counties; Cape Breton I. (Victoria county). *N.B.*—(Whitehouse, 1948). *Que.*—Covey Hill; St. Hyacinthe; Hull; Wakefield; Laurentian Mts. ("Newaygo"); Mont Tremblant; Saguenay River (Cap Jaseux). *Ont.*—Essex, Kent, Middlesex, Norfolk, Oxford, Waterloo, Huron, Peel, York, Prince Edward, Bruce, Simcoe, Victoria, Peterborough, Hastings, Lennox and Addington, Frontenac, Leeds, Carleton, Lanark and Renfrew counties; Muskoka, Parry Sound, Nipissing (incl. Algonquin Park), Sudbury, Thunder Bay, Cochrane and Kenora districts. *Man.*—Victoria Beach; Cowan. *Sask.*—Kingsmere, Prince Albert National Park. *Alta.*—Near Flatbush. *B.C.*—Chilliwack; Hope; Harrison Bay; Osoyoos; Oliver; Christina Lake; Salmon Arm; Horsefly; Vancouver I. (Langford; Shawnigan, Saanich and Nanaimo districts).

Field notes. Although not one of the commonest species of *Libellula* that frequent the ponds and marshy bays in agricultural districts, *L. julia* often occurs in multitudes about the more acid waters of the Canadian Shield. At Go Home Bay, Georgian Bay, where it is very abundant, we found the full-grown nymphs in the bog ponds and swampy bays on our arrival in early May until about the end of the month or the first week in June.

In the decaying organic matter at the bottom of such swampy waters, where other species of *Libellula* were rare or absent, *L. julia* seemed to find ideal conditions. At De Grassi Point, Lake Simcoe, it was abundant in the sluggish, weedy waters at the outlet of the small stream discharging into Lake Simcoe, but it was not observed to emerge elsewhere.

In dredgings made at Go Home Bay on May 31, 1912, and during the following week, full-grown nymphs were common. Imagos were first noticed on June 7 and had become abundant by June 13. They continued so for about a month, their numbers dwindling during the second half of July until 30th, when the last individual was noted. A few specimens emerged in the laboratory during the second half of June.

After emergence, *L. julia* wanders into the open woods where it takes short flights, frequently settling on rocks, logs or the bare ground, after the manner of *Gomphus*. They soon return to the waters where they emerged, or places of a similar kind and often appear in prodigious numbers. According to Whedon (1914) this species (as *L. exusta*) is "gregarious, as many as fifteen or twenty alighting on the same spot; it is also inquisitive and many were caught that actually alighted inside the net as it was being carried."

In fine still weather the males may be seen chasing each other swiftly and erratically over the water. They are conspicuous objects on account of the white pruinosity of the thoracic dorsum and the broad segments of the abdomen.

Near Flatbush, Alberta, an adult of this species was taken in the process of consuming an adult *Leucorrhinia proxima* (Pritchard, 1964).

Libellula exusta Say. (Pl. 25: 2, p. 168)

Libellula exusta Say, J. Acad. Phila., 8: 29, 1839.
Leptetrum exustum: Kirby, Cat. Neur. Od., p. 26, 1890.
Ladona exusta: Needham, Bull. N.Y. State Mus., 47: 529, 1901.
Libellula exusta exusta Ris, Coll. Zool. Selys, 11: 258, 1910.

The smallest of our species in the genus *Libellula*. Similar to *L. julia*: the thoracic dorsum has a creamy white humeral line on each side but does not become pruinose in the adult male, as it does in *L. julia*.

Male. Easily recognized when mature by the absence of the thoracic pruinose area which is replaced by a pair of pale yellowish stripes, straight and parallel-sided and about half a millimetre wide, separated by a median brown area and bounded laterally by black, which grades into dark brown on the sides.

Face of adult dark brown, including labrum and labium; legs black. Abdomen above grayish to white pruinose, only the last three or four segments blackish. Wings hyaline with dark basal markings relatively larger than in *julia*, the anterior streak of the fore wing extending to the second antenodal, with a spot (sometimes separated) on the antenodal, the posterior streak of the fore wing extending about half-way along the cubital space. On the hind wing the two stripes are merged into a triangular area, which extends in front to the fourth or fifth antenodal and behind to the base of the triangle or somewhat beyond.

Genitalia small, lodged in a basin-like cavity, the inner hook of each hamulus (pl. 25: 2, p. 168) straight but sharply bent at the extreme tip and directed posteriorly; more slender than in *julia*.

Female. Resembles *julia* except that the two lobes of the vulvar lamina are connected mesially by a small plate.

Venation. Anx ♂ 13–14/10–11, ♀ 12–13/9–10; pnx ♂ ♀ 9–10/9–10; cross-veins under pt. 3 or 4.

Measurements. Total length ♂ 36.5–39.5, ♀ 36.0–38; abd. ♂ ♀ 22–24; w.abd. ♂ 3.5–4.5, ♀ 3.0–3.8; h.w. ♂ 29–31, ♀ 29.0 (NW: ♂ ♀ 31–33); h.f. ♂ 6.0–6.6, ♀ 5.5–6.5; w.hd. ♂ ♀ 6.0–6.6; pt. ♂ 3.0–3.5, ♀ 3.5–4.0.

Nymph. Apparently unknown: published references to *L. exusta* have sometimes been given in mistake for *L. julia*.

Habitat and range. Lakes and ponds. From N.S. s. through the New England States to N.C. and Ga.; also La.

Distribution in Canada. N.S.—Annapolis, Digby and Halifax counties. *N.B.*—Charlotte county (Chamcook).

Field notes. In Canada (Nova Scotia) adults have been seen between June 3 and July 11. In the New England States the flight period has been given as May 19 to July 28 (Needham and Westfall, 1955).

Libellula lydia Drury. (Pl. 24: 2, p. 164; pl. 38: 3, 5, p. 254; pl. 41: 3, p. 264)

Libellula lydia Drury, Ill. Nat. Hist., 1: 112, 1770.
Libellula trimaculata De Geer, Mém. Ins., 3: 556, 1773.
Plathemis trimaculata: Hagen, Syn. Neur. N. Amer., p. 149, 1861.
Plathemis lydia: Kirby, Trans. Zool. Soc. London, 12: 288, 1889.

A stout-bodied dragonfly with a pronounced sexual dimorphism, the male having the broadest dark stripes across the wings of any of the *Libellula* group in Canada whereas the female is unlike the male but very similar to the female of *L. pulchella*, although smaller.

Male. Labium on middle parts dark brown, fading towards the periphery into much paler brownish orange; labium also somewhat paler than other parts of the face. Hair of face very short, dark gray; dorsum of head, including vertex and occiput very dark brown; rear of head polished dark brown with two yellow spots on each side close to the eye margin. Thorax dark brown with two oblique yellowish stripes on each side, each stripe bright yellow in front, where it is bordered with black but fading behind into dull grayish; anterior half of thorax clothed with rather long, dark-brown hair, posterior part around bases of wings with very short gray hair. Legs brown, grading distally on the tibiae and tarsi to black. Wings (pl. 24: 2, p. 164) hyaline, each wing crossed by a very broad dark brown stripe approximately between nodus and pterostigma and bearing also a much narrower dark stripe extending approximately from base to triangle. The larger stripes on hind wing are bordered behind with white. Abdomen of tenerals dark yellowish brown with a lateral segmental series of yellow spots enclosed in black, the first spots on seg. 3, dorsolateral in position, the others lateral, passing distad to seg. 9. In adult males the abdomen is white pruinose and makes a striking contrast with the wide black wing bands. Wings relatively short, only slightly longer than abdomen, somewhat wider than in most species of *Libellula*.

First abdominal sternum bearing a pair of large porrect processes, each with a slight spiral twist suggesting the horns of a cow.

Hamuli consisting of a broad lateral lobe and a mesal hook separated by a shallow round notch. Abdomen widest near the base, tapering steadily to the anal apps. which are relatively small.

Female. Wing pattern very similar to that of *Libellula pulchella*, i.e., with three dark spots on each wing: (1) a basal spot from base to fourth or fifth antecostal, that of fore wing touching or nearly touching the triangle; that of hind wing covering the triangle; (2) a large nodal spot, widest on costal margin, narrowed about the middle and usually covering two-thirds or more of the wing's width, sometimes crossing the entire width of one or both pairs of wings; and (3) an apical spot.

Abdomen slightly shorter than that of male with parallel or slightly convergent margins from segs. 2 to 7 or 2 to 8 with the margins strongly arcuate, 9 and 10 rapidly narrowing with apps. very short and slender. Vulvar lamina rectangularly bilobed, the lobes rounded and somewhat raised. Dorsal surface of abdomen dark brown with a series of somewhat oblique subtriangular yellow spots, edged with black, the spots distinct on 2 to 7, indistinct on 8, absent on 9 and 10.

Venation. Anx ♂ 12–14/9–12, ♀ 11–14/10–12; pnx ♂ 9–11/10–11, ♀ 9–12/9–12; cross-veins under pt. ♂ 3–4/3–5.

Measurements. Total length ♂ 42–47, ♀ 37–41; abd. ♂ 23–29, ♀ 23–25; w.abd. ♂ 4.5–5.8, ♀ 4–5; h.w. ♂ 29–35, ♀ 32–33; h.f. ♂ ♀ 5–6; w.hd. ♂ ♀ 6–7; pt. ♂ ♀ 4.5–5.0.

Nymph. Smooth but not shiny; vesture scanty except on the legs where it is fairly dense, especially on the tibiae. Head (pl. 38: 3, p. 254) transversely quadrate, about one-half as long as wide, the sides subparallel and, though somewhat convex, head is not wider behind the eyes than it is in front (contrary to current descriptions), being nowhere wider than at level of the eyes, which are small but prominent; posterior outline of head somewhat emarginate. Antennae rather short, the segments more nearly uniform in length than usual, their relative lengths being about 10:10:15:11:11:12:14. Labium with premental setae about 8, the 5 outermost on each side being the longest; distal margin of median lobe of prementum strongly crenate, a single spiniform seta between crenations; palpal setae 10; distal margin of palpus with 7 or 8 crenations, these being more deeply cut than is usual in the Libellulidae. (The setal formula probably varies widely: an exuvia associated with an emerged adult from Toronto, Ontario possesses 12 premental setae, in an unbroken series, on each side, and 7 palpal setae plus a basal setella.) Pronotum as wide as the head, both broadly rounded, on lateral and posterior margins. Legs short and sturdy, particularly the hind pair. Abdomen (pl. 41: 3, p. 264) ovate, dorsally strongly convex, widest at segs. 5 and 6, tapering with curved sides from seg. 7 to 10; seg. 10 half as wide as 9; dorsal hooks on segs. 3 to 5 sometimes with a lower hook or a knob on 6; lateral spines on 8 and 9, each about one-fifth as long as the respective segment's lateral margin (the spine included), the direction of which they continue; epiproct with basal width about two-thirds of length; cerci half as long as epiproct; paraprocts a little longer than epiproct.

Length 21–24; abd. 11–14; w. abd. 5.5–6.8; h.w. 5.6–6.0; h.f. 4.4–5.0; w. hd. 4.5–4.8.

Habitat and range. Ponds and puddles, both clean and cowtrodden; also quiet pools on streams and sheltered corners on lake shores. Recorded from all the United States except the Dakotas, Mont., Oreg., Nev. and Ariz. In the last two of these States it is replaced by *Libellula (Plathemis) subornata* (Hagen). In Canada the distribution of *L. lydia* is divided, its being known from N.S., N.B., Que. and Ont. in the e. and s.w. B.C. in the w. It will certainly be found also in P.E.I. and very probably in e. Man.

Distribution in Canada. N.S.—Halifax county. *N.B.*—Charlotte county. *Que.*—Covey Hill; Knowlton; St. Hyacinthe; Lanoraie; Aylmer; Hull; Wakefield; Alcove; Mont Tremblant Park; Nominingue. *Ont.*—Essex, Kent, Elgin, Middlesex, Norfolk, Oxford, Brant, Haldimand, Welland, Waterloo, Halton, Huron, Peel, York, Ontario, Prince Edward, Simcoe, Hastings, Lennox and Addington, Leeds, Prescott, Carleton, Lanark and Renfrew counties; Muskoka, Parry Sound, Nipissing, Algoma, Thunder Bay and Kenora districts. *B.C.*—Vancouver; Cloverdale; Rosedale; Cultus Lake;

Vedder Crossing; Osoyoos; Oliver; Boundary; Vancouver I. (Sooke; Florence Lake; Goldstream; Victoria; Saanich district; Ladysmith; Long Lake; Nanaimo district; Forbes Landing).

Field notes. The flight period of *L. lydia* in Ontario is May 28 to August 25. Full-grown overwintered nymphs begin to transform about the end of May. Our earliest record of emergence is May 28, 1944, at one of the exposed ponds near Snelgrove, Ontario. Here we noticed teneral males and many exuviae, perhaps including some of other species of *Libellula*, together with many *Leucorrhinia intacta*. These emerged in part at least a week or so earlier. Tenerals are frequent about the first week in June, although the colour pattern has not yet matured by then. At this time they are flying half a mile or more from the water where they emerged and may be seen in clearings or along the edges of woods or on roads through the woods. They usually fly a few rods and come to rest on the ground or on a log or a rock. They seldom rest on bushes or foliage as do *L. pulchella* and *L. quadrimaculata*.

At De Grassi Point on June 3, 1933 they were abundant and were flying in open places near the slow stream from which they had emerged. Their colour pattern was still not quite mature.

By the middle of June, however, or earlier in some seasons, the males are very conspicuous with their white bodies and broad black wing bands. On sunny days they chase one another over the water and seldom come to rest. About this same time, and for a month or more later, ovipositing females may be seen. The females oviposit generally without the aid of the male, sometimes in clean water and sometimes in ponds that have been trampled and befouled by cattle. On July 21, 1937 we observed a female of this species ovipositing in a low trodden pond in water about an inch deep. She kept within about three inches of the surface, striking the water with the abdomen from behind and knocking drops about a foot forward. The rate at which these strokes were produced was somewhat more than one per second although doubtless varying with temperature. It appeared to be perfectly rhythmical and was performed in one spot for a period of a minute or more. While the female is thus engaged in oviposition she will sometimes be seized and carried away by a male, or sometimes two males will scuffle before one of them will succeed in seizing the female. On the whole, however, the female of *L. lydia* is not molested much while ovipositing.

A detailed account of the reproductive behaviour of this species has been made by Jacobs (1955) from studies made near Bloomington, Indiana.

Libellula luctuosa Burmeister. (Pl. 24: 3, p. 164; pl. 40: 1, p. 260)

Libellula luctuosa Burmeister, Handb. Ent., 2: 861, 1839.
Libellula basalis Say, J. Acad. Phila., 8: 23, 1839.

A species so distinctly marked that it can scarcely be mistaken for any other. The single large basal dark area of each wing, which, in the adult male, contrasts with the intermediate white area that separates the dark base from the hyaline apex, is unique among our species of libelluloid dragonflies. The wings are rather large for the body, making the insect appear larger than it really is.

Male. General colour dark brown; labrum almost black, labium paler, brown, particularly toward the lateral angles, vertex very prominent, eye-seam short; occiput large, thinly clad with tawny hairs; dorsal parts and rear of head all nearly black with scanty dark brown pubescence. Thorax very dark above, a little paler and more bronzy on the sides, without pale stripes or spots. Legs very nearly black. Wings (pl. 24: 3, p. 164) relatively large, the width at nodus of hind wing a little more than one-third of the wing's length. Colour dark brown from base of each wing to nodus, or nearly thereto, the distal boundary of the dark area curved toward the base and reaching anal area. From nodus, about half way to pt., or a little farther, is an opaque white area beyond which the wings are hyaline except for the pt. and venation, which are black. Abdomen dark yellowish brown, shading into black along the middle.

Head rounded above, the rounded area bounded below by a transverse ridge and divided by a median furrow. Abdomen, without appendages, about two-thirds as long as hind wing, moderately narrow, tapering from seg. 3 to the end.

Anal apps. about as long as seg. 9.

Female. Paler, yellow brown; dorsum and head also yellow brown. The vertex and occiput somewhat darker than frons. Thorax yellow brown with carina and a little wider space paler. Wings with dark basal area not extending to the nodus, that of fore wing less open in posterior half, distal part of anal loop sometimes beyond the dark area.

Abdomen not very broad for a *Libellula*, sides of segs. 4 to 7 nearly straight and parallel. Seg. 8 nearly as wide as 7 but with slightly convex sides. Vulvar lamina turned vertically downwards, about one-fourth as long as the sternal area of 9, with a free edge slightly bilobed and emarginate.

Venation. Anx ♂ 13–19/12–15, ♀ 14–17/10–12; pnx ♂ 13–16/13–17, ♀ 10–13/11–14; cross-veins under pt. 3–4 (occasionally 2).

Measurements. Total length ♂ 42–48, ♀ 38.0–42.0; abd. ♂ 25.5–28.0, ♀ 24.0–27.0; w.abd. ♂ ♀ 4.0–4.5; h.w. ♂ 36.0–39.5, ♀ 33.0–36.0; h.f. ♂ 6.0–6.5, ♀ 5.5–6.0; w.hd. ♂ 6.7–7.5, ♀ 7.0–7.4; pt. ♂ ♀ 4.5–5.0.

Nymph. Smaller than usual for a *Libellula*, smooth and somewhat shiny, hair sparse, eyes rather large and more prominent than in *L. pulchella* or *L. quadrimaculata*. Postocular lateral margins well rounded, broadly curved, passing smoothly into the posterior margin, which is mesally somewhat concave. Antennae slender with relative lengths of segments approximately 2:2:3:2:3:3:3. Folded labium extending posteriorly to beginning of mesosternum; distal margin of median lobe not crenate; premental setae 10 to 11, the outer 5 or 6 longer than the others palpal setae 7; distal margin of palpus with about 10 low crenations of which those closer to the movable hook are the more prominent and rounded, most of the others being straight; intervening notches bearing 3 or 4 fairly stout spiniform setae in a graded series. Legs of moderate length, the length of the hind femur slightly greater than the width of the head. Abdomen (pl. 40: 1, p. 260) ovate and transversely rather strongly convex; dorsal hooks on segs. 4 to 8 all well developed and cultriform, the largest ones on 5 and 6, but none vestigial; lateral spines on 8 and 9, those of 8 slightly larger, about one-fifth as long as the segmental margin the spine included, that of 9 about one-sixth as long as its margin; epiproct with a distinct median ridge, its basal width a little more than one-half its length, slightly acuminate; cerci nearly one-third length of epiproct.

Brown, slightly variegated, with pale yellowish; head with two oblique dark brown spots extending mesad from the eyes. Sides of thorax pale, femora pale with three dark annuli, a sub-basal, an ante-apical, and an apical annulus. Wing-sheaths dark brown. Abdomen irregularly variegated with dark and paler brown.

Length 20.0–20.5; abd. 12.0–13.0; w. abd. 6.5–7.0; h.w. 5.0–5.5; h.f. 5.0–5.5; w. hd. 5.4–5.5.

Habitat and range. Ponds, small lakes and marshes. Ga. to N. Mex. and Tex., n. to Me., N.S., s. Que., s. Ont., Mich., Minn. and S. Dak. Also Mexico: Chihuahua.

Distribution in Canada. N.S.—Hants county (Mt. Uniacke). *Que.*—Fairy Lake, Ironside and Meach Brook (all near Hull); Rigaud. *Ont.*—Essex, Kent, Lambton, Norfolk, Oxford, Brant, Haldimand, Welland, Wentworth, Waterloo, Huron, Wellington, York, Northumberland, Prince Edward, Simcoe, Hastings, Lennox and Addington, Frontenac, Leeds and Carleton counties; Manitoulin and Algoma districts.

Although common in southern Ontario and reported farther north from Manitoulin Island, the known distribution of *L. luctuosa* elsewhere in Canada is restricted to the vicinity of Hull, Quebec and Hants county, Nova Scotia. We would expect to find it also in Quebec south of the St. Lawrence River.

Field notes. Adults have been encountered from mid-June until the end of July in Quebec (Robert, 1963) and until the end of August in Ontario. In Ontario emergence was recorded on June 26 (Carleton county) but occurs earlier than this at Point Pelee.

When we first began to observe and collect Odonata about 1901, *L. luctuosa* was fairly common in the vicinity of Toronto and even at De Grassi Point, Lake Simcoe. This may have been an unusual year, for *luctuosa* had hitherto been a stranger to us and became relatively scarce again for many years. With the general northward shifting of southern fauna, it has now become common again in York county. In fact it is now by no means rare even at the latitude of Lake Simcoe. It is not yet, however, as plentiful and generally distributed as are *Libellula pulchella, L. lydia,* or *L. quadrimaculata.* Its occurrence in Hants county, Nova Scotia, however, was the most unexpected event concerning its distribution that we had ever heard of.

Needham and Heywood (1929) considered *L. luctuosa* to be a common species in the Mississippi Valley: "It flies rather steadily over the ponds, resting occasionally on reed tips. It is not very hard to capture. In the cool of the evening adults may be found hanging by their feet to the sloping twigs of nearby shrubbery. The female usually oviposits unattended." On the bank of a small glacial lake north of Toronto, where *L. luctuosa* was emerging, we found many exuviae clinging to the long grass just as Needham had described. The exuviae appear small in comparison with the body (without the wings) of the adult insect. In Mississippi nymphs collected by Bick (1950) were from still, mud-bottomed ponds or from borrow pits.

Regarding oviposition in *L. luctuosa*, Miss Ferguson (1940) describes the process as observed in Dallas county, Texas, on June 22, 1937: "... they were very abundant at Bluff View. This was the only time the males proved easy to capture. On this date oviposition was observed. By making regular dips into the water at the narrow part of the creek, the female deposited eggs. When dipping, it flew steadily about six inches high, then suddenly a male came down, joined the female, and both rose high into the air. After a few seconds, still at the same elevation they parted. At this habitat several *Libellula luctuosa* flew higher and more slowly than did either *Plathemis* or *Pachydiplax* ... Two days after the eggs were deposited they had changed in color from white to brown, and in about five days hatching began."

Libellula forensis Hagen. (Pl. 24: 4, p. 164)

Libellula forensis Hagen, Syn. Neur. N. Amer., p. 154, 1861.

A western species closely related to *L. pulchella* and partly replacing this species on the Pacific Coast, although the ranges of the two species overlap considerably. *L. forensis* is easily known by the absence of the dark tips of the wings seen in *pulchella*.

Male. Head dark olive brown, labrum with lateral yellow spots, frons above and vertex darker, hair of face not very dense; rear of head black with two yellow spots on each side and a fringe of pale gray hair between the coloured area and the central cavity. Pterothorax orange brown above, darker on the sides, blackish over the mesopleural suture, the colours subdued by dense grayish hair; epimera each with a clear yellow oblique stripe, each stripe divided into an oval wider spot in front and below, and a narrow crooked streak behind, which may be separated or wanting. Legs black. Pattern of each wing (pl. 24: 4, p. 164) consisting of two dark brown, nearly black, areas, one a narrower stripe extending from the base to a distance a little beyond the triangle and a broader stripe across each wing from nodus towards pt. but not reaching it. The apices are without dark spots. In mature males opaque white areas alternate with the black ones. In the hind wing the white areas between basal and nodal stripes extend around the basal spot to the anal area. Abdomen dark brown in tenerals, with a lateral yellow stripe that breaks up caudad into segmentally arranged spots, which are widest and most distinct in the middle parts (segs. 4 and 5). In adults the abdomen becomes gray pruinose and the yellow spots less distinct.

Hind wing about one-third longer than the abdomen (excluding apps.); abdomen tapering at about the same rate from seg. 3 to seg. 10; seg. 5 nearly square; segs. 1 to 4 wider than long; segs. 6 to 9 longer than wide.

Sup. apps. with the apices more divergent and much blunter than in *pulchella*. Inf. app. triangular with convexly arcuate sides.

Female. Similar to male in colour pattern, the wings of adult females even having the white areas as in the males, and differing from *pulchella* in this regard. Abdomen relatively a little shorter and blunter. Seg. 8 and tergum somewhat widened, the margin being expanded and slightly rounded. Vulvar lamina as in *pulchella*.

Venation. Anx ♂ 13–17/11–13, ♀ 14–17/10–14; pnx ♂ 11–13/11–13, ♀ 9–14/10–14; cross-veins under pt. ♂ 3–4/3–4, ♀ 2–3/2 (NW: ♂ ♀ sometimes 5).

Measurements. Total length ♂ 46–50, ♀ 44–47; abd. ♂ 28–31, ♀ 27–29; w.abd. ♂ 5.0, ♀ 5.0–5.5; h.w. ♂ 35–41, ♀ 39–41; h.f. ♂ ♀ 7.0–8.0; w.hd. ♂ ♀ 8.0–8.5; pt. ♂ 4.0–4.2/4.0–4.8, ♀ 3.8–4.7/4.0–4.7.

Nymph. Similar to that of *L. pulchella* but differing in certain characters of the dorsal hooks and lateral spines. Form of head including eyes as in *L. pulchella*. Antennae slender with relative lengths of segments approximately 1:1:2:1:1.5:2.5:2. Folded prementum with basal hinge reaching middle of mesocoxae; premental setae 10 (in groups of 6 and 4) or 11 (5 and 6); palpal setae 5 to 7, usually 6; marginal setae on proximal half of outer margin of palpus; distal margin of palpus with 10 or 11 low crenations separated by distinct notches from each of which projects one relatively large spiniform seta with 2 successively smaller ones next to it. Postocular part of head with a sparse tuft of coarse dark brown setae; similar long setae also borne by lateral thoracic lobes and along dorsolateral ridges of pterothorax. Abdomen ovate, widest at segs. 5 and 6; segments with a short, closely placed fringe of setae along hind margin and with longer, darker setae arranged along this fringe at intervals, with groups along the mid-dorsal line of longer and more closely arranged setae that surround the dorsal hooks where these are present, namely on segs. 4, 5 and 6, with a vestige on 7; each dorsal hook (spine) is covered with short pale spinules, surrounded by the much longer and darker setae; lateral spines present on abd. segs. 8 and 9, those of 8 being about one-fifth as long as the lateral margin of that segment, including the spine, those of 9 being about one-sixth as long as the margin of that segment; epiproct as long as mid-dorsal length of segs. 9 + 10, not acuminate; cerci two-thirds as long as epiproct, with slender sharp apices; paraprocts about one-third longer than epiproct.

Length 25–27; abd. 15–16; w.abd. 7.8–8.5; h.w. 7–7.8; h.f. 6.3–7.

Habitat and range. Lakes and ponds. Calif. n. to B.C. and e. to Mont., Nebr. and Ariz. Also known from Wash., Oreg., Nev., Utah and Colo. There is a doubtful record from Wis.

Distribution in Canada. B.C.—Harrison Bay; Hope district; Cultus Lake; Chilliwack; Osoyoos; Oliver; Christina Lake; Penticton; Kaslo; Salmon Arm; Vancouver I. (Prospect Lake; Langford Lake; Shawnigan Lake; Victoria; Saanich district; Goldstream; Sahtlam; Nanaimo district).

In Canada *L. forensis* is limited to southern and southwestern British Columbia including Vancouver Island, where it is abundant; eastwards it has been found as far as Penticton district.

Field notes. L. forensis has a long flight period, adults having been encountered from June 3 to September 16 (Whitehouse, 1941).

At Chilliwack, British Columbia, it occurred in a moderately large, deep, stagnant slough bordering the Golf Club (Walker and Ricker, 1938). Musser (1962) found nymphs in a pond that had a bottom consisting predominantly of black ooze, and profuse aquatic vegetation around the edge. Whitehouse (1941) remarks that on occasion the nymphs travel some distance from the water to emerge: "I took one at Beaver Lake, Vancouver, on June 27th 1935, at 10 a.m., scrambling across the path in full sunshine, and making good time of it."

This species has been referred to as the western equivalent of *L. pulchella* (Needham and Heywood, 1929), although Whitehouse (1941) encountered both species flying at the same marsh at Penticton, British Columbia, in mid-July.

Libellula pulchella Drury. (Pl. 24: 5, p. 164; pl. 25: 5, p. 168; pl. 40: 2, p. 260)

Libellula pulchella Drury, Ill. Nat. Hist., 1: pl. 48, fig. 5, 1773.
Libellula pulchella: Hagen, Syn. Neur. N. Amer., p. 153, 1861.

PLATE 26

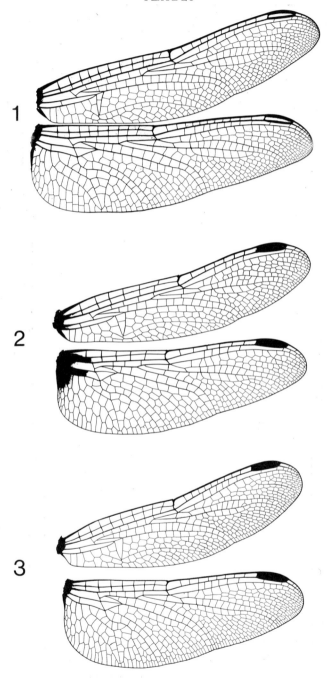

A large species frequenting ponds, the mature males being easily recognized by the alternate arrangement of black and white wing spots. Females and juvenile males lack the white spots but have similar dark markings. The females of *L. lydia* are similarly marked but smaller.

Male. Face dark brownish olivaceous, labium paler, yellowish towards base, distally brownish; labrum nearly black with broad distal margin paler, brownish; frons dorsally and laterally light yellow; rear of head black with a yellow spot above and a longer marginal stripe below. Thorax moderately dark olivaceous above with fairly thick pale gray hairs; epimera with broad and very oblique paler stripes which are light yellow below and in front; pale translucent bluish gray below and behind, widening and less distinct caudad; legs black, brown at base, centre brown. Wings (pl. 24: 5, p. 164) hyaline with three large, dark brown spots. The basal spots elongate, widening from base to the transparent distal ends, including in their area the arculus and triangles; middle spots subtriangular, wider than basal spots, enclosing the nodus but almost entirely distad of nodus, about half way to pt.; apical spots from the distal ends of pt. to apices of wings. In mature males there are large white spots in the intervals between black spots and in the anal angle. Abdomen nearly uniform brown, three-fourths as long as hind wing, tapering caudad evenly to seg. 4 but somewhat more rapidly distad; width of abdomen at seg. 2 (excluding anal apps.) about five-eights of length.

Entire abdomen in tenerals brown, the basal four segments and ultimately almost the entire abdomen becoming grayish pruinose.

Female. Wings with dark spots like those of male but white spots absent. Basal wing spots somewhat narrower, nodal spots a little smaller. Abdomen somewhat shorter and less tapering, almost truncate, brown with a yellow-green lateral stripe, which is submarginal in front, marginal behind. Vulvar lamina (pl. 25: 5, p. 168) with each lateral half rolled into a kind of scroll.

Venation. Anx ♂ 15–19/12–14, ♀ 15–20/13–15; pnx ♂ 11–14/12–15, ♀ 11–13/11–15; crossveins under pt. 3–4(5)/3–4(5).

Measurements. Total length ♂ 49–52, ♀ 43–51; abd. ♂ 30–32, ♀ 28–31; w.abd. ♂ 4.9–5.2, ♀ 4.9–5.7; h.w. ♂ 40.0–42.5, ♀ 39.0–42.0 (NW: ♂♀ 42–46); h.f. ♂ 6.0–7.0, ♀ 5.0–6.0; w.hd. ♂ 8.2–8.7, ♀ 8.0–8.9; pt. ♂ 5.1–5.7/5.2–5.7, ♀ 5.1–6.0/5.1–6.0.

Nymph. Similar to nymph of *L. quadrimaculata* but larger and sometimes darker. Head about twice as wide as long, anterior margin of frons straight or barely convex; eyes not very prominent; postocular margins of head broad and rounded, hind margin nearly straight, posterior half of surface moderately hairy. Antennae as long as head, segments with relative lengths about as follows: 1:1.5:.75:1.2:1.5:1.5:1. Premental setae 13 to 15; palpal setae 8 or 9. Abdomen (pl. 40: 2, p. 260) ovate, widest at seg. 5 or 6, curving and narrowing evenly to seg. 10, which is about one-half as long as the mid-dorsal length of 8; dorsal hooks on segs. 4 to 7, short and inconspicuous, each surrounded by a group of rather more than a dozen setae, largest on 4 and 5, very small on 7, slightly incurved; lateral spines on segs. 8 and 9, slender and nearly continuing the curve of the lateral margin of these segments and forming about one-seventh of the lateral length of each segment (including the spine); epiproct a little longer than mid-dorsal

---

PLATE 26

Wings of Libellulidae: (1) *Erythemis simplicicollis*; (2) *Pachydiplax longipennis*; (3) *Erythrodiplax berenice*.
(1–3 from Needham and Westfall, 1955.)

length of segs. 9 + 10, its base about three-fifths of its length; paraprocts about same length; cerci one-half as long, straight and acute.

General surface of body smooth, setae sparse, extending little more than one-half of a segment-length (including intersegmental membrane), longest and nearest together in middle.

Length 24–27; abd. 15–19; w. abd. 7.5–9; h.w. 7; h.f. 6; w. hd. 6–6.2.

Habitat and range. Ponds, marshy borders of lakes or bays and slow streams, chiefly in calcareous soils. Fla. to Tex., Nev. and Calif., n. to N.S., N.B. and Me., s. Que., s. Ont., Mich. and Minn. to Man. and Sask.; also from Iowa, Mo. and Nebr., w. through Colo. and Utah to Oreg., Wash. and B.C. There is a wide gap in the distribution of *L. pulchella* between Man. and B.C. from the Columbia River westward. Farther south, however, between Colo. and Calif., this species is transcontinental.

Distribution in Canada. N.S.—Hants county (Mt. Uniacke). *N.B.*—Sunbury and York counties. *Que.*—Covey Hill; Hemmingford; Vaudreuil; Hull; Wakefield; Alcove; Nominingue; Abitibi county. *Ont.*—Essex, Kent, Lambton, Elgin, Middlesex, Norfolk, Oxford, Brant, Haldimand, Welland, Wentworth, Waterloo, Halton, Huron, Wellington, Peel, York, Prince Edward, Bruce, Simcoe, Hastings, Leeds, Prescott, Carleton and Renfrew counties; Muskoka, Parry Sound, Nipissing, Algoma, Thunder Bay and Kenora districts. *Man.*—Treesbank; Stockton; Aweme; Onah; Selkirk; Husavick; Winnipeg Beach; Victoria Beach. *Sask.*—(Whitehouse, 1948). *B.C.*—Osoyoos, n. through Oliver, Peachland and Vernon to Salmon Arm and Revelstoke.

Although *L. pulchella* is one of the commonest of dragonflies in southern Ontario and in the St. Lawrence Valley, its main area of distribution is in the United States, where it is almost ubiquitous, except in the drier parts of the west. Its absence or rarity on the Precambrian region of Canada is noteworthy and it is also decidedly rare in New Brunswick and Nova Scotia. It is unknown in Newfoundland and the Northwest Territories.

Field notes. Libellula pulchella is a typical pond dragonfly, as indicated by its greater abundance about ponds than elsewhere, particularly ponds in open places, fully exposed to the sun. The outlets of slow streams are also favourite haunts of this species for they are often essentially similar to ponds as habitats of Odonata.

In Ontario, *L. pulchella* is noticeably more abundant in the regions underlain by sedimentary rocks than in the granitic areas. Thus, when one travels from an agricultural district to the Canadian Shield this species immediately becomes scarcer. It appears to avoid peaty waters, though breeding to some extent in the pond-like outlets of slow streams with an abundant emergent flora. We believe that in such places adults are likely to emerge much later than are those from the sun-warmed ponds. We have occasionally taken *L. pulchella* about streams in late September whereas the usual flight period is nearly finished by the end of August.

Emergence begins in Ontario during late June or the first week of July, sometimes before the end of May. Such early individuals come from the shallow ponds that are soon sufficiently warmed by the sun to stimulate emergence. Most of them are out by the middle of June and many are by this time fully mature, the males appearing very conspicuous from the pattern of alternate dark and white spots on their wings.

The males fly swiftly up and down the ponds, not patrolling the shore very regularly, if at all, but frequently driving away other dragonflies. We recall a visit to a pond-like expansion of a small stream near Niagara, Ontario, where *Gomphus villosipes* was in flight. In the usual *Gomphus* fashion, *villosipes* was often on the point of settling on a projecting stone, when a big male *pulchella* would sweep down upon it and drive it away. In their usual flights over a pond male *pulchella* will make a long swoop of ten or more yards, then come to an abrupt halt, hover for a few seconds a few feet along the water, then dash off at quite a different angle. When flying over a marsh or field they often come to rest on a long branch or a rush and they then illustrate well the habit which is prevalent in the Libellulidae of returning to the same support when driven away from it.

Mating occurs during most of June and July and also oviposition, the female performing this function alone. She usually picks a spot with plenty of submerged vegetation. Descending from a height of a few inches she strikes the water with the abdomen rhythmically at the rate of nearly two strikes per second for sometimes a minute or more. Frequently ovipositing females are disturbed by males. On June 29, 1948 we observed such a female, which was constantly pursued by a male and was prevented from depositing eggs repeatedly in one spot after their usual practice. It was compelled to fly ahead while ovipositing.

In August the number of *pulchella* in flight drops off considerably and by the end of this month the flight period is nearly over. But, as we have already noted, occasional fresh individuals are seen in September and these apparently come from the cooler water of streams. Our latest record from Ontario is September 28, and Whitehouse (1941) gives the latest record from British Columbia as September 29.

Libellula semifasciata Burmeister. (Pl. 24: 6, p. 164; pl. 40: 3, p. 260; pl. 41: 4, p. 264)

Libellula semifasciata Burmeister, Handb. Ent., 2: 862, 1839.

A yellowish brown, rather small species, the wings with brown basal streaks, large yellow-brown nodal spots and apical cross-bands. Found in Canada only in southern Ontario.

Male. Ochraceous and brown, not very hairy. Face without dark markings, pale orange brown to deep salmon without dark markings, palest on labium, deepening above to red with red-brown hairs on frons; vertex and occiput ochraceous, vertex with darker brown hairs, occiput with paler hairs behind, ocelli narrowly surrounded with black; rear of head dull brownish with a pale fringe of tawny hairs. Prothorax brown, darkest on sides and posterior lobe, without elevation of pronotal parts, hairless. Pterothorax ochraceous with two oblique pale stripes which are clear and distinct below but sometimes faint above and behind; hair of thorax dull ochraceous, not very dense; legs pale brown at base, deepening toward apices of femora; tibiae and tarsi black. Wings (pl. 24: 6, p. 164) tinted with brownish yellow, most deeply in proximal fourth; each with a dark brown streak from base to triangle, more distinct on hind wing; a large spot over the nodus, halfway across the wing and a dark stripe across apex of wing including pt., becoming clearer at the extreme apex. Most of the main veins light red. Abdomen ochraceous, darkening caudad from about seg. 6 to the end.

Female. Similar to male in colour pattern except in a few details. Face pale yellow to green, without the reddish tint of the male. First five abdominal segments almost entirely ochre yellow; 6 to 10 with a black median stripe, narrowest on 6, 7 to 10 black with a marginal yellow area as in the male but more distinct on account of the wider segments. Wings similar to those of male except in the width of the apical stripe, which narrows rearward, leaving a clear or almost clear apical area.

Abdomen tapering evenly as far as end of 7, then widening a little on 8, and narrowing again on 9 and 10; anal apps. very small; vulvar lamina consisting of two short broad processes, widely separated and not projecting beyond the tergal margins.

Venation. Anx ♂ 14–17/11–13, ♀ 15–17/10–12; pnx ♂ 9–12/9–13, ♀ 8–11/10–12; cross-veins under pt. ♂ 2–5/3–5, ♀ 3–5/4–5.

Measurements. Total length ♂ 40–45, ♀ 37–43.5; abd. ♂ 25–28, ♀ 23–26; w.abd. ♂ 4.2–4.9, ♀ 4.4–5.0; h.w. ♂ 34–37, ♀ 32–38 (NW: ♂ ♀ 31–37); h.f. ♂ 5.5–6.1, ♀ 5.2–6.0; w.hd. ♂ 7.1–8.0, ♀ 7.3–7.8; pt. ♂ ♀ 5.0–5.3.

Nymph. The two exuviae which we have studied are a dull dark brown and lack a distinct pattern. They are slightly smaller than *L. quadrimaculata*, but similar to that species in general form, differing in the larger and more curved dorsal abdominal hooks and longer lateral spines of abd. segs. 8 and 9 (pl. 40: 3, p. 260). They are not very hairy in appearance.

Head measured from anterior margin of frons to posterior margin of head about twice as wide as long; eyes slightly more prominent than those of *L. quadrimaculata*. Antennae slender, relative lengths of segments being approximately 3:4:6:4:5:6:5; frons bearing a moderate fringe of setae chiefly from near the front margins; postocular area also moderately hairy; lateral margin somewhat oblique, passing into hind margin in a broad curve; hind margin slightly concave. Labium with 11 to 13 premental setae, the 7 or 8 outermost setae being the longest; palpal setae 7 or 8; distal margin of palpus with about 12 low crenations, each with 4 or 3 spiniform setae in a steeply graded series. Legs rather slender with tibiae setiferous. Abdomen (pl. 41: 4, p. 264) not hairy in appearance, ovate, greatest width between segs. 6 and 7, at seven-twelfths of length; wing-sheaths in natural position reaching a little beyond base of seg. 6; dorsal hooks on abd. segs. 4 to 8, very short and stubby on 4, increasing in size rapidly, largest on 8, somewhat curved and acute on 6, 7 and 8 and each projecting a little over the succeeding segment; lateral spines on 8 and 9, each about one-fifth as long as the lateral margin of the respective segment (the spine included); tips of spines on 8 strikingly incurved, those of spines on 9 less so; epiproct about twice as long as wide; paraprocts very little longer; cerci nearly three-fifths as long as epiproct.

Length 20–24; abd. 12–12.2; w. abd. 6.5–7.0; h.w. 5.5; w. hd. 5–5.6.

Habitat and range. Forest brooks and marshy bays, more rarely ponds. Fla. to Tex. and n. through the Atlantic and middle States to Me., N.Y., s. Ont., Mich. and Wis.

Distribution in Canada. Ont.—Essex (Pt. Pelee; Pelee I.), Kent, Norfolk, York, Ontario and Simcoe counties.

Known only from southern Ontario where it is locally common in the Lake Erie counties and has been observed and taken during a single season as far north as De Grassi Point, Lake Simcoe. It was formerly not uncommon at Scarborough, east of Toronto.

Field notes. In Ontario this is the first *Libellula* to appear in the spring. Our earliest recorded date is May 25 (Lake Simcoe) and our latest July 15 (Point Pelee). In the New England States it has been encountered until August 16 (Needham and Westfall, 1955).

At Scarborough, Ontario, I first found this species along a small brook in a partly wooded ravine. The discovery of this species at De Grassi Point in 1936 for the first time in over thirty years' collecting was surprising. They flew in an open wood in company with large numbers of *L. quadrimaculata* from which they were difficult to distinguish at a short distance. This is well out of the normal range of this species.

Needham and Heywood (1929) record that adults fly "in swift dashes, and in long sweeping curves."

Libellula incesta Hagen. (Pl. 22: 3, p. 148; pl. 41: 5, p. 264; pl. 45: 1, p. 276)

Libellula incesta Hagen, Syn. Neur. N. Amer., p. 155, 1861.
Holotania incesta: Kirby, Cat. Neur. Od., p. 29, 1890.
Libellula vibrans incesta Ris, Coll. Zool. Selys, 11: 270, 1910.

A long slim species, which in the teneral state is pale yellow or greenish yellow, broadly striped lengthwise with brown, but which soon darkens and finally becomes an almost uniform blue-black in both sexes.

Male. Teneral coloration pale yellow striped with brown. Face including entire frons yellowish olivaceous; labium and labrum reddish brown; vertex and occiput light brownish, hair of face and dorsum of head sparse and rather short; rear of head pale yellowish. Thoracic dorsum rich yellowish brown with a median furrow and the ante-alar sinuses light yellow; also a median inter-alar stripe yellow; sides of thorax mostly pale greenish yellow, this area expanding rearward and edged below with deep brown. Coxae and trochanters pale yellowish, the legs otherwise dark brown. Wings hyaline, venation and pt. dark brown; other brown markings usually absent or reduced to a slight apical tinge. Abdomen yellow with a median dark brown stripe and narrow lateral margins, the median stripe widening as far as seg. 5 or 6 and covering all of segs. 8 and 9.

In the mature male the face has changed from yellow to a deep shining olivaceous brown. Only the labium and sides of the mandibles remain a dull yellow. Dorsal part of head shining black; rear of head black with two pale yellow spots on each side. Sides of thorax dark olivaceous, finally becoming blue-black. Dorsum of pterothorax bluish black, scantily beset with fine black hairs; legs black, wings hyaline with venation and pt. black and usually with no other spots except, sometimes, a narrow smoky apical margin; pt. black, 6 × 8 mm.

Frons not very prominent, with an antero-dorsal roughened area bounded below by a sharp transverse ridge, which curves upward on each side, the rough area with a broad smooth groove extending from anterior to dorsal surface in front of the vertex, which is elevated in front with two anterolateral angles. Pronotum with middle lobe flat, with a semicircular hind margin. Abdomen long, slender, half as wide as thorax at seg. 2, parallel-sided from seg. 1 to end of 5, thence tapering to apex or, in some individuals, from seg. 2.

Sup. apps. twice as long as seg. 10, slightly arched from base to middle, meeting at about three-fourths of their length, apices slightly upcurved, bluntly pointed inf. app. three-fourths as long as sup. apps. Cerci slightly longer than seg. 10, very slender.

Female. Similar to male in coloration and proportions except that the abdomen is stouter and is only very slightly tapered from segs. 2 to 7; 8 laterally expanded, with lateral margins rounded; seg. 10 about half as long as 9. Apices of wings usually dusky beyond pt., a small nodal spot sometimes present. Vulvar lamina a semitubular trough formed from the free margin of the eighth sternum, bent downwards, its lateral walls swollen and rounded behind. It bounds the genital orifice in front.

Venation. Anx ♂ ♀ 15–17/12–14; pnx ♂ 12–16/12–17, ♀ 12–14/12–15; cross-veins under pt. 4–5.

Measurements. Total length ♂ 47–55, ♀ 47–51; abd. ♂ 30–34, ♀ 31–35; w.abd. ♂ 2.5–3.6, ♀ 3.0–5.5; h.w. ♂ 38.0–40.0, ♀ 40.0–45.0 (NW: ♂ ♀ 36–42); h.f. ♂ 6.0–7.0, ♀ 6.0–6.5; w.hd. ♂ ♀ 7.5–8.0; pt. ♂ ♀ 5.5–6.5.

Nymph. Only two exuviae of *L. incesta* were available for the study of the nymph. They are somewhat smaller than average-sized *L. pulchella* but similar in general form: reddish brown with no distinct pattern; body not very hairy.

Head about 2.5 times as wide as long; eyes of average prominence at the anterolateral angles; labrum truncate, its sides convexly curved; posterolateral corners of head passing into the posterior margin by a broad, strongly concave curve. Labium with apex of median lobe forming an angle of 120°, its margins smooth; crenations of distal margin very low, its alternating notches bearing groups of usually 3 to 5 spiniform setae of which the longest in a group is nearly one-half as long as an average crenation, the next about one-half as long as the longest, the others rapidly diminishing to minute "dots"; premental setae 8 or 9, the outer 5 much longer than the inner 3 or 4, which tend to be successively smaller; palpal setae 5; movable hook nearly as long as the adjacent palpal seta; outer margin of palpus with a basal series of about 15 very short setae. Prothorax slightly narrower than head; mesothorax about as wide as head. Legs stout, femora with long but sparse hairs, tibiae with lateral surfaces very hairy. Hind wing-sheaths reaching base of seg. 6, where the width slightly exceeds one-half the length, somewhat tectate in segs. 7 to 9; dorsal hooks on segs. 4 to 8 (pl. 41: 5, p. 264), the largest on 7 and 8, arising on posterior half of the segment and on 6 to 8 extending on posterior half of each segment and beyond base of following segment; lateral spines on segs. 8 and 9, the length of each spine about one-fifth of the lateral margin of the segment that bears it; spines of both segments slightly incurved, tapering to a fine point; epiproct (pl. 45: 1, p. 276) about as long as mid-dorsal length of segs. 9 + 10, its base about two-thirds of its length, somewhat acuminate, very slenderly pointed and distinctly decurved at tip; paraprocts slightly longer than epiproct; cerci about one-third as long as paraprocts.

Length 21; abd. 11.5; w. abd. 6.5; h.f. 6.

Habitat and range. Ponds, small marshy lakes and slow streams. Most of the United States e. of the Rocky Mts. and s. of the northern parts of the Great Lakes' States; Fla. to Tex. and Okla., n. through the Atlantic States to Me. and N.H., w. through s. Ont., Mich., Wis. and middle States, N.Y., Pa., Ohio, Ky., Tenn. and Mo.

Distribution in Canada. Ont.—Essex, Norfolk, York, Simcoe, Victoria, Lennox and Addington, Frontenac and Leeds counties; Muskoka district (Kahshe Lake; Georgian Bay; Dwight, Lake of Bays).

The distribution of *L. incesta* in Ontario appears to be divided into two main areas, i.e. a southwestern area along Lake Erie, and an eastern area approximately along the southern margin of the Canadian Shield, from Seebright in Simcoe county, and Bobcaygeon (Victoria county) to the Rideau Lakes and the St. Lawrence River in Leeds county.

Field notes. This species flies from the third week in June to the second week in August. The recorded flight period in Michigan is somewhat longer—from the end of May to the end of August (Kormondy, 1958).

We have found this insect to be most common over clear quiet waters, such as lakes and wider streams, but not in marshy places. It appears to be absent from Lake Simcoe, which is underlain by stratified limestone, but is common on Kahshe Lake, Muskoka district, a short distance north of Simcoe county, but in the typical granitic region of the Canadian Shield. On one occasion we saw many *L. incesta* on Kahshe Lake but have failed to catch a single specimen since. When disturbed they would always fly from the water up over the rocky banks to the top branches of the button bushes, where they were well out of each of the net; but in other situations they are not especially difficult to capture.

While locally not uncommon, this is one of our rarer species of *Libellula* and is, so far as we know, confined in Canada to the Province of Ontario. To judge from its wide east-west distribution in this province, however, it is probable that *L. incesta* will be discovered in southern Quebec and, quite possibly, also in Nova Scotia. In the vicinity of Washington, D.C., Donnelly (1961) writes of it as "one of the most common pond dragonflies."

Libellula vibrans Fabricius. (Pl. 40: 4, p. 260; pl. 41: 6, p. 264; pl. 45: 2, p. 276)

Libellula vibrans Fabricius, Ent. Syst., 2: 380, 1793.
Libellula vibrans: Ris, Coll. Zool. Selys, 11: 268, 1910.

Similar to *L. incesta* but larger, edges of wing-tips beyond pt. with black pigment and nodal spots small but clearly defined; pt. black. Space between Sc and R + M black from base to second or third antenodal cross-vein.

Male. The coloration of tenerals (dried specimens) is as follows: eyes reddish brown; vertex darker brown; labrum clear light yellow; labium straw yellow to gray; rear of head on each side with a short yellow stripe between two black stripes. Prothorax and dorsum of pterothorax reddish brown with a clear-cut median yellow stripe. Pleura next to dorsum pale yellowish to gray with a large dark brown anterior spot, often divided into two spots and another smaller spot in the second lateral suture. Legs yellow or brown at base and distally, including a part of each femur; tibiae and tarsi black. Interalar area of thorax yellow. Abdomen with yellow lateral marginal stripe extending entire length of abdomen; median area dark brown to black.

The coloration of mature males is as follows: face greenish white or pale green; labrum dull white, labium pale yellowish with inner margins and middle lobe blackish. Transverse frontal groove and vertex blackish with pale margins, with angular points; facial hair brown, short, fine and sparse, vertex dark blue and green; occiput blackish above, yellowish brown below. Rear of head black above with a yellow postoccipital spot. Prothorax black, little elevated; pterothorax dorsally pale pruinose gray with rather thin gray hair, the sides yellowish fawn with a crooked black mesepisternal suture at base of which is a large triangular black spot extending rearward

PLATE 27

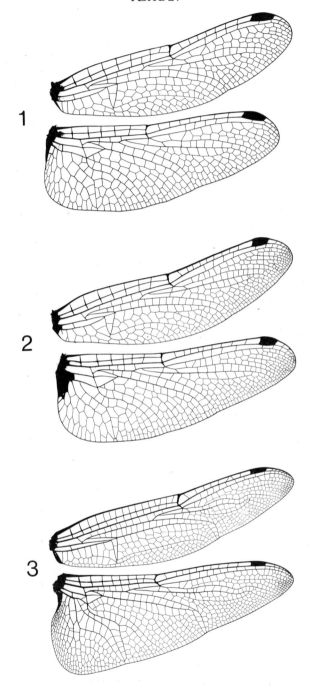

over the mesepimeron and a smaller black spot below the spiracle. Mesonotum pruinose gray; metanotum with little or no pruinosity. Legs brownish at base, distally darkening.

Abdomen very slender for a *Libellula* and tapering at a steady rate from a width of nearly 4 mm. to barely over 2 mm.

Sup. apps. slender, parallel, with apices slightly divergent; viewed laterally they curve downwards from the bases and become larger distally.

Female. Coloration as described above for the teneral male, except that on the abdomen (which is broader) the lateral yellow margin is much wider and divided into segmental areas. Seg. 8 is laterally expanded, the outcurving margins with their thickened rims being suggestive of the abdominal expansions so general in both sexes of the Gomphidae.

Venation. Anx ♂ 20/20, ♀ 16/17; pnx ♂ 16/16, ♀ 19/19; cross-veins under pt. 3–4.

Measurements. Total length ♂ 56–62, ♀ 52–58; abd. ♂ ♀ 35–37; w.abd. ♂ 4.0–4.6, ♀ 3.7–4.0; h.w. ♂ 43–45, ♀ 40–48 (NW: ♂ ♀ 48–51); h.f. ♂ ♀ 5–6; w.hd. ♂ 8.6–9.2, ♀ 8.1–9.2; pt. ♂ 6, ♀ 7–8.

Nymph (description from Needham and Westfall (1955)). Body not particularly hairy in appearance. Premental setae 11, with the 5 outermost being longest; palpal setae 5; dorsal hooks (pl. 40: 4, p. 260; pl. 41: 6, p. 264) on abd. segs. 4 to 8, the largest on 8; lateral spines on segs. 8 and 9, their tips diverging slightly from the outline of the abdomen, each spine about four-tenths of the lateral length of the respective segment, including the spine; cerci less than one-half as long as paraprocts; epiproct (pl. 45: 2, p. 276) straight or only very slightly decurved at extreme tip.

Length 21.5–27; abd. 13.0–15.0; w. abd. 7.0–8.0; h.w. 7.5–8.5; h.f. 5.5; w. hd. 5.5–6.0.

Habitat and range. Marshes and standing water. Fla. to Tex., n. to Mass., Pa., Ohio, Ind. and Ill. to Wis. and s. Ont.

Distribution in Canada. Ont.—Essex county (Pt. Pelee).

Known here only from a single mature male taken at Point Pelee, June 21, 1951 by Mr. Eric Thorn.

Field notes. It seems likely that in Ontario *L. vibrans* will be encountered mainly in June, the month in which captures have been recorded also from the nearby States of Ohio (Kellicott, 1899) and Indiana (Montgomery, 1950). In the southern United States it flies from April 1 to December 8 (Needham and Westfall, 1955).

Although this may be only an adventive species on Point Pelee and Pelee Island, the fauna and flora of these localities provide much evidence that they are somewhat richer in southern species than are the other parts of Ontario adjoining Lake Erie.

In North Carolina *L. vibrans* is, according to Brimley (1903), rather common in summer "over marshes and standing water," being the "largest and most sluggish of the *Libellulas*."

PLATE 27

Wings of Libellulidae: (1) *Sympetrum vicinum*; (2) *Leucorrhinia intacta*; (3) *Pantala flavescens*. (1–3 from Needham and Westfall, 1955.)

Genus **Erythemis** Hagen

Syn. Neur. N. Amer., p. 168, 1861. Type species *Libellula peruviana* Rambur.

This is a genus of moderate-sized, clear-winged dragonflies of mostly southerly distribution; only one species occurs in Canada, where it has a very limited range.

The head is of medium size or small, the eye-seam short, and the frons without an anterior ridge; the vertex is moderately high, with the apex slightly emarginate. The hind lobe of the prothorax is as wide as the other lobes or wider, the posterior and lateral margins having long hairs; the legs are robust and spiny, the femora of the males having two ventral rows of small regular teeth, and the tibiae very long black spines. Accessory genitalia of the males are small, and the hamuli biramous; the lateral margins of segment 8 of females are not expanded, the vulvar lamina being a more or less erect plate or trough. The wings (pl. 26: 1, p. 180) are long, moderately broad, and hyaline or in some species (though not ours) brown or flavescent; the triangles of the fore and hind wings are at about the same level; the fore-wing triangle is narrow, the anterior side being much shorter than the proximal side, and two-celled; the triangle of the hind wing and all supratriangles are free; the sectors of the arculus are joined in a long stalk; the arculus lies between the first and second antenodals; the basal IR3 and Rspl veins are separated by one row of cells; vein R3 has a single smooth curve.

The nymph is stocky with prominent, green and brown eyes, and strong spiny legs. It lacks dorsal hooks or lateral spines on the abdomen and is distinguished from other libellulids (in Canada) by the strongly decurved paraprocts.

The genus contains eight or nine species, most of which inhabit some part of the Neotropical region; five or six occur in North America depending on whether *E. collocata* Hagen is a subspecies of *E. simplicicollis* or a species distinct from it. In either case the record of *collocata* from British Columbia is almost certainly erroneous: all specimens that we have examined so far from this Province are *simplicicollis*. Thus a single species, *E. simplicicollis*, is known from Canada, being quite common in southern Ontario and southwestern British Columbia. (Since these words were written Paulson (1970) has stated that *E. simplicicollis* does not occur west of the Great Plains and that all records of it from such westerly areas should be assigned to *E. collocata*. This statement was based on determinations made by several authorities using diagnostic features recognized by Gloyd (1958). Clearly the status of populations in British Columbia needs to be resolved.)

Bick (1941) has described morphological changes that occur during nymphal development in *E. simplicicollis*.

Erythemis simplicicollis (Say). (Pl. 1: 2, p. 6; pl. 25: 3, p. 168; pl. 26: 1, p. 180; pl. 45: 3, p. 276)

Libellula simplicicollis Say, J. Acad. Phila., 8: 28, 1839.
Mesothemis simplicicollis: Hagen, Syn. Neur. N. Amer., p. 170, 1861.
Erythemis simplicicollis: Calvert, Trans. Amer. Ent. Soc., 20: 265, 1893.

A medium-sized dragonfly with clear wings and a nearly uniform green to gray-blue body colour, the abdomen having segmentally arranged dark brown spots chiefly behind the middle. The males are pruinose blue when mature. This species is widely distributed in the United States.

Male. General colour of teneral grass-green with dark brown legs and abdominal markings. Labrum and labium pale yellow, the labium often distally greenish; clypeus and frons pale green with a very sparse clothing of minute brownish hairs; vertex green; rear of head with many pale greenish spots on a black background. Prothorax and pterothorax green (drying to yellow) almost without dark spots, except along the ante-alar ridges, lower margins of epimera and upper parts of pleural sutures. Legs dark brown, becoming distally black with outer surface of fore femora and most of coxae and trochanters of all the legs paler brown. Wings hyaline without dark spots except the anal triangles, which are mostly brown with a pale basal spot. Pt. pale fawn. Abdomen green with dark brown spots on segs. 3 to 10, chiefly behind the transverse carina of each segment, the spots appearing successively more crowded caudad. The last two or three segments wholly dark brown; anal appendages pale yellow.

After the teneral stage the body colour changes from green to gray-blue, this colour spreading from the abdomen forward until the body is entirely gray-blue except the legs, the dark spots of the abdomen being also included in the bluish area, and the head alone retains the teneral coloration.

Vertex of head quadrate, about twice as wide behind as in front, eyes meeting at a point; occiput about as long as vertex. Abdomen nearly as long as hind wing, segs. 1 and 2 a little enlarged, 3 slightly constricted, the remaining segments parallel-sided to end of 8, narrowing beyond 8, triquetral with sharp dorsal and lateral carinae.

Sup. apps. nearly as long as seg. 9, straight, cylindrical with acute apices, convergent; inf. app. truncate sub-triangular with the usual upward curvature.

Female. Colour pattern similar to that of the teneral male and remaining so throughout life. Abdomen scarcely wider than that of male, constricted at base of seg. 4, widening again slightly to the end of 8; seg. 10 very small. Vulvar lamina a blunt, angular spout-like process, nearly as long as seg. 9, with margin rounded at base, projecting (in dried specimens) perpendicularly to axis of abdomen, this position probably being that which is used in oviposition.

Venation. Anx ♂ ♀ 9–11/7–9; pnx ♂ ♀ 7–10/7–11; rows of cells subtended by Rspl and Mspl 1; cross-veins under pt. 2 (occasionally 1).

Measurements. Total length ♂ 41–46, ♀ 40–43; abd. ♂ 27–30, ♀ 27–29; h.w. ♂ 31–33, ♀ 30–34; h.f. ♂ ♀ 6.5–7.0; w.hd. ♂ ♀ 6–7; pt. ♂ 3.5–4.0, ♀ 4.0–4.3.

Nymph. Short, compact, smooth, with nearly uniform pale green coloration. Head with labium prominent and bluntly angular in front; eyes rather large, rounded and very prominent, directed anterolaterally, lateral margins of head behind eyes straight in front, posterolateral angles rounded, posterior margin straight, somewhat excavate. Antennae slender, relative lengths of segments approximately 4:5:12:10:10:12:12. Prementum of labium about as broad as long; median angle of median lobe somewhat greater than a right angle, smooth; palpi with distal margins smooth, without crenations. Prothorax without lateral processes, spiracular plates very high, elevated in exuviae. Legs robust, not hairy, hind pair much the largest, the hind tibia longer than hind femur. Wing-sheaths smooth, width of hind pair at nodus three-sevenths of their length. Abdomen ovate, very compact and convex, widest at seg. 6, but widths of 5 to 7 nearly equal, rapidly narrowed from 8 to 10, the lateral margin widely curved; dorsal hooks and lateral spines absent; abd. seg. 8 less than one-half as long as 9; anal appendages all decurved, the paraprocts (pl. 45: 3, p. 276) most strongly so; cerci slightly longer than 10, three-fourths as long as epiproct, which is wider than long, and with an acute apex. Colour uniform or nearly so; femora sometimes with basal and ante-apical dark annuli.

Length 15–17; abd. 9; w. abd. 5–6; h.w. 5.5; h.f. 5–6; w.hd. 5. (Measurements incorporating those of Needham (1901).)

Habitat and range. Lakes, ponds and slow streams. In the United States from the Atlantic to the Pacific (with the possible exception of the Dakotas); in Canada s. Que., s. Ont. and s. B.C.

Distribution in Canada. Que.—Near Hull; Rigaud. *Ont.*—Essex, Kent, Norfolk, Oxford, Welland, Wentworth, Waterloo, Huron, York, Prince Edward, Simcoe, Leeds, Lanark and Carleton counties. *B.C.*—Vancouver; Chilliwack; Vancouver I. (Goldstream; Florence Lake; Pike Lake; Long Lake; Newcastle).

Field notes. The recorded flight period is from June 1 to July 25 in British Columbia and from June 19 to September 5 in Ontario. Tenerals have been encountered in British Columbia (Vancouver) in early July (Whitehouse, 1941) and in Ontario (Leeds county) on June 21 (Walker, 1941b). In Ontario it is most abundant during July; records for late August and early September are from Point Pelee, where it is a regular constituent of the aggregations of Anisoptera that appear there then (Root, 1912; Corbet and Eda, 1969).

From its habit of resting on the bare ground and its green colour this insect is easily mistaken for an *Ophiogomphus*. It also squats on logs or trash along the lake shore, particularly in open areas. Or it may perch on weeds and dart out to catch insects of various kinds.

"Diptera form the bulk of its food, many *Chrysops* justly perishing in this way. They have also been observed to kill and eat butterflies (*Pamphila*), moths and dragonflies (*Lestes vigilax* and *Argia violacea*). On several occasions and at different localities males have been seen going through manoeuverings which are hinted at frequently by some of the *Libellulas* but seem to have been perfected only by this species. Two males are necessary for the performance. They flutter motionless, one a few inches in front of the other, when suddenly the rear one will rise and pass over the other, which at the same time moves in a curve downwards, backwards and then upwards, so that the former position of the two is just reversed. These motions kept up with rapidity and regularity give the observer the impression of two intersecting circles which roll along near the surface of the water" (Williamson, 1900).

Variations. The few specimens of *E. simplicicollis* from British Columbia recorded by Whitehouse as race *collocata* (on the present writer's authority) we now refer to the typical variety, which is now known from the west as well as the east (but see remarks above and Paulson (1970)).

Genus **Pachydiplax** Brauer

Verh. Zool.-Bot. Ges. Wien, 18: 368, 728, 1868. Type species *Libellula longipennis* Burmeister.

A monotypic genus of wide distribution through almost the entire United States and in adjacent parts of Mexico, the Bahamas, Bermuda and a little of the extreme south of Ontario and British Columbia. The single species is easily recognizable, having no very close resemblance to any other genus. It was regarded by Ris (1909) as one of the *Sympetrum* series of genera, although its appearance is not suggestive of *Sympetrum*: it is somewhat larger and has a pale olivaceous thorax striped with brown, and a short abdomen. The blue males might be mistaken in life for *Erythemis simplicicollis* but the smaller size and striping of the thorax are distinctive features.

Pachydiplax is unlike any other libellulid genus in our fauna in the colour pattern in which the brown thoracic stripes on a pale green background are combined with a pruinose blue abdomen in the male and with longitudinal green abdominal spots, segmentally divided in the female. The most useful venational character (pl. 26: 2, p. 000) is the presence of only a single cross-vein behind the pterostigma at the extreme distal end, the only other cross-veins in this space being the four postnodals, the nearest of which stands at a distance of about four normal cells away from the cross-vein. Vein R_3 is not undulate.

The nymph is strikingly patterned and rather like a large *Leucorrhinia* without dorsal hooks, e.g., *L. borealis*, but is more depressed, smoother and with a broader head and abdomen and longer lateral spines. A dark ridge runs mesad from the mesoposterior part of the eye. The lateral spines on segment 9 are twice as long as those on 8. The epiproct is much shorter than the paraprocts.

Pachydiplax longipennis (Burmeister). (Pl. 26: 2, p. 180; pl. 38: 4, p. 254; pl. 44: 1, p. 272)

Libellula longipennis Burmeister, Handb. Ent., 2: 850, 1839 (♀).
Mesothemis longipennis: Hagen, Syn. Neur. N. Amer., p. 172, 1861.
Pachydiplax longipennis: Calvert, Trans. Amer. Ent. Soc., 20: 265, 1893.
Pachydiplax longipennis: Calvert, Trans. Amer. Ent. Soc., 25: 66, 1898.

A species of variable, usually moderate, size, being somewhat smaller than *Erythemis simplicicollis*, and easily recognized by the striped thorax of both sexes and the blue pruinosity of mature males and old females, which is most intense on the abdomen but spreads forward on to the thorax.

Male. Labium and face greenish white; labrum pale yellow. Antennae slender, 7-segmented. Facial hair dark but scanty except on frons where it is moderate. Vertex and large spot on frons bright metallic blue; eye-seam as short as occiput, rear of head black. Prothorax blackish, the hind margin an upright crest bearing a fringe of whitish hair like that which forms a moderate cover for the pterothorax, the latter conspicuously striped with pale green, or olivaceous with dark brown. Dorsum brown with a pair of short, narrow, straight stripes, diverging forward and a narrow transverse greenish stripe immediately in front of the ante-alar sinuses. Sides of thorax pale green, crossed obliquely by three brown stripes, including the humeral stripe which is broadest and has a strongly undulate front margin coinciding with the humeral suture. Legs black, femora becoming pruinose. Wings each with a basal transparent brown spot, that of the hind wing with three short dark brown streaks on the costal vein and the subcostal and cubito-anal space. There are also two large yellow-brown clouds occupying most of the post-nodal areas of each wing from nodus to distal ends of pt. Underside of thorax pale yellow. Abdomen dark brown, becoming pruinose blue when mature. Segs. 1 to 3 light yellow below.

Length of abdomen five-sixths of hind wing, little enlarged at base, scarcely constricted beyond but diameter smallest between segs. 4 and 5 and only slightly larger beyond 6.

Anal apps. about as long as segs. 9 + 10, nearly straight and parallel as viewed from above, with apices acute; in lateral view moderately arcuate but with apices curved upward, angulate below at distal third. Inf. app. reaching a point between inferior angle and apices of sup. apps. Inner branch of hamulus as broad as long, with short hook abruptly inturned.

Female. Similar to male in general form and colour pattern, with a slightly shorter and wider abdomen and with paired dorsal and lateral longitudinal yellow stripes divided into segmental pairs of spots, the lateral spots about twice as wide as the dorsal spots; last spots on seg. 9 and very small. Wing spots frequently absent but sometimes developed as in males. Abdomen somewhat shorter than in the male with the last four or five segments depressed and slightly expanded. The colour pattern is similar except in some details of the abdomen. On all but the last segments there is a dorsal pair of longitudinal yellow stripes and a wider lateral yellowish stripe on each side. The lateral stripes are widest in front and decrease in size caudad.

Venation. Anx ♂ ♀ 6/5; pnx ♂ ♀ 6–7/6.

Measurements. Total length ♂ 36–41, ♀ 35–40; abd. ♂ 22–36, ♀ 22–24; h.w. ♂ 28–30, ♀ 29–32 (NW: ♂ ♀ 30–42); h.f. ♂ 6–7, ♀ 5–6; pt. ♂ 3.5–4.2, ♀ 4.0–4.3.

Nymph. Head twice as wide as long, the eyes prominent, directed more laterally than those of *Erythemis simplicicollis*, much as in *Sympetrum*; dark ridge running mesad from mesoposterior part of eye. Antennae slender, their length about three-fifths that of the head. Premental setae 10 to 13; palpal setae 9 or 10; movable hook long and slender; crenations on distal margin of palpus shallow, each with 2 or 3 long spiniform setae. Wing-sheaths extending to base of seg. 6; legs slender, moderately long, the hind pair, when straightened out, being about half as long as the entire body. Abdomen long-oval, being longer than in *Sympetrum* or *Leucorrhinia*, somewhat depressed, widest at seg. 6 but scarcely changing in width from the anterior margin of 6 to the posterior margin of 8, smooth and without dorsal hooks; segs. 8 and 9 with lateral spines, the tips of those on 8 slightly divergent, in length about one-fourth of that segment's lateral margin, including the spine; the spines of 9 subparallel, their length a little more than one-third of that segment's lateral margin, and equal to or greater than its mid-dorsal length; epiproct one-fourth longer than its basal width; cerci nearly three-fourths as long as the basal width of the epiproct; paraprocts about one-fourth longer than epiproct, somewhat tapering throughout their length.

Length 18.1–21.0; abd. 11.8–12.1; w. abd. 6.2–7.5; h.w. 5.6–5.8; h.f. 4.8–5.2; w. hd. 5.0–5.5.

Habitat and range. Lakes, ponds and slow streams throughout almost the entire United States, extreme southern Canada (Ont., Man. and s.w. B.C.), Mexico (Coahuila, Baja California), Bermudas and Bahamas.

Distribution in Canada. *Ont.*—Essex, Kent, Elgin, Norfolk, Haldimand, Welland, Wentworth, York, Prince Edward, Simcoe, Frontenac and Leeds counties. *Man.*—Lac du Bonnet. *B.C.*—Vancouver I. (Waugh Creek; Goldstream Creek; Victoria; Saanich district; Nanaimo district; Newcastle I.).

This is an abundant dragonfly in the counties along Lake Erie and locally along Lake Ontario, particularly on the sandy areas of Prince Edward county. Fifty years ago it was common in High Park, Toronto, breeding in the small lake known as Grenadier Pond. It is now extinct there. During the season of 1933 a male was taken at the stream near De Grassi Point, Lake Simcoe, fifty miles north of Toronto, but this stream is now destroyed as a breeding place for dragonflies.

Field notes. In Canada adults of *P. longipennis* have been taken from early June until early September. In Ontario an early capture was on June 11, and tenerals were seen in Prince Edward county on June 19 (Walker, 1941b). At Point Pelee, where it is common in summer (Root, 1912), it has been seen as late as September 5. In Michigan records extend from May 31 to August 19 (Kormondy, 1958), and in Indiana it has one of the longest flying seasons among libellulids: from the last week in April until the last week in September (Montgomery, 1945).

We visited Pelee Island for a few days in June, 1959 and found this insect to be common there. On the more sheltered parts of the beach of Lake Erie where there is a fairly open stand of herbaceous vegetation many individuals come to rest horizontally on the upper side of the leaves and about a foot or fifteen inches above the ground. Various observers have also noted *Pachydiplax* resting with the abdomen curved upwards between the folded wings.

According to Whitehouse (1941) this species and *Sympetrum illotum* are much alike in form and behaviour. "They both delight in perching, with wings drooped, at the extreme tip of a dry twig; from which point of vantage they make short sallies to catch a fly or to rush at any other species invading their chosen territory. They are both given to hovering and are thus easy to net."

Needham and Westfall (1955) write that the females "are less in evidence. Except when foraging or ovipositing, they rest on trees back from the shore. When ovipositing over open water, they have an odd habit which we have not observed in other dragonflies: they do not rise and descend again between strokes of the abdomen against the surface of the water, but fly horizontally close to the surface and from time to time strike downward with the abdomen alone, presumably washing off eggs. In the midst of vegetation, however, they fly down and up again, as do other species." Egg masses observed by Montgomery (1941) almost always separated after sinking a few inches, but with less violence than characterizes the "explosion" of the egg masses of *Perithemis tenera*.

Adult recognition of the male is partly on the bluish-white pruinescence on the abdomen, and that of the female on the series of yellow spots on the abdomen (Johnson, 1962a).

Nymphs collected by Bick (1950) in Mississippi were from standing water with a mud bottom, in ponds, borrow pits or creek pot holes. In Utah Mrs.

Musser (1962) found nymphs among aquatic vegetation along the edge of a spring-formed pond.

Genus **Erythrodiplax** Brauer

Verh. Zool.-Bot. Ges. Wien, 18: 368, 722, 1868. Type species *E. fusca* (Rambur).

This is a large Neotropical genus of some fifty species: about six species inhabit the southern United States and one of these has spread northeastward along the Atlantic coast (including Nova Scotia) and has once been recorded from Quebec, probably as an adventive. The genus therefore hardly justifies inclusion in a treatise on Canadian Odonata. *Erythrodiplax* is, however, quite nearly related to our familiar genus *Sympetrum*, also being a member of the libellulid tribe Sympetrini.

Most species of *Erythrodiplax* are a little smaller than the average *Sympetrum*, although a few are larger than any of the North American species of the latter genus. They are more varied in colour than are *Sympetrum* and undergo a greater colour change during adult maturation. Although the face is pale in tenerals it may become metallic black, shining brown or red in old individuals (Needham and Westfall, 1955); brownish black, bluish black and green are other colours that appear in some of the numerous species.

The wings (pl. 26: 3, p. 180) are rather broad; the fore-wing triangle is narrow and the hind-wing triangle is usually clear; the arculus is a little nearer to the second antenodal cross-vein than to the first. The venation is variable between and within species. It would be out of place to deal with this subject further here, since Canada is virtually outside the geographical range of *Erythrodiplax*. An exceptionally detailed and comprehensive monograph of the genus has been prepared by Borror (1942).

The nymphs are short, blunt posteriorly, and smooth, somewhat resembling certain *Sympetrum*. If dorsal abdominal hooks are present they are poorly developed or may be represented by shallow prominences sometimes bearing tufts of setae. Lateral spines are present on segments 8 and 9 but are small.

Erythrodiplax berenice (Drury). (Pl. 25: 4, 6–9, p. 168; pl. 26: 3, p. 180)

Libellula berenice Drury, Ill. Nat. Hist., 1: 114 (pl. 48, fig. 3), 1770.
Diplax berenice: Hagen, Syn. Neur. N. Amer., p. 178, 1861.
Micrathyria berenice: Karsch, Berl. Ent. Zeits., 33: 371, 1889.
Trithemis berenice: Kirby, Cat. Neur. Od., p. 19, 1890.
Erythrodiplax berenice: Calvert, Occ. Pap. Bost. Soc. Nat. Hist., 7: 36, 1905.

A slender clear-winged dragonfly with a body coloration that varies a great deal with age. The nymphs characteristically inhabit brackish water.

Male. In tenerals the general colour is black and yellow, the appearance being striped. Face equally black and yellow; labrum yellow with a black front border; sides of postclypeus and frons yellow, with front blackish; frons blue-black with a large yellow spot on either side above in front, and a small yellow spot on either side near the ventrolateral corners; vertex metallic blue except for a small yellow spot on the dorsal side. Thorax striped with yellow and black, a mid-dorsal black stripe being divided by yellow of carina; on the sides black stripes begin in sutures and spread and coalesce at their ends to form a big 'N' mark over the humeral area and a coarse meshwork of black farther back; sparse white hair on underside of thorax. Legs black. Abdomen black with large yellow lateral spots on terga of segs. 3 to 7 or 4 to 7 (pl. 25: 8, p. 168). Hamuli with outer branches flat, obliquely truncate and usually straight (pl. 25: 6, 7, p. 168).

The body colour darkens progressively with age so that fully mature individuals may be entirely blackish blue. The wings remain hyaline.

Female. Generally similar to male with these exceptions. In tenerals the abdomen (pl. 25: 9, p. 168) is yellowish or yellowish brown with the sutures, segs. 3 to 7 below and along the sides just above the lateral carina, the posterior half of 7, and all of 8 and 9, black. With increasing age the light areas darken and eventually the colour pattern is obscured. Usually a small brownish basal spot in each wing extending almost to arculus and caudad a little beyond the membranule; old individuals may be diffusely clouded with brown to just beyond the nodus, or across the basal area, or both. The large mid-dorsal spots on abd. segs. 3 to 7 may remain yellow or even become blood-red.

The ovipositor is composed of the projecting sternite of seg. 8.

Venation. Anx ♂ ♀ 7–11.5 (usually 8.5)/6 (usually)–8; pnx ♂ ♀ 6–10 (usually 8)/6–10 (usually 8); rows of cells subtended by Rspl 1, occasionally 2; CuP in hind wing distinctly separated from anal angle of triangle (Borror, 1942).

Measurements. Total length ♂ 30.5–34.1, ♀ 31.4–33.0; abd. ♂ 21.2–23.1, ♀ 21.0–22.9; h.w. ♂ 24.6–27.1, ♀ 24.3–26.0 (NW: ♂ ♀ 21–26); h.f. ♂ 4.0–5.3, ♀ 4.5–5.1; w.hd. ♂ 5.0–5.5, ♀ 5.0–5.3; pt. ♂ 2.8–3.9/2.8–3.9, ♀ 2.9–3.5/2.9–3.8.

Nymph (description mainly from Garman (1927)). Head abruptly narrowed behind prominent eyes; posterolateral corners of head smoothly rounded, the posterior margin with 6 longitudinal setose ridges separated by smooth areas. Premental setae 10 to 13, the 6 to 8 outer setae being longer than the inner ones; palpal setae 9 or 10; 11 to 12 crenations on distal margin of palpus, each bearing 3 or 4 small spiniform setae. Pronotum of thorax with a low mesial setose prominence. Abdomen with a faint brown dorsal stripe on each side; dorsal hooks absent, represented on segs. 4 to 9 only by prominences each bearing a conspicuous tuft of thick setae; small, straight lateral spines on segs. 8 and 9, that on 9 barely as long as seg. 10; epiproct about four-fifths as long, and cerci about two-thirds to three-quarters as long, as paraprocts.

Length 17.0–17.4 (Needham and Westfall, 1955: 14); abd. 10.0–10.8; w.abd. 5.7–6.4; h.w. 4.7–5.7; h.f. 4.3–4.6; w.hd. 4.3–4.7. (Measurements from Dennis R. Paulson based on specimens from Florida.)

Habitat and range. Brackish coastal marshes. Que. s. along the Atlantic coast to Fla., w. to Tex. and Mexico, Baja California; Bahamas.

Distribution in Canada. N.S.—Argyle, Yarmouth county (August 6, 1957, 1 ♂ 1 ♀, D.C. Ferguson). Que.—East Bolton (July 16, 1911, 1 ♀, A. F. Winn).

There is one record from the Thousand Is., St. Lawrence River, but the particular island happened to belong to New York State and not to Ontario, Canada. It occurs in Maine and may therefore be expected from the adjacent Canadian Province of New Brunswick.

Field notes. In Canada adults have been encountered on July 16 and August 6. The recorded flight period in New England is from June 21 to

August 28 (Garman, 1927) and in Massachusetts (Cape Cod) from June 30 to August 6 (Gibbs and Gibbs, 1954).

E. berenice is typically an inhabitant of brackish water. On the central Gulf Coast Wright (1943) found adults and nymphs in both fresh and brackish parts of the coastal marshes, but their abundance increased with the salinity. Females have been seen ovipositing in ponds with a salinity of 24 to 37 and 55 to 68 percent sea water, and nymphs have been obtained from those where the corresponding salinity values were 55 to 68, 64 to 77 and 157 to 170 (Pearse, 1932; in Wright, 1943).

Variations. Some females may have a rather distinct, diffusely bordered, brownish spot in each wing just beyond the nodus, several cells in diameter (Borror, 1942).

Genus **Sympetrum** Newman

Ent. Mon. Mag., 1: 511, 1833. Type species *Libellula vulgata* Linné.
Diplax Charpentier, Lib. Eur., p. 12, 1840.

This is a genus of more than fifty species, most of which are natives of Europe and Asia, particularly eastern Asia. Sixteen species are recorded from North America, including two which Needham and Westfall (1955) placed in a new genus, *Tarnetrum*, a name whose generic value has been questioned by Gloyd and Wright (1959) and Kormondy (1958, 1960). We would agree with these authors in limiting *Tarnetrum* to the rank of a subgenus and, for reasons given below, we would include in this subgenus the species hitherto known as *Sympetrum madidum* Hagen. It has two rows of cells in the space between IR$_3$ and Rspl (as has *Tarnetrum*) and the genitalia, both male and female, resemble those of *Tarnetrum* rather than *Sympetrum*. The absence of an accessory transverse carina on abdominal tergite 4 we regard as relatively unimportant.

Adults of *Sympetrum* are rather small and slender, the total length of most Canadian representatives lying between 30 and 40 mm., and usually reddish (except *S. danae*, which is black and yellow).

The head is rounded and the frons low; the vertex is high, transversely oval and slightly crescentic or bilobed. The prothorax has a bilobed hind margin with long hairs; the pterothorax is slightly compressed, thinly clothed with hair and often has pale blotches or stripes in the teneral stage though usually lacking this pattern in the mature adult. The legs are slender and lack conspicuous spines. The wings (pl. 27: 1, p. 188) are generally hyaline, often flavescent, and in the American fauna seldom have definite stripes or patterns. The antenodal cross-veins are incomplete; there is a single cross-vein under the pterostigma; vein R$_3$ is smoothly curved. The abdomen is enlarged on segments 1 and 2 and on the base of 3, constricted slightly on 3, and then widens again nearly to the posterior end. Both males and females are best defined according to the form of the genitalia – the hamuli of males and the vulvar lamina of females.

These are mainly pond species that frequent meadows and swamps near shallow, standing water. The flight is neither swift nor strong and is frequently interrupted by periods of rest. In warm weather they usually rest on foliage or tall grasses several feet above the ground; at low temperatures they tend to rest on bare ground or to sun themselves on rocks, walls or tree-trunks. The name of the genus probably derives from this habit of basking (often on stones) which in turn is a reflection of the time of flight, which is by day, and chiefly in late summer and autumn. The normal time for emergence is from late June to early August.

The nymphs are small, slender, mottled with green and brown, and live amongst submerged plants. They closely resemble nymphs of *Leucorrhinia* but are unpigmented ventrally. Nymphs of *Sympetrum* may be characterized as follows. The head is rather large, the width across the eyes being little less than that of the abdomen; the head is not more than twice as broad as long; the eyes are moderately to decidedly prominent, the lateral margins being very oblique and curving into the straight, posterior margin with no indication of an angle. The prementum of the labium is scarcely longer than wide; there are 13 to 15 premental setae (on each side); the palpal setae usually number 9 to 11 but sometimes up to 14; the distal margins of the palpi have very low crenations; the spiniform setae are in groups of two to four; one seta is much the longest; the movable hook is slender and rarely more than one-half as long as the distal margin. The abdomen is broadest at segment 6 and narrowed more abruptly posteriorly than anteriorly; dorsal hooks are never present on segments 1 to 3 nor on 9 and 10, and are usually shorter than the segments that bear them; lateral spines occur on segments 8 and 9, or on 9 only, or are absent and, if present, are usually shorter than the lateral margin of the segments that bear them (excluding the spine); the epiproct is acuminate but little longer than wide; the cerci are about one-half as long as the paraprocts; the paraprocts are decidedly longer than the epiproct.

The North American species of *Sympetrum* show affinities among themselves that are most clearly detected by the form of the accessory genitalia of the male, particularly the hamuli.

In the basic libellulid plan only one pair of hamuli is present, this being the pair that is homologous with the posterior hamuli of the Aeshnoidea. These hamuli are bifid and their two limbs are known as the outer and inner branches from their relative positions. The inner branch is a hook or bears a hooked extremity, whereas the outer branch is not hooked but is usually broader and suggests a sheath for the hook. This appears to be the plan in all species of *Sympetrum* as well as in many other genera of Libellulidae. But in a number of the commonest North American species of *Sympetrum*, the hamular branches have acquired a closer union with each other, the outer branch being fixed to the end of the common stem and the inner branch having a position beside the outer branch and opposing it like the movable joint of a crab's claw.

Five of the commonest species of *Sympetrum* in North America have these and other peculiarities which indicate a closer relationship with each other than with any other species: the association of the two hamular branches, the inner branch being much smaller than the outer branch and separated from its partner by a single suture; the two branches lying close to each other on the same plane as far as one surface is concerned; and their apices curving towards one another like a pair of forceps instead of being independent. On the strength of these characteristics we are tentatively applying the term "forcipate" to the species of this group.

Females of the forcipate group of *Sympetrum* differ from more typical species of this genus in the form of the vulvar lamina (the supra-anal plate of Needham and Westfall (1955)). In typical *Sympetrum*, e.g. *S. costiferum* or *S. danae*, the vulvar lamina consists of a posterior prolongation of the sternal margin of abdominal segment 8, which in some species lies horizontally whereas in others it is more or less bent downwards or forwards at various angles; in the forcipate group the vulvar lamina consists of a small thickened plate in the middle of segment 8, which is partly divided by a median groove, flanked on each side by a ridge which terminates behind in a pair of pointed apices. Thus both kinds have very distinct characters whereby they may be placed in their proper group.

Three species remain outside these two categories. They are the two species assigned by Needham and Westfall (1955) to *Tarnetrum*, and the aberrant *S. madidum*, which in some respects appears to be intermediate between the subgenera *Sympetrum* and *Tarnetrum*.

A synopsis of the main characters of the three categories of *Sympetrum*, as we have proposed them, is given below.

Section 1. An American group, given generic rank as *Tarnetrum* by Needham and Westfall (1955), and in our view characterized mainly by the form of the male hamuli. We have added *S. madidum* which agrees with the two other members of this Section in having two rows of cells between IR_3 and Rspl in the fore wing. Large robust species with abdomen not, or scarcely, constricted near the base. Hamuli with short wide branches, the outer branch conical and saccular, the inner branch a short curved hook. Genital lobe small. Vulvar lamina a pair of angular sternal folds. Three Canadian species, known also from the United States except Alaska.

corruptum

illotum

madidum

Section 2. The group to which the type species of the genus (the European *S. vulgatum* (Linnaeus)) belongs. Body smaller and more slender than in

Section 1, with a slight but distinct constriction immediately following the basal abdominal segments. Wings with a single row of cells between IR₃ and Rspl in the fore wing. Wings (in species of our fauna) generally hyaline or with more or less flavescence, which in some species covers most of the hind wing from base to nodus. Branches of male hamuli longer than in Section 1, the outer being wider and more leaf-like and the inner more slender, and terminating in a minute hook. Genital lobe much larger than in Section 1, usually forming a broadly rounded lobe, but sometimes (rarely) much longer than wide. Vulvar lamina very variable. Superior anal appendages dorsally convex when viewed laterally, without a prominent ventral tooth, and with the apices not tilted. Five Canadian species, known also from the United States; one recorded also from Alaska.

vicinum
costiferum
semicinctum
occidentale
danae

Section 3. A group of North American species in which the predominant colour is blood-red in the males, the females being often olivaceous but also frequently red. Both sexes more or less marked with black or brown, especially along the sides of the abdomen; the legs commonly black or dark brown. Face often with distinctive colour such as yellow, greenish white or reddish. Body small or medium-sized with the constriction of the abdomen as in Section 2. Wings with a single row of cells between IR₃ and Rspl in the fore wing. Wings usually hyaline but often slightly to strongly flavescent. The main stem of each hamulus projects from the genital fossa and is strongly bent before terminating in the two hamular branches. These lie close together and do not overlap each other as in Section 2. The outer branch broader than the inner, and the tips of the two branches bent so as to appear forcipate, approaching one another like a pair of forceps or pliers. The inner branch is slightly hooked but has no minute hook at the anterior end, as do the hamuli of the other Sections. Vulvar lamina consists of two angular or cylindrical ridges that arise from the sternum of segment 8 on each side of the genital opening; these two lobes usually nearly meet mesially but curve away from each other anteriorly and posteriorly. The genitalic differences excepted, this group closely resembles Section 2. Five Canadian species, known also from the United States; one recorded also from Alaska.

ambiguum
rubicundulum
internum
obtrusum
pallipes

The extent to which these subgeneric groupings are reflected in the morphology of the nymphs remains to be determined. At present the nymphs of species in Section 1 (except *S. madidum* if Needham's (1904) supposed determination proves to be correct) stand apart from other North American *Sympetrum* in lacking dorsal hooks on the abdomen, and in the extreme reduction or absence of lateral spines on abdominal segment 8. No characters have yet been detected in final-instar nymphs that enable species of Sections 2 and 3 to be distinguished; on the contrary, from her thorough study of seven such species, Musser (1962) found it impossible to distinguish reliably between *S. obtrusum* and *S. pallipes* (both Section 3), and between *S. danae* and *S. internum* (Sections 2 and 3, respectively). Such differentiating characters, should they exist, may be revealed by detailed examination of the early instars (see Nevin, 1930; Corbet, 1951; Gardner, 1951; Trottier, 1969). Indeed, Trottier's (1969) analysis of certain morphological features in instars II to V of *S. vicinum* (Section 2), and *S. rubicundulum* and *S. obtrusum* (both Section 3) supports the subgeneric grouping that we have proposed on the basis of the adults.

Key to the Species of Sympetrum

Characters used in this key refer only to mature adults; for these the key can serve as a useful guide but determinations will usually need to be checked against the illustrations of the hamuli (pl. 31, p. 218) and anal appendages (pl. 30, p. 212) of the male and the vulvar laminae (pl. 31–33, pp. 218, 224 and 228) of the female.

Adults

1 Two rows of cells between IR$_3$ and Rspl in fore wing   (Section 1) 2

One row of cells between IR$_3$ and Rspl in fore wing   4

2 Tergite of seg. 4 with a transverse carina in addition to the usual apical carina (pl. 29: 1, p. 208)   3

Tergite of seg. 4 lacking an additional transverse carina *madidum* (p. 211)

3 Antenodal cross-veins in fore wing 7, in hind wing 5; sides of thorax with two white stripes   *corruptum* (p. 205)

Antenodal cross-veins in fore wing more than 7, in hind wing more than 5; sides of thorax with two white spots   *illotum* (p. 209)

4 Sup. apps. with denticles but no prominent ventral tooth (e.g. pl. 30: 8, p. 212); vulvar lamina emarginate, not bifid at posterior margin (e.g. pl. 32: 4, p. 224)   (Section 2) 5

Sup. apps. with a prominent ventral tooth (e.g. pl. 30: 10, p. 212); vulvar lamina bifid at posterior margin (e.g. pl. 33: 3, p. 228)   (Section 3) 9

5 Face black; sides of thorax yellow and black (pl. 29: 5, p. 208)
danae (p. 222)

Face not black; sides of thorax not yellow and black, usually red or
brown 6

6 Wings tinged with yellow over basal half 7

Wings not tinged with yellow or, if so, only along costal margin or at
extreme base 8

7 Thorax with a black streak in front of spiracle and narrow lines of black
in lateral sutures (pl. 29: 3, 4, p. 208) *occidentale* (p. 220)

Thorax without these markings (pl. 29: 2, p. 208) *semicinctum* (p. 219)

8 Tibiae entirely yellow *vicinum* (p. 213)

Tibiae striped with black *costiferum* (p. 215)

9 Tibiae yellow externally 10

Tibiae black externally 11

10 Abdomen with black markings; face white; branches of hamuli widely
separated at tips (pl. 31: 9, p. 218); vulvar lamina as in pl. 33: 1 (p. 228)
ambiguum (p. 226)

Abdomen with yellowish brown markings; face yellow; branches of
hamuli narrowly separated at tips (pl. 31: 13, p. 218); vulvar lamina as
in pl. 33: 5 (p. 228) *pallipes* (p. 234)

11 Male 12

Female 14

12 Gap separating branches of hamuli shallower than width of hamulus
at tip (pl. 31: 12, p. 218); face white *obtrusum* (p. 232)

Gap separating branches of hamuli about as deep as width of hamulus
at tip (pl. 31: 10, 11, p. 218); face yellow or red 13

13 Median margin of outer (dorsal) branch of hamulus forming a blunt
tooth before meeting the lateral margin, thus making this branch
appear bidentate in apical view (pl. 31: 11, p. 218) *internum* (p. 230)

Median margin of outer branch of hamulus not forming such a tooth
and so not appearing bidentate in apical view (pl. 31: 10, p. 218)
rubicundulum (p. 227)

14 Vulvar lamina swollen, fused along median line for nearly its whole
length (pl. 33: 2, p. 228) *rubicundulum* (p. 227)

Vulvar lamina not swollen, not fused along median line for the posterior third of its length 15

15 Apices of bifurcate vulvar lamina distinctly divergent at posterior
margin (pl. 33: 3, p. 228) *internum** (p. 230)

Apices of bifurcate vulvar lamina with margins parallel or convergent
(pl. 33: 4, p. 228) *obtrusum* (p. 232)

*It is often very difficult to distinguish females of *S. internum* and *S. obtrusum*.

Nymphs

Nymph unknown   *madidum* (p. 211)
1 Dorsal hooks absent   2
  Dorsal hooks on segs. 5 to 7 at least   3
2 Total length about 21 mm.; cerci about one-half as long as paraprocts;
  palpal setae (on each palpus) 13 or 14; premental setae (on each side) 15
  to 18, usually 16 or 17   *corruptum* (p. 206)
  Total length about 18 mm.; cerci about two-thirds as long as paraprocts; palpal setae 9; premental setae about 13   *illotum* (p. 210)
3 Dorsal hook on seg. 8   4
  No dorsal hook on seg. 8   5
4 Dorsal hook on abdominal seg. 4   7
  No dorsal hook on seg. 4   *rubicundulum* (p. 229)
5 Dorsal hook on seg. 4; lateral spines on seg. 9 at least three-eighths
  lateral length of seg. 9 (including spine) (pl. 39: 3, p. 258)
                                               occidentale (p. 221)
  No dorsal hook on seg. 4 (though present on segs. 5 to 7); lateral spines
  on seg. 9 less than three-eighths length of seg. 9   6
6 Lateral spines on seg. 9 about one-fourth lateral length of seg. 9 (including the spine) (pl. 42: 4, p. 266)*   *danae* (p. 223)
  Lateral spines on seg. 9 about one-fifth length of seg. 9 (pl. 42: 6, p. 266)
                                               internum (p. 231)
7 Lateral spines on seg. 9 extending posteriorly beyond tips of cerci by at
  least one-third of length of cercus   8
  Lateral spines on seg. 9 not extending posteriorly beyond tips of cerci
  or, if doing so, then by less than one-third of length of cercus   9
8 Lateral spines on seg. 8 scarcely twice, and those of seg. 9 two and
  one-half times, as long as their widths at base; outer margins of lateral
  spines on seg. 9 distinctly incurvate; cerci one-half as long as the
  paraprocts; paraprocts not acuminate (pl. 42: 3, p. 266)
                                               semicinctum (p. 220)
  Lateral spines on seg. 8 more than twice, and those of seg. 9 three times,
  as long as their widths at base; outer margins of lateral spines on seg. 9
  nearly straight; cerci less than one-half as long as the paraprocts;
  paraprocts apically acuminate (pl. 42: 1, p. 266)   *vicinum* (p. 214)

*Musser (1962) was unable to distinguish reliably between *danae* and *internum*. Caution
should be exercised when using this couplet for diagnosis.

PLATE 28

Wings of *Sympetrum occidentale*: (1) *S.o. occidentale*; (2) *S.o. fasciatum*.

PLATE 28

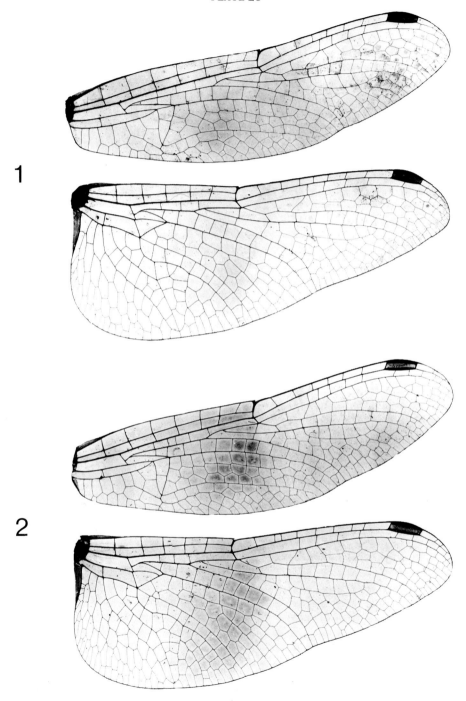

1

2

9 Lateral spines on seg. 9 more than one-third lateral length of seg. 9 (including the spine) (e.g. pl. 42: 2, p. 266)   10

Lateral spines on seg. 9 less than one-third length of seg. 9   11

10 Dorsal hook on seg. 8 extending posteriorly over about one-half mid-dorsal length of tergite of seg. 9 (pl. 42: 5, p. 266; pl. 43: 6, p. 270)

ambiguum (p. 226)

Dorsal hook on seg. 8 extending posteriorly only barely beyond the anterior margin of tergite of seg. 9 (pl. 42: 2, p. 266; pl. 43: 4, p. 270)

costiferum (p. 216)

11 Lateral spines on seg. 9 nearly one-third lateral length of lateral margin of seg. 9 (including the spine) (pl. 43: 2, p. 270); dorsal hooks well developed, slender and very acute, that on seg. 7 nearly as long as the mid-dorsal length of seg. 7 and that on seg. 8 usually more than one-half mid-dorsal length of seg. 8 (pl. 43: 9, p. 270)   *pallipes* (p. 235)

Lateral spines on seg. 9 about one-fourth length of seg. 9 (pl. 43: 1, p. 270); dorsal hooks less developed, stouter and distinctly curved, that on seg. 7 distinctly less than mid-dorsal length of seg. 7 and that on seg. 8 rarely one-half mid-dorsal length of seg. 8 (pl. 43: 8, p. 270)

obtrusum (p. 233)

Section 1

Sympetrum corruptum (Hagen). (Pl. 29: 1, p. 208; pl. 30: 1, p. 212; pl. 31: 1, p. 218; pl. 32: 1, p. 224)

Mesothemis corrupta Hagen, Syn. Neur. N. Amer., p. 171, 1861.
Diplax corrupta: Selys, Ann. Soc. Ent. Belg., 28: 43, 1884.
Sympetrum corruptum: Calvert, Trans. Amer. Ent. Soc., 20: 264, 1893.
Tarnetrum corruptum: Needham and Westfall, Drag. N. Amer., p. 546, 1955.

A large, robust, light reddish species with pink venation and a colour pattern unlike those of our other species of *Sympetrum*.

Male. Labium olivaceous, labrum, ante- and post-clypeus pinkish brown; frons deep yellow or orange in dried specimens, often bright red in life, especially in the more prominent parts; vertex duller red, occiput darker, brownish; hair, where present on the head, very scanty. Rear of head dark brownish, the lateral margins orange brown, interrupted by two pale yellowish spots.

Prothoracic crest of hair inconspicuous, brownish gray; general colour of pterothorax pale brownish or pinkish brown in immature to nearly mature individuals, marked by two oblique narrow whitish stripes which, with the coming of maturity, largely disappear, leaving only two small round yellow spots near the level of the metaspiracle, each spot edged below with black. Legs with basal segments and extensor surfaces of all other segments red, flexor surfaces black. Wings pale brownish pink, nearly hyaline, without spots, venation pink.

Abdomen light orange and gray-brown with irregular blackish markings along the sides, best defined along segs. 6 to 8 where they enclose a series of elongated segmental areas, which appear as fairly definite spots; segs. 8 and 9 also with well defined black dorsal spots, each covering the length of its segment. Abdomen cylindrical and evenly tapering.

Sup. apps. (pl. 30: 1, p. 212), seen from above, appear straight; they may be close together or separated, the apices being divergent and acute; in lateral view, dorsally straight, ventrally concave in basal third, convex beyond. Genitalia of seg. 2 very inconspicuous, the genital lobe rounded but very small. Hamuli (pl. 31: 1, p. 218) saccular, outer branch scarcely defined, bluntly pointed, inner branch a short, sharply curved hook.

Female. Very similar to male except in the usual differentiating characters, such as the stouter abdomen and the lack of any constriction near the base. Vulvar lamina (pl. 32: 1, p. 224) consisting of a pair of obtuse angular folds formed of the two angles of the divided projection of the 8th sternum.

Venation. Anx ♂ ♀ 7/5; pnx ♂ ♀ 6–7/6–7.

Measurements. Total length ♂ 37–40, ♀ 38–43; abd. ♂ 23–25, ♀ 25–26; h.w. ♂ 27.0–29.0, ♀ 29.0–33.0; h.f. ♂ ♀ 6.0–7.0; pt. ♂ 3.0, ♀ 3.0–3.4.

Nymph (description compiled from Needham (1903), Wright (1946b) and Musser (1962)). This nymph, and that of *S. illotum* have a more conspicuous pattern on the abdominal terga than do any other *Sympetrum* except *S. occidentale fasciatum.*

Width of head twice its length, anterior margins straight; eyes directed slightly forward of transverse axis; posterolateral corners of head sloping behind the eyes in a wide curve to the almost straight posterior margin. Premental setae about 14 to 18 (usually 16 or 17), the 6th or 7th from the outside being the longest, the outer 9 on each side being longer and thicker than the very short inner ones; palpal setae 13 to 15; distal margin of palpus smooth, without crenations. Abdomen broadly ovate, widest at seg. 6, the middle width about two-thirds of the length; dorsal hooks absent; lateral spines of seg. 8 absent or, if present, less than one-tenth the lateral length of that segment (including the spine); lateral spines on seg. 9 very small, about one-eighth or less the lateral length of that segment; epiproct as long as segs. 9 + 10 at the mid-dorsum; cerci about one-half as long as paraprocts; epiproct extending slightly beyond tip of paraprocts. Dorsal surface of head with 6 longitudinal lines of setae behind the transverse suture. Dorsum of thorax fuscous, divided by a narrow longitudinal pale line, the sides with some ill-defined pale markings. Legs pale, the femora sometimes with 3 dark annuli; tibiae less distinctly annulate. Abdomen with two longitudinal dorsal fuscous stripes extending caudad to the end of seg. 10.

Length 17–19; abd. 11; w. abd. 6.0–6.5; h.w. 6.5; h.f. 6.5 (Needham, 1903: 4.5); w. hd. 5.8 (Needham, 1903: 4.5).

Habitat and range. Ponds or slow streams, most commonly in more or less arid, sandy or gravelly situations, such as beach ponds, but by no means confined to such conditions. Mexico s. to Honduras; generally distributed throughout the western United States becoming rarer eastward and not reaching the New England States. Southern Canada from B.C. to s. Ont. Also in extreme eastern Asia and the Sea of Ochotsk.

Distribution in Canada. B.C.—Vancouver; Harrison Bay; Hatzic Lake; Cultus Lake; Chilliwack district; Hope; Osoyoos; Oliver; Penticton; Peachland; Creston; Cranbrook; Wasa; West Kootenay; Kaslo; Ainsworth; Kamloops. *Alta.*—Banff; Laggan (51°26′N., 116°11′W.); Red Deer. *Sask.*—Maple Creek; Swift Current; Regina; Katepwa Lake; Hamton; Redberry Lake; Kinkerdines Lake; Prince Albert; Little Fish Lake. *Man.*—Treesbank;

Stockton; Aweme; Onah; Winnipeg; Victoria Beach; The Pas. *Ont.*—Essex, Kent, Haldimand, York, Durham, Grey and Simcoe counties; Muskoka, Thunder Bay and Kenora districts.

Field notes. "On July 14, 1912, this species appeared on the low, sandy, eastern end of the Giant's Tomb Island (Georgian Bay, Ontario). The island is divided here by a narrow channel, close to which, on the outer side, is a shallow pond or lagoon. It was about the margins of the channel and lagoon, especially the former that *Sympetrum corruptum* was observed. They were flying about from place to place, sometimes hovering over one spot, sometimes settling for a moment on the wet sand. They were so shy that it was almost impossible to get within striking distance, and more than an hour of patient effort was spent before one was secured. Two males and one female were all that were taken, all fully mature and in good condition" (Walker, 1915).

"Very abundant everywhere over roads, lawns, ponds, on beach and shallow parts of the lake itself (Lake Winnipeg, Manitoba) and on muddy or rocky shores. All were mature, some much worn, many pairs in tandem over ponds and the lake itself. Many pairs ovipositing over open water, occasional pairs in cop." (First author, field notes for June 28, 1931.)

It has been often noted that *S. corruptum* emerges over a long period and that there appear to be two peaks of emergence (1) from June to early July, and (2) in August and September. In Manitoba the recorded flight period at Aweme is June 7 to September 8, whereas at Victoria Beach, Lake Winnipeg, adults were "all old and mature" at the end of June (Walker, 1933b). Similarly in British Columbia emergence was witnessed at Kaslo during the first week in June and at Hatzic Lake on August 8 (Whitehouse, 1941). Whitehouse considered that the two peaks he recorded at these localities and elsewhere in British Columbia were indicative of two "broods" and that the first brood began during the first week of June and the second during the last week in July, flight continuing until about the middle of August. The earliest and latest recorded dates for this species are June 7 (British Columbia, Manitoba, Ontario) and October 1 (Ontario), although it is seldom that adults are encountered beyond mid-September. In parts of the United States this species is said to overwinter, but it is unlikely to do so in any part of Canada.

Young adults of both sexes have two oblique parallel whitish stripes on the sides of the thorax. As they approach maturity the lower end of each stripe becomes a bright yellow round spot bordered below with black. The white stripe becomes increasingly less distinct and finally, in the fully mature insect, only the yellow spots, edged below with black, remain. Pairing may begin, however, before the white stripes disappear.

According to Needham and Westfall (1955) the long season of flight sometimes lasts over winter and, in the north, adults are sometimes taken in

PLATE 29

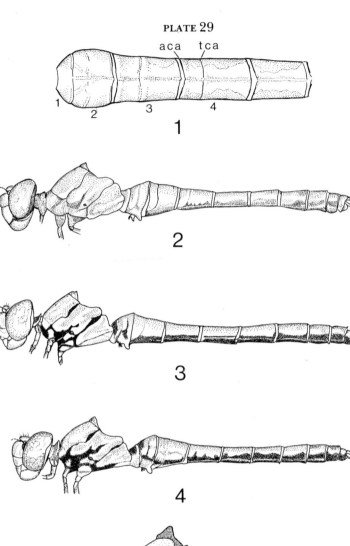

aca   tca

1 2 3 4

1

2

3

4

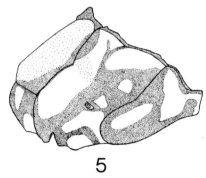

5

early spring. It is quite probable that from some part of the western states overwintered individuals of *S. corruptum* may enter southern British Columbia in the spring, as *Anax junius* does, or even in June in other parts of Canada, the earliest Canadian records of *S. corruptum* being of adult insects taken in June.

In the western United States Kennedy (1915a) regarded *S. corruptum* as an extremely adaptable species of dragonfly, being "found in a greater variety of environments than any other."

Sympetrum illotum (Hagen). (Pl. 30: 2, p. 212; pl. 31: 2, 14, p. 218)

Mesothemis illota Hagen, Syn. Neur. N. Amer., p. 172, 1861 (♂ from northern California).
Diplax illota: Selys, Ann. Soc. Ent. Belg., 28: 43, 1884.
Sympetrum illotum: Kirby, Cat. Neur. Od., p. 17, 1890.
Tarnetrum illotum: Needham and Westfall, Drag. N. Amer., p. 548, 1955.

Somewhat smaller than *S. corruptum* and very unlike a typical *Sympetrum* in appearance, the abdomen being broader, and parallel-sided, suggesting a small *Libellula* in appearance.

Male. Face uniform brown, hairs sparse and short, inconspicuous. Vertex somewhat low and slightly bilobed in front; eye-seam very short; occiput large, slightly longer than vertex; eyes of dried specimens darker than the face.

Pterothorax covered with dense reddish brown hair without dorsal stripes but sides with two oblique narrow pale yellow stripes, only the lower ends of which are conspicuous, rounded and marked below with black; all but these rounded spots disappear with maturity. Legs reddish brown. Abdomen rather dark red, almost hairless except dorsally on segs. 1 and 2. Wings flavescent from base to beyond the nodus, each with one or two dark brown basal streaks extending from base to about third antenodal cross-vein.

Abdomen dark brownish red, almost hairless, except for a small tuft on the dorsum of seg. 2. Abdomen flat and parallel-sided as far as seg. 7, the last three segments tapering rather abruptly.

Anal apps. (pl. 30: 2, p. 212) red. Hamuli (pl. 31: 2, p. 218) similar to those of *corruptum* but with outer branch more elongated.

Female. Generally similar to male. The scooplike vulvar lamina (pl. 31: 14, p. 218) is emarginate with the edge rolled outwards like the lip of a water jug; it reaches but does not cover the palps of the 9th sternite and projects beyond the posterior margin of seg. 8 by about one-half the length of that segment.

Venation. Anx ♂ ♀ 9–10/6–7; pnx ♂ ♀ 6–9/9–10.

PLATE 29

Sympetrum: (1) *S. corruptum*, anterior segments of abdomen, dorsal view; (2–4) head, thorax and abdomen of males, left lateral view: (2) *S. semicinctum*; (3) *S. occidentale fasciatum*; (4) *S.o. occidentale*; (5) *S. danae*, thorax of male, left lateral view.
(2–4 after Walker, 1951a.)
(Abbreviations: aca—apical carina; tca—transverse carina; 1, 2, 3, 4—abdominal segments.)

Measurements. Total length ♂ 38–40, ♀ 36–37; abd. ♂ 24–26, ♀ 23–25; h.w. ♂ ♀ 26.5–29.0; h.f. ♂ ♀ 6–7; pt. ♂ 2.3–2.5, ♀ 2.5–2.8.

Nymph (description from Byers (1972b)). Head shaped as in *S. corruptum*. Premental setae about 13, the 4th or 5th from the outside being the longest; palpal setae 9; distal margin of palpus smooth, without crenations. Abdomen without dorsal hooks; lateral spines on seg. 8, if present, extremely minute, about one-tenth as long as lateral margin of that segment (including the spine); lateral spines on seg. 9 larger, one-third to one-half as long as mid-dorsum of that segment; anal appendages short, about as long as mid-dorsum of segs. 9 + 10; epiproct slightly shorter than paraprocts; cerci about two-thirds as long as paraprocts. Abdominal pattern as in *S. corruptum.*
Length 18; abd. 10; w. abd. 6.5; h.w. 6; h.f. 5; w. hd. 5.

Habitat and range. In the United States *S. illotum* ranges from Calif. and Nev. n.e. to the Yellowstone National Park and n. through Oreg. and Wash. Southward it is distributed through Mexico, incl. Baja California into South America as far as Argentina.

Distribution in Canada. This is very limited, being almost exclusively from Vancouver I. (Victoria; Shawnigan Lake; Nanaimo district; Lost Lake; Florence Lake; Courtenay; Forbes Landing; Newcastle I.). The only record from mainland British Columbia is that of Whitehouse (1941) who took a single specimen at Beaver Lake, Stanley Park, Vancouver.

Field notes. The earliest recorded date for *S. illotum* is May 20 (Vancouver Island, Lost Lake) and the latest is the third week in August. "Its extremely early appearance is in striking contrast to that of most species of *Sympetrum*" (Walker, 1927). Most captures in Canada have been between mid-June and mid-July.
In Canada this dragonfly has been met with in numbers only on Vancouver Island where Whitehouse (1941) observed its behaviour. Like *Pachydiplax longipennis* it often perches with wings drooped at the extreme end of a dry twig, making sallies from there to feed or to investigate another dragonfly. Oviposition often occurs, or at least begins, in tandem, the eggs being "dipped in, an inch or so apart, where open spaces occur between the reeds." "The male not infrequently . . . relinquishes his grasp, when the female . . . continues her duties alone." Kennedy (1917c) records that near Auburn, California a pair of *S. illotum* copulated on the wing and then rested for half a minute in copula on a branch, after which they flew in tandem over the water, making tentative dives from a height of about two feet. "After a half minute they dropped two inches above the water when with a swinging motion the female dipped her abdomen in the water about 30 times, after which they made a sudden upward flight and separated, each to seat itself on a twig."
We have been impressed by the handsome appearance of this species when mature. As Whitehouse (1941) observes: "What more striking contrast in nature than a number of bright scarlet *illotum* flying over green reeds!"

Sympetrum madidum (Hagen). (Pl. 30: 3, p. 212; pl. 31: 3, p. 218; pl. 32: 2, p. 224)

Libellula madida Hagen, Syn. Neur. N. Amer., p. 174, 1861.
Sympetrum madidum: Osburn, Can. Ent., 16: 194, 1905.
Sympetrum madidum: Ris, Coll. Zool. Selys, 13: 679, 1911.

A large red *Sympetrum* with whitish thoracic stripes; body and wings red in life but deep yellow when dried; and abdomen slender and distally tapering, with a slight but distinct constriction near the base.

Male. Labium and labrum pale greenish yellow; face, including postfrons and vertex, dark yellow to olivaceous; rear of head brownish with gray hairs, and lateral margin yellowish, each sixth divided by two dark stripes.

Thorax covered with olivaceous hairs, the dorsum of the pterothorax bearing two whitish inconspicuous stripes on each side, widest below. Legs pale at base but beyond trochanters dark brown or nearly black. Wings strongly stained with yellow, especially at the anal angle and along a wide costal margin. Pt. light brown.

Abdomen (when dried) yellowish to brown, with an inconspicuous brown streak along the midlateral line on each side, and irregular cross-bars, segmentally arranged. These areas red in life.

Sup. apps. (pl. 30: 3, p. 212) about as long as seg. 9. Viewed from above they appear parallel at base with the apices divergent, and in lateral view slightly arched, the inferior surface with an oblique angle at about the distal third, the middle third with 6 to 8 very minute denticles. Genitalia small, genital lobe slightly stalked, broadly rounded distally; hamuli (pl. 31: 3, p. 218) horizontal with branches little developed, the inner branch represented by a short hook, the outer branch much broader but little differentiated from the stalk.

Female. Resembles male except that abdominal colour, including anal apps. is yellowish rather than red, and that there are black markings on sides of segs. 8 and 9.

Venation. Anx ♂ 8–9/5–6, ♀ 7–8/5–6; pnx ♂ 7–8/7–10, ♀ 6–9/6–9.

Measurements. Total length ♂ ♀ 38–39; abd. ♂ ♀ 26; h.w. ♂ ♀ 27–30 (NW: ♂ ♀ 28–31); h.f. ♂ ♀ 7; pt. ♂ ♀ 3.

Nymph. Unknown (1972). We have been unable to confirm the identity of the material described as *S. madidum* by Needham (1904).

Habitat and range. Shallow lakes and ponds. A northern and western species. From N.W.T. s. through B.C. along coastal States to Calif.; e. to Man. in n. and Mo. in s.

Distribution in Canada. N.W.T.—Ft. Smith; Great Slave Lake (delta of Slave River; Ft. Resolution; Wool Bay; Yellowknife); Ft. Simpson; Norman Wells. *B.C.*—Chilliwack; Hope; Princeton; Aspen Grove; Osoyoos; Creston; Cranbrook; Hat Creek; Salmon Arm; Kamloops; Jesmond; 70-mile House; Bridge Lake; 100-mile House; Tatla Lake. *Alta.*—Waterton Lakes; Lethbridge. *Sask.*—Piapot; Maple Creek; Markinch; near Regina; Indian Head; Elbow; Liberty; s. of Loverna; Watrous; Vonda; Scott; Cudworth; Rutland Station; Basin Lake; Battleford; Prince Albert. *Man.*— Aweme; Portage la Prairie; Cormorant Lake.

PLATE 30

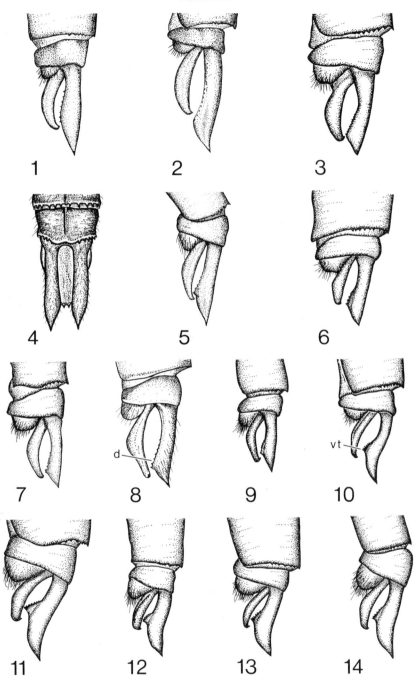

Field notes. Although a young male has been caught as early as June 7 on Vancouver Island (Whitehouse, 1941) most adults are encountered during the second half of June and in July. In British Columbia *S. madidum* flies until the first week in September. The latest record elsewhere in Canada is of a male from Great Slave Lake, Northwest Territories, captured on August 27. At Great Slave Lake specimens taken in July were all rather young (Walker, 1951b).

On Vancouver Island Whitehouse (1941) found this species emerging from a reedy pool in early June and remarked that specimens on the Island where usually heavily laden with red mites.

In the Yakima Valley, Washington, Kennedy (1915b) found *S. madidum* to be peculiar in that males and females were seldom caught in the same place except when teneral.

Section 2

Sympetrum vicinum (Hagen). (Pl. 27: 1, p. 188; pl. 30: 4, 5, p. 212; pl. 31: 4, 15, p. 218; pl. 32: 3, p. 224; pl. 42: 1, p. 266; pl. 43: 3, p. 270)

Diplax vicina Hagen, Syn. Neur. N. Amer., p. 175, 1861 (types).

A slender, thin-legged species of small to medium size, yellow when young, and becoming almost entirely red when mature. Its flight period is in late summer and autumn when in certain localities it may be extremely abundant.

Male. Face light orange brown, grading into bright red on the frons, sometimes becoming red all over except on the labium, which is dull yellowish or brownish; vertex and occiput uniform reddish. Hair of thorax light reddish brown; sides of pterothorax yellowish or olivaceous until fully mature, becoming uniform reddish with no black markings. Legs at maturity red; tibiae entirely yellow. Wings hyaline with yellow to deep orange at the extreme base. Pt. dull brown. Abdomen slender, cylindrical, nearly hairless, stoutest at seg. 2, smallest at 4 and proximal end of 5, increasing in diameter to 8.

Sup. apps. (pl. 30: 4, 5, p. 212) as long as seg. 8, in profile appearing dorsally straight and horizontal. Ventral profile moderately concave in proximal two-thirds, meeting at distal third in an obtuse angle, with 3 or 4 minute denticles. Genital lobe long-oval, widening very little beyond the base, more than twice as long as broad. Hamuli (pl. 31: 4, p. 218) nearly erect, the two branches more slender than usual, the outer branch much wider than the inner branch, which terminates in a minute hook.

Female. General form and colour as in male. Vulvar lamina (pl. 31: 15, p. 218; pl. 32: 3, p. 224)

PLATE 30

Sympetrum—anal appendages of males, left lateral view (except 4, dorsal view): (1) *S. corruptum*; (2) *S. illotum*; (3) *S. madidum*; (4) *S. vicinum*; (5) *S. vicinum*; (6) *S. costiferum*; (7) *S. semicinctum*; (8) *S. occidentale*; (9) *S. danae*; (10) *S. ambiguum*; (11) *S. rubicundulum*; (12) *S. internum*; (13) *S. obtrusum*; (14) *S. pallipes*.
(1, 2 after Needham and Heywood, 1929; 3, 5–7, 9–14 after Needham and Westfall, 1955.)
(Abbreviations: d—denticles; vt—ventral tooth.)

with an angular median protrusion and arcuate lateral expansions, forming a ventrally directed scoop that becomes very wide and conspicuous after oviposition.

Venation. Anx ♂ ♀ 6.5/5–6.5; pnx ♂ ♀ 6–7/6–7.

Measurements. Total length ♂ 32.5–35.0, ♀ 26.0–32.0; abd. ♂ 21–23, ♀ 18–23; h.w. ♂ 23–26, ♀ 20–24; h.f. ♂ 4.7–5.1, ♀ 4.5–5.1; pt. ♂ 2.1–2.5, ♀ 2.0–2.3.

Nymph (description compiled from Needham (1901) and Garman (1927)). The unusually prominent eyes, and large dorsal hooks and long lateral spines on the abdomen (pl. 42: 1, p. 266; pl. 43: 3, p. 270) make this species easy to recognize; among the known North American *Sympetrum*, only *S. semicinctum* has larger dorsal hooks.

Head nearly twice as wide as long, the posterolateral corners rounded and with rather long heavy setae. Premental setae 11 to 13, the 5th from the outside the longest; palpal setae 9 or 10; distal margin of palpus with very shallow crenations, each with 1 large and 1 small spiniform seta; movable hook exceptionally long and slender. Dorsal hooks on abd. segs. 4 to 8, well developed, those on 7 and 8 the largest, arched, robust and each extending posteriorly well beyond the suture of the succeeding segment; long lateral spines on segs. 8 and 9, those on 8 about one-third that segment's lateral length (including the spine), those on 9 nearly one-half the segment's lateral length similarly measured, both pairs of spines gently curved but in general direction parallel; epiproct two-thirds as long as paraprocts; cerci less than one-half as long as paraprocts; relative lengths of cerci, epiproct and paraprocts: 4:6:10. Labial palpi conspicuously spotted on ventral (i.e. anterior) surface; median dark band on metathorax and anterior half of abdomen; abdominal segments dark-banded with lighter spots on posterolateral angles of segs. 1 to 9; lateral spines on segs. 8 and 9, and anal appendages, black-tipped.

Length 12–15; abd. 8; w. abd. 5; h.w. 4; h.f. 4.5; w. hd. 4.5.

Habitat and range. Slow streams and permanent ponds. Eastern and middle United States, s. to Colo., Miss. and Ga.; and eastern Provinces of Canada. Also in Wash. and extreme western B.C.

Kennedy (1917a) called this a northern transcontinental species, a term that is somewhat misleading in two senses. First, its north-south range on the Atlantic coast of the United States is from Maine to Florida and about midway across the continent it is from Minnesota to eastern Texas. Thus "northern" is hardly applicable to the United States and of course still less so to Canada; second, records west of the Mississippi are so sporadic that, with the exception of Washington and southwestern British Columbia, we are inclined to believe that the main range of *S. vicinum* is restricted to the eastern half of the continent with a few scattered outliers westward to Colorado, Kansas, while a distinct western population, perhaps of small extent, is formed along the coast of Washington and British Columbia.

Distribution in Canada. N.S.—Annapolis and Halifax counties. *N.B.*—Queen's county. *Que.*—Hull; Cawood; Kazabazua; Lac Ste. Marie; Mont Tremblant Park; Isle of Orleans. *Ont.*—Essex, Kent, Lambton, Middlesex, Oxford, Haldimand, Peel, York, Dufferin, Ontario, Bruce, Grey, Simcoe, Peterborough and Carleton counties; Muskoka, Parry Sound, Nipissing (Algonquin Park), Algoma, Sudbury and Timiskaming districts. *B.C.*—Cultus Lake; Chilliwack; Vancouver district; Vancouver I. (Loon Lake; Prospect Lake; Burnaby Lake; Langford Lake; Wellington; Saanich district; Nanaimo district; Forbes Landing).

Field notes. This is not only the latest species of *Sympetrum* to emerge in eastern Canada but also the latest of all the Odonata. Individuals of other species may sometimes emerge at later dates but *Sympetrum vicinum* only begins to appear as a teneral in southern Ontario during the last week of July. The great majority are first seen in August and the fully coloured adults are most abundant about the middle of this month. They fly throughout September and sometimes well into October, and on one occasion we saw a few on November 3. In Philadelphia, Pennsylvania, an adult has been seen as late as December 1 (Calvert, 1926).

It is the habit of *S. vicinum* to rest at a greater height than other species of *Sympetrum* do—on bushes, tall herbs and grasses— whereas other species rest near the ground. On August 25 hundreds of mature *S. vicinum* were resting in grassy clearings among pines at Long Point, Norfolk county, Ontario. The following morning adults "were perched on grass stems 0.3–0.7 m. above the ground," the air temperature (at a height of 1 m.) being 13–14°C. The first shafts of sunlight touched them at 6.10 a.m. (about half an hour after sunrise) and they first flew about three minutes later. "Then they flew only a few metres, often upwards, so as to alight on the insolated foliage of nearby pine trees, 2–3 m. above the ground." By 6.35 a.m., when the air in the sun was about 28°C, many were flying spontaneously (Corbet and Eda, 1969).

S. vicinum appears to be entirely or at least normally an inhabitant of permanent waters. It may be seen ovipositing in pairs on the banks of slow streams or ponds that are normally or at least usually permanent. The species is peculiar among our species of the genus in that it not only oviposits in tandem but the female strikes the surface of the water alternately with another object projecting from the water, usually a wet part of the stream bank. The alternating movements are not regular: the ovipositing female may strike the bank twice before touching the water again.

We have not been able to discover whether the eggs are deposited in the water or on the bank, but since water is usually where the eggs of *Sympetrum* are placed we would postulate that this is their destination in *S. vicinum*. In this connection it may be noted that there is a larger space for holding the eggs during oviposition in *S. vicinum* than in our other species of *Sympetrum*.

Sympetrum costiferum (Hagen). (Pl. 30: 6, p. 212; pl. 31: 5, p. 218; pl. 32: 4, p. 224; pl. 42: 2, p. 266; pl. 43: 4, p. 270)

Diplax costifera Hagen, Syn. Neur. N. Amer., p. 174, 1861.

Sympetrum costiferum: Needham, Bull. N.Y. State Mus., 47: 522, 1901.

Young individuals of both sexes are light yellow with the thorax somewhat darkened above and marked with darker brown along the sutures. The legs are also marked lengthwise with light yellow and dark brown or black, and the wings in tenerals of both sexes have conspicuous broad yellow costal borders.

Male. When mature of an almost uniform brownish red, the dark markings on the thoracic sutures having disappeared. Legs with extensor surfaces of about the same red colour, the flexor surfaces dark brown from bases of femora to tips of tarsi. Wings gradually lose the conspicuous yellow costal margin to become nearly or quite uniform yellowish, almost hyaline; mating may begin, however, while the costal margins are still conspicuous. Abdomen of mature adult red with lateral carinae black, narrowed on seg. 4 and widening again very little following 5 and 6, slightly constricted between 8 and 9.

Sup. apps. (pl. 30: 6, p. 212) slightly longer than seg. 9, in dorsal view straight, approaching each other at an angle of about 20°, the apices sharp, but not meeting; in ventral view sigmoid with a short series of minute denticles; inf. app. two-thirds as long as sup. apps., deepest at about mid-length. Genitalia inconspicuous; stem of hamulus (pl. 31: 5, p. 218) short and wide, outer branch twice as wide as inner branch, overlapping it at base; inner branch a curved hook; genital lobe constricted at base, distally broadly rounded, somewhat hairy, nearly as wide as long, and with a rounded free margin.

Female. Face pale greenish white to buff; labrum greenish. Golden or reddish stain in costal strip of wings broadens in some females until it covers most of hind wing. Vulvar lamina (pl. 32: 4, p. 224) does not project ventrally as in *vicinum*.

Venation. Anx ♂ 6.5/5, ♀ 6.5–7.5/6–7; pnx ♂ 5–7/5–8, ♀ 5/6–8.

Measurements. Total length ♂ 34–37, ♀ 31–36 (1 ♀ 40); abd. ♂ 23–25, ♀ 22–26; h.w. ♂ 26–27, ♀ 25–26 (NW: ♂ ♀ 25–28); h.f. ♂ ♀ 5–6; pt. ♂ ♀ 3.

Nymph. Differs from *S. ambiguum* in that the dorsal hook on abd. seg. 8 (pl. 43: 4, p. 270) barely projects beyond the anterior margin of tergite 9. Eyes a little less prominent, and movable hook shorter and thicker than in *S. vicinum.* Body larger, with longer, less curved and more slender dorsal hooks, and with longer lateral spines than in *S. obtrusum.* Prementum proportionately wider at base than in *S. obtrusum* and *S. pallipes.*

Ventral surface of prementum without brown spots; base of prementum excluding hinge, one-third as wide as prementum at its widest point; premental setae 13 to 18, usually 14 or 15; 8 to 10 setae in the stronger, outer group, the 5th from the outside usually being the longest of all; palpal setae usually 11, sometimes 10 or 12; distal margin of palpus with very shallow crenations, each with one long spiniform seta, and sometimes a second, short one. Dorsal hooks on abd. segs. 4 to 8, all narrow and arched, longest on 7 or 8, where they measure about three-fifths of the mid-dorsal length of those segments; lateral spines on segs. 8 and 9 (pl. 42: 2, p. 266), following the outline of the abdomen; lateral spine on 8 one-fifth the lateral length of that segment (including the spine), the spine on 9 one-third to two-fifths the lateral length of that segment; cerci a little more than one-half the length of the epiproct, which is itself about two-thirds as long as the paraprocts. Lateral surface of pterothorax with incomplete brown stripe. Legs of uniform colour. Abdomen with light brown stripe on either side of mid-dorsum, darkest on seg. 9; dorsal spots medium brown.

Length 18; abd. 11; w.abd. 6; h.w. 6; h.f. 8.5; w.hd. 5. (Measurements from Musser (1962), except the last which is from Needham (1901).)

Habitat and range. Reedy marshes bordering ponds, especially shallow, sandy or gravelly ponds; occasionally in bogs (Robert, 1953; Kormondy, 1958). Maritime Provinces and Que. to B.C., s. through the New England States and N.Y., Ohio, Mich., Wis., Mo., Nebr. to Nev., Calif., Oreg. and Wash.

Distribution in Canada. Nfld.—Gander; Bay of Islands. *N.S.*—Annapolis, Colchester, Halifax, Kings, Pictou and Yarmouth counties; Cape Breton I. (Inverness county). *N.B.*—Charlotte and York counties. *Que.*—Longueuil; Joliette; Mont Tremblant Park; Nominingue; Isle of Orleans; Saguenay River

(Cap Jaseux); Ellis Bay; Anticosti I.; Abitibi county. *Ont.*—Lambton, Huron, Peel, York, Dufferin, Bruce, Simcoe, Peterborough, Hastings, Leeds, Carleton and Renfrew counties; Muskoka, Parry Sound, Manitoulin, Nipissing (Algonquin Park), Algoma, Sudbury, Timiskaming, Thunder Bay, Cochrane and Kenora (incl. Patricia Portion) districts. *Man.*—Treesbank; Stockton; Aweme; Onah; Carberry; Winnipeg; Westbourne; Riding Mountain National Park; Blue Lake. *Sask.*—Swift Current; Moosomin; Regina; Markinch; Hamton; Big Manitou Lake; Saskatoon; Jackfish Lake; Prince Albert; Midnight Lake. *Alta.*—Red Deer; "Settler" (presumably Stettler); Jasper Park. *B.C.*—Vancouver district; Cultus Lake; Chilliwack; Penticton; Peachland; Vernon; Okanagan Falls; Okanagan Landing; Kootenay Landing; Crawford Bay; Invermere; Brisco; Golden; Salmon Arm; Kamloops; Clinton; Tête Jaune; Vancouver I. (Langford Lake; Wellington Lakes; Victoria; Loon Lake; Gabriola I.; Prospect Lake; Forbes Landing). *N.W.T.*—Wood Buffalo Park (Little Buffalo River; Bear River); Great Slave Lake (Outpost I.; Ft. Resolution).

Field notes. Among Canadian species of *Sympetrum*, this is one of the latest to appear, and it has a long flight period—from the second week in July until mid-September or (in the west) October or even occasionally November. Emerging or teneral adults have been seen in Ontario (Go Home Bay) and Manitoba (Stockton) in mid- and late July, in British Columbia (Cultus Lake and Vancouver Island) in early and mid-August, and in the Northwest Territories (Great Slave Lake) in late August. In Quebec it appears about the beginning of August (Robert, 1963).

Adults are most numerous in August. The capture of a female in Stanley Park, Vancouver, on November 1 (Whitehouse, 1941) is exceptionally late. Outside British Columbia late recorded dates include September 12 (Bear River, Northwest Territories), September 13 (Moosomin, Manitoba) and September 16 (Talon Lake, Nipissing district, Ontario). At Go Home Bay, in the Muskoka district, Ontario, in 1912 adults were "still common" on September 11 (Walker, 1915).

In its choice of habitat *S. costiferum* appears to be the *Sympetrum* that is most tolerant of saline waters. In Georgian Bay, Ontario, it has been found breeding in a shallow, sand-bottomed lagoon (Walker, 1917b); and in the Cap Jaseux region of the Saguenay River, Quebec, Fernet and Pilon (1970b) regarded *S. costiferum* as one of the five species of dragonfly characteristic of flooded sand-pits (sablières), the others being *Lestes unguiculatus, Sympetrum internum, Pantala flavescens* and *P. hymenea.*

Whitehouse (1941) states that the male usually accompanies the female while ovipositing, and that adults "delight in resting on earth paths . . ." W. E. Ricker captured a male flying over a patch of low rushes and horsetails in very shallow water on the edge of a quiet part of a stream near North Bay, Ontario, in early September. In the Yakima Valley, Washington, where *S. costiferum* is the most abundant species of the genus around alkaline ponds,

PLATE 31

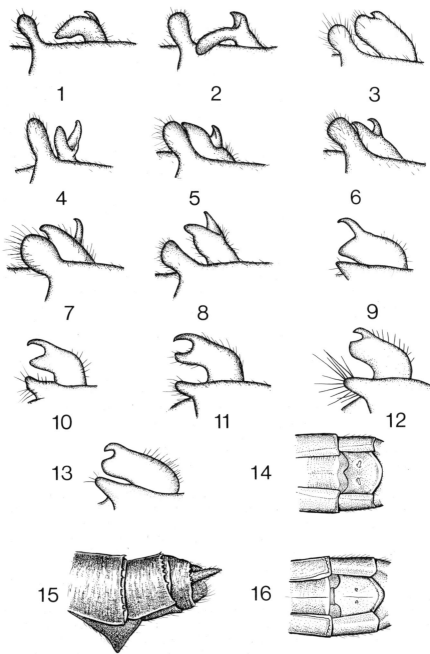

Kennedy (1915b) once saw thousands of adults perched on a telephone wire for a distance of a mile, and all facing the same way.

Sympetrum semicinctum (Say). (Pl. 29: 2, p. 208; pl. 30: 7, p. 212; pl. 31: 6, 16, p. 218; pl. 42: 3, p. 266)

Libellula semicincta Say, J. Acad. Phila., 8: 27, 1839.

Diplax semicincta: Hagen, Syn. Neur. N. Amer., p. 176, 1861.

Sympetrum semicinctum: Walker, Ent. News, 62: 153–158, 1951a.

A small brown species with wings dark yellow on basal half. The coloured area larger and darker on the hind wing than the fore wing. In mature individuals the only pale yellowish areas are the small marginal spots on the rear of the head and two ill defined spots on the sides of the pterothorax, which often disappear in old males.

Male. Face rather dark olivaceous, sometimes uniform, sometimes with labium and labrum much paler or bright orange. Rear of head brown around centre, the lateral margin with three dark spots alternating with whitish ones. Prothorax with reddish brown crest; pterothorax medium orange brown varied with ill defined paler areas on the epimera and dark areas around the coxae but no clear-cut black lines on any of the sutures. Legs dark brown to black; trochanters and posterior face of fore femora reddish brown. Wings with a basal yellow area and an apical hyaline zone. The basal area of the fore wing is pale yellow, about as far as the second cell beyond the triangle; the hind wing is coloured as far as the nodus or one cell beyond, the proximal half being medium yellow, and the distal half darker, this shade extending in a broad arc to the anal triangle. Abdomen reddish brown on the basal three segments, becoming distally red in the male with black lateromarginal spots on segs. 3 to 6. Last two or three segments with an elongate dark spot on each.

Sup. apps. (pl. 30: 7, p. 212) slightly arched, without a prominent ventral tooth; apex not raised above the general dorsal level; apices acute. Length of inf. app. five-sixths that of sup. app.

Female. Generally resembles male in colour, differing in these respects. Basal yellow area of fore wing deeper in colour, more extensive, more sharply delimited, and reaching to within 2 or 3 cells of nodus; hind wing more uniform yellow. On the abdomen a wide black band covers the ventral half of each side, this being double on segs. 1 to 3, and there being small mid-dorsal black triangles on 7, 8 and 9; seg. 1 is black across the dorsum.

Venation. Anx ♂ 6.5–7.5/6–7, ♀ 5.5–6.5/5–6; pnx ♂ 6–7/6–7, ♀ 5–6/6–7.

Measurements. Total length ♂ 24.5–31.0, ♀ 24.0–29.0; abd. ♂ 15.0–19.5, ♀ 15.0–19.0; h.w. ♂ 18.5–22.5, ♀ 18.0–22.0; h.f. ♂ 5.5, ♀ 4.5; pt. ♂ 1.9, ♀ 1.9–2.1.

PLATE 31

Sympetrum—hamulus, left lateral view: (1) *S. corruptum*; (2) *S. illotum*; (3) *S. madidum*; (4) *S. vicinum*; (5) *S. costiferum*; (6) *S. semicinctum*; (7) *S. occidentale*; (8) *S. danae*; (9) *S. ambiguum*; (10) *S. rubicundulum*; (11) *S. internum*; (12) *S. obtrusum*; (13) *S. pallipes*.
Sympetrum—vulvar lamina, ventral view (except 15, left lateral view): (14) *S. illotum*; (15) *S. vicinum*; (16) *S. semicinctum*.
(1, 2 after Needham and Heywood, 1929; 3, 4, 6, 8–11, 13 after Needham and Westfall, 1955; 5 after Robert, 1963.)

Nymph. The nymph is clearly distinguishable from all other known North American *Sympetrum* nymphs by the unusually broad bases to the lateral spines on abd. segs. 8 and 9 (pl. 42: 3, p. 266), and by the direction of these spines which gives the nymph a superficial resemblance to that of *Pantala.* Similar in general form to nymph of *S. vicinum* but larger and more robust.

Eyes very prominent. Premental setae about 12, of which the 5th from outside is longest; palpal setae 9; dorsal hooks on abd. segs. 4 to 8, those on 4 and 5 slightly curved and shorter than the segments that bear them; dorsal hook on seg. 6 about as long as its segment; those on 7 and 8 longest, slightly curved, and blunter than in most other species of *Sympetrum*, the hook on 7 being about as long, and that on 8 being about one-half as long, as the respective segments that bear them; lateral spines on seg. 8 about one-third of the segment's lateral length (including the spine); lateral spines on 9 much larger, almost one-half as long as the segment's lateral length and very broad at base; epiproct about three-fourths as long as paraprocts; cerci about one-half as long as epiproct.

Length 14–15 (Needham and Westfall, 1955).

Habitat and range. Spring-fed ponds and marshes; becoming rarer in populated districts. N.C. and Tenn. n.e. to N.B. and n.w. to Mich.

Distribution in Canada. N.S.—Annapolis, Colchester and Halifax counties. *N.B.*—Charlotte, Queen's and York counties. *Que.*—Covey Hill; Knowlton; upper St. Lawrence River; Richelieu River; St. Hilaire; Hull; Wakefield; Kazabazua. *Ont.*—Middlesex, Welland, Peel, York, Ontario, Bruce, Simcoe, Leeds, Carleton and Lanark counties; Muskoka, Parry Sound, Nipissing (incl. Algonquin Park), Algoma, Rainy River and Kenora districts.

Field notes. In Canada adults of *S. semicinctum* have been encountered between June 19 (Ontario, Niagara Falls) and September 30 (Nova Scotia, Halifax county). It is a late species: in Ontario it is most often seen in August (Walker, 1941); in the Laurentides, Quebec, its flight period seldom begins before August and lasts until October (Robert, 1963). The record for June 19 is unusually early.

"Although generally distributed, *S. semicinctum* appears to be everywhere a rather scarce species, never occurring in large numbers as most of the species of its genus do. This is probably due, in part at least, to its type of breeding place, which is chiefly marshy spots on the course of small, spring-fed streams" (Walker, 1951a). In the vicinity of Go Home Bay, Muskoka district, Ontario, specimens were taken in open marshes adjoining shallow bays and creeks.

In August, 1941 we watched a pair ovipositing in tandem at the outlet of Baillie Lake, Annapolis county, Nova Scotia: the female was striking the water but not the bank of the stream.

Little is known of the biology of this dragonfly.

Sympetrum occidentale Bartenev (Pl. 28: 1, 2, p. 204; pl. 29: 3, 4, p. 208; pl. 30: 8, p. 212; pl. 31: 7, p. 218; pl. 32: 5, p. 224; pl. 39: 3, p. 258)

Ris, Coll. Zool. Selys, 13: 692, 1911 (no name).

Sympetrum semicinctum occidentalis Bartenev, Univ. Izviestija Varsava, 46: 1, 1915.

Sympetrum occidentale: Walker, Ent. News, 62: 156, 1951a.

A western species closely related to *S. semicinctum* but usually larger, and distinguishable by the clear yellow face (particularly in young adults) and by the clearly defined black markings on the sides of the thorax. Three sub-species have been recognized (Walker, 1951) two of which occur in our region: *S.o. occidentale* and *S.o. fasciatum*.

Male. Face light yellow (sometimes darkening with age); front half of labrum and top of vertex yellow. Sides of thorax initially clear yellow with definite black markings, giving this species an appearance very different from that of *semicinctum*; with age the yellow parts may become darkened to a dull ochraceous, greenish or grayish and the black markings are then less conspicuous, though always discernible. The latter comprise markings on the pleural sutures and an oblique black stripe in front of the spiracle. Trochanters and posterior surface of fore femora yellow. On abdomen the black latero-marginal spots on segs. 3 to 6 form a more definite and continuous stripe than in *semicinctum*; this stripe is occasionally divided anteriorly (on segs. 3 and 4) into a mid-lateral and ventro-lateral part.

Anal apps. (pl. 30: 8, p. 212) as in *semicinctum*. The outer branch of the hamuli (pl. 31: 7, p. 218) tends to be longer and more slender than in *semicinctum* but this character is not constant.

S. o. occidentale. This species, when old and darkened, is most like *semicinctum* in appearance; this is evident in the wing pattern as well as the dark body colour and relative obscurity of the black thoracic markings, which however are always discernible. Yellow area of wings (pl. 28: 1, p. 204) rarely darkened distally and, if so, not forming a transverse band crossing the two wings on each side.

S. o. fasciatum. Easily recognized by the wing pattern (pl. 28: 2, p. 204) in which the dark yellow area forms a stripe across the middle of each wing, the area next to the body being lightly tinted whereas the lateral areas are clear. The pale yellow parts of the thorax and abdomen turn greenish gray in old individuals; the black markings on the sides of the thorax tend to be less heavy than in the other subspecies. The body is usually somewhat less robust than that of *S. o. occidentale* but the abdomen is relatively a little longer; these differences however are not constant enough to be useful as a taxonomic character.

Female. Resembles male in colour except that the black lateral stripe on the abdomen is normally (as distinct from occasionally) divided anteriorly (on segs. 3 and 4) into a mid-lateral and ventro-lateral part. This, however, is not characteristic of *occidentale*, being found not infrequently in females of *semicinctum*.

S. o. occidentale and *S. o. fasciatum*. See *Male*.

Venation. S. o. fasciatum. Anx ♂ 6.5/5, ♀ 5.5–6.5/5–7; pnx ♂ 6–7/7–8, ♀ 5–7/6–8. *S. o. occidentale.* Anx ♂ 6.5/5(6), ♀ 6.5–7.5/5(5.5); pnx ♂ 5–7/6–9, ♀ 5–6/5–7.

Measurements. S. o. fasciatum. Total length ♂ 30–37, ♀ 30–34.5; abd. ♂ 18–23, ♀ 20–23; h.w. ♂♀ 23–27; h.f. ♂♀ 5.0; pt. ♂ 1.9–2.0, ♀ 2.2–2.5. *S. o. occidentale.* Total length ♂ 32.5–40.0, ♀ 31.0–35.0; abd. ♂ 20.0–23.5, ♀ 20.0–24.0; h.w. ♂ 25–28, ♀ 24–26; h.f. ♂ 5.0–6.0, ♀ 4.4–5.2; pt. ♂ 1.9–2.6, ♀ 2.0–2.3.

Nymph (pl. 39: 3, p. 258). The nymph of *S. occidentale fasciatum* has been described by Musser (1962); that of *S. o. occidentale* is unknown (1972); and that of *S. o. californicum* features in a key to *Sympetrum* nymphs prepared by Smith and Pritchard (1971). *S. o. fasciatum* differs from *S. semicinctum* (a closely related species) in several respects, notably: *S. o. fasciatum* has the lateral spines on abd. seg. 9 small and inwardly curving (compare pl. 39: 3 and 42: 3, p. 266); it has no dorsal hook on abd. seg. 8 and its dorsal hooks are all small and low, none extending beyond the suture lines of the succeeding segments, whereas (according to Needham (1901)) in *S. semicinctum* the dorsal hooks on segs. 4 and 5 are less than one-third as large as those on segs. 6 to 8; indeed Needham (1901) illustrates the dorsal hooks on segs. 7 and 8 as being much larger than in Musser's drawing of *S. o. fasciatum*. The following description refers to *S. o. fasciatum*, and is from Musser (1962).

Posterior margin of head straight. Ventral surface of prementum with brown spots; premental setae 13 to 15; palpal setae 10; distal margin of palpus with very shallow crenations, each bearing 1 long and 1 short spiniform seta. Dorsal hooks on abd. segs. 4 to 7 as specified above, largest hook on 7, about one-fourth that segment's mid-dorsal length; lateral spines on segs. 8 and 9 with lateral margins nearly straight but following curvature of abdominal margin; spines on 8 each about one-third and on 9 two-fifths the lateral length of their respective segments (including the spine). Lateral surface of thorax divided in half by brown stripe, darkest on mesothorax; femora and tibiae each with 3 brown annuli (proximal 2 faint); wing-sheaths brown basally. Abdominal pattern comprises a median triangle with base touching posterior margin on each of segs. 7 to 10, a light brown area on each side of a triangle; dorsal puncta brown, area between lateral and dorsal puncta dark brown on seg. 9; lateral scars edged in brown, especially along their posterior margins; mesial and lateral surfaces of paraprocts fringed with golden setae.

Length 16; abd. 9.5–10.0; w. abd. 5; h.w. 5; h.f. 5; w. hd. 5.

Habitat and range. A western species ranging from B.C. and Alta. s. to Calif. and Utah. The three geographic groups that we regard as subspecies (Walker, 1951a) are distributed as follows: *Sympetrum o. occidentale*, B.C. and Oreg. (New Bridge); *S. o. fasciatum*, Alta. and Utah (Grand Canyon); and *S. o. californicum*, Calif. and Nev. The species probably occurs in a wide variety of habitats: Walker (1927) found exuviae of *S. o. occidentale* at a grassy roadside ditch at Penticton, B.C., and Whitehouse (1941) saw large numbers ovipositing in a large bay at the side of the Harrison River near its mouth; and Musser (1962) found nymphs of *S. o. fasciatum* in sloughs, swamps and ponds with muddy bottoms in Utah.

Distribution in Canada. S. o. occidentale. B.C.—Cultus Lake; Chilliwack; Harrison Bay. *S. o. fasciatum. Alta.*—Suffield.

Field notes. S. o. occidentale flies mainly in July and August, its season in British Columbia extending from early July to late September.

S. o. fasciatum has been recorded in Canada (Alberta) in the first week of August. Three males and two females were taken flying over the ditches along a prairie road in company with *S. internum*; the ditches had probably contained water earlier in the season (Walker, 1952).

Sympetrum danae Sulzer. (Pl. 29: 5, p. 208; pl. 30: 9, p. 212; pl. 31: 8, p. 218; pl. 32: 6, p. 224; pl. 42: 4, p. 266; pl. 43: 5, p. 270)

Libellula danae Sulzer, Abgekürzte Gesch. Ins., p. 169, 1776.
Libellula scotica Donovan, Brit. Ins., 15: 523, 1811.
Diplax scotica: Evans, Brit. Lib., p. 27, 1845.
Sympetrum scoticum: Wallengren, Ent. Tijdskr., 15: 246, 1894.

A small northern species of black and yellow coloration, remarkable for the absence of any trace of red, a colour that is present to some extent in all our other species of *Sympetrum*. It is also exceptional in being the only circumboreal species of its genus.

The mature male is almost entirely black, whereas the young male and mature female have a very similar pattern of black and yellow. Immature adults of both sexes are similar in their black and yellow pattern but, as maturity approaches, the yellow of the males changes gradually to brown and finally to darker brown or black. Since the yellow and black pattern remains in the female throughout life, we shall describe the female first, the pattern of the immature male being quite similar to that of the female.

Female. Face light yellow, labium and labrum narrowly edged with black; vertex pale brown; rear of head dark brown. Sides of thorax with two large subtriangular yellow spots with two or three smaller yellow spots between them, the light spots separated by black or dark brown interstices. Wings hyaline with a trace of gold at extreme base. Abdomen with a series of pairs of dorsal yellow spots, each pair at first covering the entire dorsum of its segment; in the female a yellow band develops from the confluence and extension of these spots. Seg. 10 and anal apps. black.

The vulvar lamina (pl. 32: 6, p. 224) projects perpendicularly to the axis of the abdomen resembling the lip of a pitcher, the margin being slightly pointed.

Male. Coloration during the teneral stage essentially as in female. With maturity the yellow spots gradually darken to a deep brown and eventually become almost or wholly black. Legs black. On the abdomen, the lateral spots of dull yellow show through the black on segs. 2 to 4 and on 8 to 9.

Sup. apps. (pl. 30: 9, p. 212) without a prominent ventral tooth. Hamuli (pl. 31: 8, p. 218) similar to those of *costiferum* but with the branches more acute, especially the inner branch. Genital lobe less constricted basally and the free end is less rounded.

Venation. Anx ♂ 6.5/5, ♀ 5–6/5; pnx ♂ 5–6/5–6, ♀ 5–7/6–8.

Measurements. Total length ♂ 29.0–31.5, ♀ 25.5–33.0; abd. ♂ 17.0–20.5, ♀ 15.0–22.0; h.w. ♂ 22–23, ♀ 21–25 (NW: ♂♀ 20–27); h.f. ♂ 5.2, ♀ 5; pt. ♂ 1.9–2.3, ♀ 1.9–2.5.

Nymph. No external characters are known by which the nymphs of *S. danae* and *S. internum* can be consistently distinguished despite the fact that the adults of these two species do not appear to be closely related.

A small nymph. Head about two-thirds as long as wide; eyes prominent, posterior margins rounded and not angular. Premental setae 12 to 15, usually 14; the 7 to 9 outer ones being longer than the others; palpal setae 10 to 12, usually 11. Dorsal hooks (pl. 42: 4, p. 266; pl. 43: 5, p. 270) on abd. segs. 5 to 7 small, that on 7 the largest, being one-seventh as long as that segment's mid-dorsal length; dorsal hook on 8 represented in most specimens only by a minute denticle; lateral spines on segs. 8 and 9, those on 8, 12 to 18 per cent, and those on 9, 25 to 28 per cent of the lateral length of their respective segments (including the spine); inner margin of lateral spine on seg. 9 almost straight; cerci about one-half as long as epiproct; epiproct nearly one-half as long as paraprocts.

Length 14.0–15.5; abd. 8.5–9.5; w. abd. 4.5–5.0; h.w. 5.0; h.f. 4.5–5.0; w. hd. 4–5. (Measurements from Musser (1962), except the last which is from Gardner (1951).)

Habitat and range. Marshy ponds, especially on peaty soils. Circumboreal; Nfld. to Mackenzie District, N.W.T. and Alaska, s. to Me., Ohio, Ky. and Ill.; Man., Sask. and Alta., and B.C. s. to Calif. Also in n. Europe and Asia to n. Japan and Kamchatka.

PLATE 32

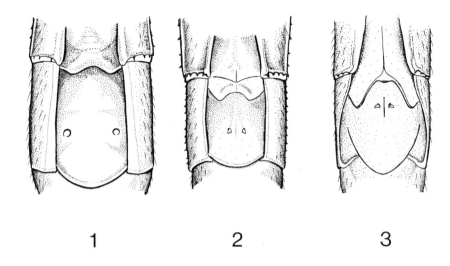

1   2   3

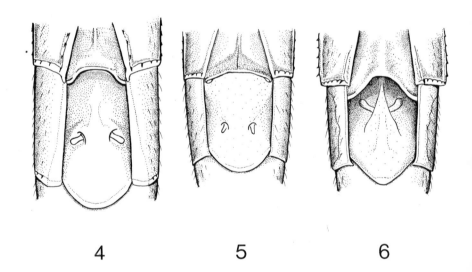

4   5   6

Distribution in Canada and Alaska. Nfld.—St. John's; Gander; Cow Head. *N.B.*—Charlotte county (St. Andrews, Bocabec Lake). *Que.*—Rigaud; Isle of Orleans; Saguenay River (Cap Jaseux); Thunder River; Anticosti I.; Nominingue; Abitibi county; near Lake Mistassini. *Ont.*—Ontario, Bruce, Simcoe and Carleton counties; Muskoka, Nipissing (incl. Timagami and Algonquin Park), Timiskaming, Thunder Bay; Cochrane and Kenora (incl. Patricia Portion) districts. *Man.*—Treesbank; Stockton; Aweme; Winnipeg; Westbourne; Neepawa; Grandview; The Pas; Wabowden; Blue Lake. *Sask.*—Moosomin; Regina; Hamton; Liberty; Big Quill Lakes; Saskatoon; Lac Vert; Redberry Lake; Prince Albert; Midnight Lake; Waskesiu Lake. *Alta.*—Banff; Red Deer; Jasper Park; Beaver Lake; Flatbush; Ft. Smith. *B.C.*—Vancouver; Chilliwack; Rosedale; Hope; Okanagan Landing; Mt. Robson; Cranbrook; Tête Jaune; Quesnel Forks; McBride; Prince George; Metlakatla; Atlin; Vancouver I. (Forbes Landing); Queen Charlotte Is. *N.W.T.*—"Salt Plain" (presumably n. of Ft. Smith); Great Slave Lake (Ft. Resolution; Gros Cap; Outpost I.; Jones Point; Yellowknife). *Alaska*—Admiralty I.; Juneau.

Although the distribution of *S. danae* in North America is not known in detail, there are enough records to indicate that it is generally distributed in the boreal region except in the Arctic.

Field notes. In British Columbia the flight period is from the third week in June to the second week in October. It begins later elsewhere in Canada, especially it seems in Quebec where adults rarely appear "before the end of August" (Robert, 1963). An early date for Ontario (De Grassi Point) is July 3, and tenerals have been seen in mid-July in Manitoba (Stockton) and Ontario (Lake Abitibi region), and in late July and early August in Ontario (Kenora district, Patricia Portion) and the Northwest Territories (Great Slave Lake). There are many August records from the Northwest Territories and prairie Provinces, and one from Gros Cap (Great Slave Lake) for September 6. At Great Slave Lake the flight period is centred on August, adults being juvenile at the beginning of that month and old at the end (Walker, 1951b).

Although the distribution of *S. danae* in North America is not known in detail, there are enough records to indicate that it is generally distributed in the boreal region, except in the Arctic. In Europe, where it is common and widespread in acid bog pools and marshes, several observers have remarked upon its biology. It was one of the first species suspected of hibernating as an adult (see Corbet, 1962) and it has been seen migrating in large numbers off

---

PLATE 32

Sympetrum—Vulvar lamina, ventral view: (1) *S. corruptum*; (2) *S. madidum*; (3) *S. vicinum*; (4) *S. costiferum*; (5) *S. occidentale occidentale*; (6) *S. danae*.
(1 after Walker, unpublished.)

the Irish coast (Moore, 1960). P.-A. Robert (1958) writes of *S. danae* in Switzerland that during the flight period mating seldom begins before 10 a.m., the males spending most of their time perched but alert and ready to fly at the slightest signal. Towards noon the females, in tandem or alone, fly to the water's edge, and with a rapid, darting flight place their eggs here and there on the water. In P.-A. Robert's experience, *S. danae* oviposits only where exposed water is present.

Section 3

Sympetrum ambiguum (Rambur). (Pl. 30: 10, p. 212; pl. 31: 9, p. 218; pl. 33: 1, p. 228; pl. 42: 5, p. 266; pl. 43: 6, p. 270)

Libellula ambigua Rambur, Ins. Neur., p. 106, 1842.
Sympetrum albifrons: Williamson, Drag. Indiana, p. 177, 1900.
Sympetrum ambiguum: Ris, Coll. Zool. Selys, 13: 689, 1911.

A rather large, clear-winged *Sympetrum*, with a black-ringed abdomen. Both sexes are easily recognized by the white face.

Male. Top of frons and vertex yellow to China blue, separated by a broad black transverse stripe. Thorax pale reddish brown in front with only narrow black margins to crest; sides olivaceous with irregular brown stripes on the three lateral sutures, these stripes joining above the legs and below the wings. Legs pale with only inner face of tibiae, leg spines, and claws, black; tibiae yellow externally. Wing bases bright red. Wings hyaline with a faint flavescence at extreme base; costa and pt. yellow but veins bordering pt. black. Abdomen red with broad and ill-defined black apical rings on segs. 4 to 9. Genital lobe and hamuli pale on seg. 2.

Sup. apps. (pl. 30: 10, p. 212) yellow, with a prominent ventral tooth. Hamuli (pl. 31: 9, p. 218) forcipate.

Female. Similar to male in colour. Vulvar lamina (pl. 33: 1, p. 228) emarginate, not bifid at posterior margin.

Venation. Anx ♂ ♀ 7.5/6; pnx ♂ 7–8/7–8, ♀ 6–7/7–8.

Measurements. Total length (NW: ♂ ♀ 36–38*); abd. (NW: ♂ ♀ 23–25*); h.w. (NW: ♂ ♀ 26–28). *These measurements include the abdominal appendages.

Nymph (description from Wright (1946a)). This nymph can be distinguished from that of *S. costiferum*, which it otherwise closely resembles, by the dorsal hook on abd. seg. 8 (pl. 43: 6, p. 270) which extends posteriorly well beyond the anterior margin of the tergite of seg. 9.

Premental setae usually 11, but sometimes 12, the 12th being one-fifth or less the length of the adjoining setae and lateral; usually the 8 outer setae are distinctly longer than the inner 3; palpal setae 9. Dorsal hooks on abd. segs. 4 to 8, robust and spine-like; strongest on segs. 6 to 8, but prominent on 4 and 5; those on segs. 7 and 8 about three-fourths the length of mid-dorsum of their respective segments; that on seg. 6 variable, usually one-half to two-thirds the length of that segment; those on segs. 4 and 5 shorter but spine-like; lateral spines on segs. 8 and 9 (pl. 42: 5, p. 266); those on seg. 8 about one-fifth the lateral length of that segment; those on seg. 9 slightly more than one-third the lateral length of that segment (including the spine), and reaching to, or just short of the tips of the cerci; lateral spines on segs. 8 and 9 with lateral margins incurved, following body outline, mesial margins straight and not incurved; cerci one-half as long as paraprocts, both acuminate at tip.

Length 15–17, usually 16; abd. 10; w. abd. 5.

Habitat and range. Ponds, including temporary bodies of water. A southern and eastern species ranging from Tex. along the coastal United States n. to Me. To the w. and n. it is found in Kans., Minn., Mich., Ont., Ohio and Pa.

Distribution in Canada. Ont.—Essex county. Known in Canada only from one male taken at Point Pelee, Sept. 5, 1964 (Corbet, 1967).

Field notes. The recorded flight period for *S. ambiguum* in Ohio is July 1 to September 10 (Borror, 1937) and in Indiana from May 30 to October 1 (Montgomery, 1941).

Very little is known regarding the biology of this species, apart from the detailed description of a favoured breeding site found by Wright (1946a) near Nashville, Tennessee. A large emergence occurred in late May from a shallow, temporary pond, about one to three feet deep at the time and slowly drying up. The bottom was mucky and covered with grass and weed roots. Grass, weeds, shrubs and small willow trees were abundant at the margin and in the water. The pond was not shaded by large trees or other vegetation. Among algal pads and weeds along the shore numbers of full-grown nymphs were collected. Exuviae were found about one to two feet from the prevailing margin of the water, usually two to four inches above the ground on grass or twigs, and almost always in the midst of large clumps of grass and weeds.

Sympetrum rubicundulum (Say). (Pl. 30: 11, p. 212; pl. 31: 10, p. 218; pl. 33: 2, p. 228)

Libellula rubicundula Say, J. Acad. Phila., 8: 26, 1839.
Sympetrum assimilatum Uhler, Proc. Acad. Phila., 1857: 88, 1857.
Diplax rubicundula: Hagen, Syn. Neur. N. Amer. p. 176, 1861.
Sympetrum rubicundulum: Scudder, Proc. Bost. Soc. Nat. Hist., 10: 219, 1866.

A slender species, generally larger than its nearest relatives, *S. internum* and *S. obtrusum*, the male predominantly red with a yellow face, the female olivaceous.

Male. Face pale brownish yellow, deepened above on the frons to reddish brown, thinly clothed with short dark brown hairs; labium pale brown; rear of head light brown. Prothorax and pterothorax red brown, covered with hair of same colour, the prothoracic lobes with a high crest; sutures without dark lines. Legs above basal segments black and a streak along the latero-ventral surface of fore femora pale drab or tawny. Wings hyaline or washed with yellow from base to a variable extent as far as the nodus or beyond; venation black or dark brown; pt. brown, paler at ends. Abdomen clear red with black lateral triangles; anal apps. red.

Abdomen widest at ridge between segs. 2 and 3, then narrowing to 4 and widening again to 6, narrowing again on 7 to 10.

Anal apps. (pl. 30: 11, p. 212) about as long as 9; as viewed from above, straight, converging but not meeting at apices, gently curved distally upward, with a ventral tooth-like prominence near the middle bearing a row of about 7 teeth on the anterior slope. Hamuli (pl. 31: 10, p. 218) bifid for about one-third of their visible length, the outer posterior branch twice as stout as inner

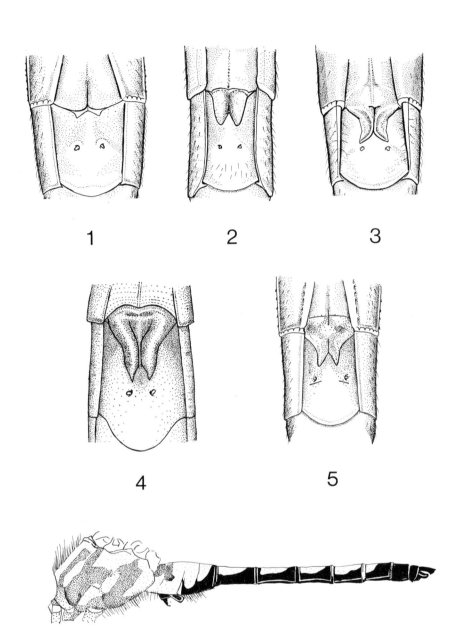

1   2   3

4   5

6

branch. Outer branch about twice as wide as the inner but blunt and a little shorter; inner branch with a small hook at tip.

Female. Similar to male in size and dimensions but differing in the stouter abdomen as well as in colour; the parts that are red or red-brown in the mature male are yellow in the young female or young male; in the mature female they are olivaceous or occasionally red as in the male. Vulvar lamina (pl. 33: 2, p. 228) tumid or "inflated," their tips short and upturned against the sternum of seg. 9.

Venation. Anx ♂ ♀ 8–9(10)/6(5 or 7); pnx♂ ♀ 8–9/8 (less commonly 9, rarely 10).

Measurements. Total length ♂ 37–38, ♀ 29–37; abd. ♂ 21–25, ♀ 25–26; h.w. ♂ 26–30, ♀ 24–29; h.f. ♂ 6.0–6.5, ♀ 5.5–6.5; pt. ♂ 2.5–2.7, ♀ 2.0–2.5.

Nymph. We have neither reared this species nor found its exuvia with the emerged adult. No satisfactory description of the nymph seems to have appeared apart from that given by Musser (1962) which accordingly is used here. This description is based on an exuvia from New York State.

Premental setae 15; palpal setae 11. Dorsal hooks on abd. segs. 5 to 8, the longest on 7, this being one-half that segment's length; lateral spines on seg. 8 12 per cent, and on seg. 9 23 per cent of the length of these segments respectively (including the spine); mesial margin of spine on seg. 9 straight.

Length 16.5; abd. 10.0; w. abd. 5.5; h.w. 5; h.f. 4.5; w. hd. 4.5.

Habitat and range. Still waters, incl. ponds, ditches, open marshes and slow streams. N.S., s.w. Que., s. Ont. and Mich., s. along the Atlantic from the northeastern United States, N.Y. and Pa. to N.C., w. through Ky. and the middle United States to Colo. and Utah.

Distribution in Canada. N.S.—Pictou county (Kormondy, 1960). *P.E.I.*— Charlottetown; Little Harbour; Souris (New Harmony Road). *Que.*—Montreal; Ft. Coulonge; "La Chaloupe River" (probably near Berthierville). *Ont.*— Essex (incl. Pelee I.), Kent, Lambton, Norfolk, Brant, Welland, Lincoln, Waterloo, Huron, Wellington, Peel, York, Simcoe and Carleton counties; Nipissing (incl. Algonquin Park) and Cochrane districts.

This is not a common species in Canada except in southern Ontario and even there it is abundant only in the Lake Erie counties and the vicinity of Niagara and probably Hamilton and London. It is not uncommon near Toronto, but fifty miles farther north, in Simcoe county, it is already much scarcer.

Field notes. S. rubicundulum flies from about the third week in June to the last week in September. Emergence has been seen in Quebec on June 13 (Robert, 1963), and tenerals encountered in Ontario (Kent county) on June

---

PLATE 33

Sympetrum—vulvar lamina, ventral view: (1) *S. ambiguum*; (2) *S. rubicundulum*; (3) *S. internum*; (4) *S. obtrusum*; (5) *S. pallipes.*
Leucorrhinia borealis: (6) thorax and abdomen of male, left lateral view.
(6 after Walker, 1943.)

23 and in Prince Edward Island (Souris) on August 1. Early and late recorded dates in Ontario are June 21 (Welland county) and September 24 (Haldimand county).

Along the Lake Erie shore at Port Rowan *S. rubicundulum* was present in abundance during June and July, 1960. Emergence did not take place from the lake itself but from the shallow pools which are found behind the sandy ridges which lie between the beach and the inland swamp. There pools contained a fairly dense growth of many shrubs. Tenerals of other species of *Sympetrum* were sometimes met with, especially *S. internum*. When our first days of collecting were spent in June, the species of *Sympetrum* were nearly all somewhat immature. *S. rubicundulum* also breeds in small ponds in the open, permanent or semi-temporary. Near Berthierville, Quebec, Robert (1963) found adults by a pool surrounded by woodland; the pool was prevented from completely drying up by the floodwater, each spring, from a neighbouring river.

According to Needham and Westfall (1955) the male accompanies the female during oviposition. "He seems to direct the course and to assist in the flight. Together the two descend to touch the water many times in rapid succession in nearly the same place; then a short flight and many more descents in another place."

A female adult taken in an orchard in Indiana in July was eating a tarnished plant bug (Montgomery, 1929).

Sympetrum internum Montgomery. (Pl. 30: 12, p. 212; pl. 31: 11, p. 218; pl. 33: 3, p. 228; pl. 42: 6, p. 266; pl. 43: 7, p. 270)

> *Sympetrum rubicundulum decisum* (Hagen): Ris, Coll. Zool. Selys, 13: 684, 1911.
> *Sympetrum rubicundulum*: Walker, Can. Ent., 49: 416–417, 1917b.
> *Sympetrum decisum*: Walker, Trans. R. Can. Inst., 23: 256, 1941b.
> *Sympetrum internum* Montgomery, Can. Ent., 75: 57, 1943a (nomenclatorial history).

Very similar to *S. rubicundulum* but on average somewhat smaller with a shorter abdomen and a reddish face in the mature adult. General colour more reddish or golden brown. The genitalia are distinctive in both sexes.

Male. Face yellowish red; labrum with a black line along front border; vertex and occiput shining brown. Thorax reddish brown clothed with hair of same colour, without definite stripes. Legs black beyond bases except for pale underside of fore femur. Wings hyaline, with a golden tinge at extreme base and generally not reaching the first antenodal cross-vein. Abdomen cherry red with a dorsal transverse black band on seg. 1, an oblique wash of black across sides of seg. 2, and large black subequal triangles on sides of segs. 4 to 8 or 4 to 9; seg. 10 and anal apps. yellow. Sup. apps. with black tips.

Sup. apps. (pl. 30: 12, p. 212) with a prominent ventral tooth. Hamuli (pl. 31: 11, p. 218) forcipate, differing in minor details of form from those of *rubicundulum*.

Female. Resembling male in colour and dimensions. Vulvar lamina (pl. 33: 3, p. 228) bifurcate with apices distinctly divergent at posterior margin, closely similar to that of *obtrusum*.

Venation. Anx ♂ 6.5–7.5/5(6), ♀ 6.5–8.5/5–6; pnx ♂ 6–8/6–8, ♀ 5–9/6–9.

Measurements. Total length ♂ 29–32, ♀ 32–33 (Chilliwack specimens: ♂♀ 36); abd. ♂♀ 21.0–22.0; h.w. ♂♀ 28.0–29.0 (NW: ♂♀ 23–27); h.f. ♂♀ 4.5; pt. ♂♀ 2.0–2.4.

Nymph. Cannot yet be consistently distinguished from the nymph of *S. danae* (p. 223). In *S. internum* (unlike *S. danae*) there is seldom if ever a minute denticle on abd. seg. 8 representing the dorsal hook (pl. 42: 6, p. 266, pl. 43: 7, p. 270), and the lateral spines on seg. 9 are usually distinctly less than one-fourth of the lateral length of that segment (including the spine).

Habitat and range. Ponds and slow shady streams. Alaska and N.W.T. to Nfld., s. throughout the Canadian Provinces to Calif., Nev., Utah, Colo., Mo., Ky. and Pa.

Distribution in Canada and Alaska. Nfld.—Gander. *N.S.*—Annapolis, Colchester, Digby, Halifax and Pictou counties; Cape Breton I. (Inverness and Victoria counties). *N.B.*—Charlotte, Gloucester, Kings, Queen's, St. John and York counties. *P.E.I.*—Charlottetown (Victoria Park). *Que.*— Covey Hill; Hemmingford; Vaudreuil; Montreal; Aylmer; Hull; Ironside; Wakefield; Alcove; Lac Ste. Marie; Mont Tremblant Park; Nominingue; Abitibi county; Godbout; Natashquan; Thunder River; Saguenay River (Cap Jaseux); Anticosti I. *Ont.*—Brant, Waterloo, Halton, Peel, York, Ontario, Bruce, Grey, Simcoe, Hastings, Leeds, Carleton, Lanark and Renfrew counties; Muskoka, Manitoulin, Nipissing (Algonquin Park), Algoma, Thunder Bay (Lake Superior), Rainy River, Cochrane (Smoky Falls; Moosonee) and Kenora districts. *Man.*—Deloraine; Stockton; Treesbank; Aweme; Onah; Winnipeg; Portage la Prairie; Westbourne; Teulon; Grandview; Dauphin Lake; The Pas; Wabowden; Gillam. *Sask.*—Maple Creek; Swift Current; Wascana Lake; Bredin; Waldeck; Indian Head; Katepwa Lake; Balgonie; Regina; Regina Beach; Lumsden; Markinch; Hamton; Davidson; Elbow; Watrous; Big Manitou Lake; Liberty; n. of Alsask; Madison; Glidden; Duro; Snipe River; Kindersley; Eston; Goose Lake; Saskatoon; Vonda; Carlton; Duck Lake; Kinistino; Lac Vert; Birch Hills; Prince Albert; Waskesiu Lake; Polwarth; Lashburn. *Alta.*—Near Waterton Lake; Cardston; Claresholm; Banff; Red Deer; Jasper Park; "Settler" (presumably Stettler); near Flatbush; Beaver Lake. *B.C.*—Vancouver (Beaver Lake); Cultus Lake; Chilliwack; Hope Slough; Indian Reserve Slough; Hope; Princeton; Kawakawa Lake; Harrison Bay; Hatzic Lake; Osoyoos; Penticton; Okanagan Landing; Peachland; Kootenay Landing; Vernon; Armstrong; Kamloops; Canal Flats; Golden; Jesmond; Canim Lake; 150-mile House; McBride; Prince George; Vanderhoof; Burns Lake. *N.W.T.*—Ft. Smith; Slave River (delta near McDonald Lake); Great Slave Lake (Outpost I.; Jones Pt.; Wool Bay; Ft. Resolution; Yellowknife; Gros Cap); Ft. Simpson; Norman Wells. *Alaska*—Chitina.

Field notes. This species has a long flight period lasting from the third week in June to the second week in October in British Columbia, and to the second half of September in the Prairies and Maritimes, where specimens have been captured as late as September 18 (Saskatchewan, Grandview) and September 30 (Nova Scotia, Halifax county). Full-grown nymphs can be found in early July and most records for emergence are for July, and mainly for the first half: representative records are July 1 (Northwest Territories, Yellowknife), July 9 (New Brunswick) and July 21 (Ontario, Moosonee). Adults are most numerous in July and August.

In New Brunswick during July, 1932 we found *S. internum* emerging from a small weedy pond in wet pasture, and also from a bog pond; large numbers of tenerals were seen in an open marsh and many adults were flying and resting in the shelter of willows at the edge of the marsh (Walker, 1933a). In Annapolis county, Nova Scotia, this appears to be by far the most abundant species of Anisoptera during late summer.

Near Cultus Lake, British Columbia, in 1936 *S. internum* was common at a partly boggy slough covering about five acres, situated in a hollow at about 300 feet elevation, and surrounded by heavy forest. Most of it was covered by tall *Scirpus* and *Typha*, with patches of *Carex*, bog shrubs, sphagnum and perhaps a third of an acre of very shallow open water (Walker and Ricker, 1938).

In the vicinity of Flatbush, Alberta, an adult of this species was preying on simuliids around an observer's head on August 27; and earlier in the month an adult *S. internum* had itself fallen prey to *Aeshna eremita* (Pritchard, 1964b).

Variations. Most adults examined from the vicinity of Cultus Lake, British Columbia have been about the usual size of eastern individuals; but two specimens collected from Chilliwack on August 11, 1936 constitute exceptions, being very large and exceeding in size any seen from eastern Canada (Walker and Ricker, 1938).

Sympetrum obtrusum (Hagen). (Pl. 30: 13, p. 212; pl. 31: 12, p. 218; pl. 33: 4, p. 228; pl. 43: 1, 8, p. 270)

Diplax obtrusa Hagen, Stett. Ent. Zeitung, 28: 95, 1867.
Sympetrum obtrusum: Kirby, Cat. Neur. Od., p. 16, 1890.

Very similar to *S. internum* but usually a little smaller than either *internum* or *rubicundulum*; abdomen a little more slender than in *internum* from which it is to be distinguished only by close examination of the male hamuli. Dark red like *rubicundulum* but lacks the pale lateral thoracic stripes.

Male. Labrum and frons yellowish white, face greenish white, bounded above with a black line which separates it from the brown vertex. In immature specimens the face and other pale parts are yellow. Rear of head medium brown. Thorax golden brown, hairiest dorsally with one or more ill-defined pale blotches on the sides in juvenile individuals, which disappear when maturity is reached. Legs black except extreme bases and the ventrolateral surfaces of the fore femora which are whitish. Wings hyaline except at the extreme bases which are golden yellow. Venation dusky; pt. smoky gray. Abdomen in life blood red when mature, yellowish to brownish red when immature, with black lateral spots on segs. 4 to 9, narrowing forward and widening backward, tending to become larger and less regular at the ends.

General proportions relatively more slender and elongate than in *internum*, nearly as in *rubicundulum*. Abdomen narrowing from middle of seg. 2 to end of seg. 3, where it is constricted to about one-half its greatest width, in the anterior half of the body, then widening slightly up to seg. 7, with another slight constriction between 8 and 9, the abdomen being widest at its posterior extremity. Anal apps. (pl. 30: 13, p. 212) scarcely, if at all, distinguishable from those of *rubicundulum* and *internum*. Hamuli (pl. 31: 12, p. 218) with very short branches, the length of each not more than one-fifth as long as the visible part of the hamulus, the outer branch wider than the length and at least twice as wide as the inner branch, which is blunt and curved. Genital lobe with straight sides, bluntly angular.

Female. Similar in size and proportions to the male except in the stouter and less constricted abdomen. Face greenish white, teneral coloration like that of male, olivaceous with sides and venter of thorax marked with large pale yellow spots of variable form; mature coloration generally olivaceous, on both thorax and abdomen, with black legs; occasionally homoeochromatic, the thorax red-brown and the abdomen red with black lateral spots. This type of coloration in our experience is quite rare. Vulvar lamina (pl. 33: 4, p. 228) similar to that of *internum*, but the two valves (or scales) are bent, rather than curved, inward, the apices being directed caudad, meeting or approaching each other at the posterior end. There is considerable variation in the form of these valves, but they are apparently always somewhat straighter and more prolonged at the posterior end than in *internum*. In *rubicundulum* the tumid form of the valves at once distinguishes them from those of the other two species.

Venation. Anx ♂ 7–8(9)/5–6, ♀ 7–8/5–6; pnx ♂ ♀ 6–7(8)/7–9.

Measurements. Total length ♂ 33–36, ♀ 30–34; abd. ♂ 23.0–24.5, ♀ 20.0–23.5; h.w. ♂ ♀ 22–26 (NW: ♂ ♀ 20–29); h.f. ♂ ♀ 5.0–5.5; pt. ♂ 1.6–1.9, ♀ 1.7–2.0.

Nymph. Musser (1962), who reared both *S. obtrusum* and *S. pallipes* in Utah, found no differences by which the two species could consistently be distinguished. As her material was more extensive than ours, we do not hesitate to accept her conclusions. However, our own observations (Walker, 1917b) appeared to be constant for the locality where they were made; these indicated that nymphs of *S. obtrusum* are, on average, somewhat smaller than those of *S. pallipes*, and that there are persistent differences as indicated in the key (p. 205).

Habitat and range. Small ponds, often temporary ones, in fields, pastures and gravel pits and occasionally marshes; also streams. N.S., N.B. and P.E.I. w. to B.C. and sparingly n. into the N.W.T. Also abundant in the middle United States s. to Ky.; s.w. to Nebr., Kans., Colo. and Utah.

Distribution in Canada. N.S.—Annapolis, Colchester and Pictou counties; Cape Breton I. (Inverness county). *N.B.*—Charlotte and Northumberland counties. *P.E.I.*—Charlottetown; Little Harbour; Souris (New Harmony Road). *Que.*—Covey Hill; Montreal; Hull; Kingsmere; Alcove; Mont Tremblant Park; Nominingue; Lanoraie; Isle of Orleans; Abitibi county; Saguenay River (Cap Jaseux); Lake St. John. *Ont.*—Essex, Kent, Middlesex, Norfolk, Brant, Lincoln, Peel, York, Dufferin, Ontario, Durham, Northumberland, Bruce, Grey, Simcoe, Peterborough, Hastings, Leeds, Carleton and Renfrew counties; Muskoka, Parry Sound, Manitoulin, Nipissing (incl. Algonquin Park), Algoma, Sudbury, Timiskaming, Thunder Bay, Rainy River, Cochrane and Kenora (incl. Patricia Portion) districts. *Man.*—Westbourne; Swan River. *Sask.*—Big Manitou Lake; Lac Vert; Waskesiu Lake. *Alta.*—Near Waterton Lake; Claresholm; near Flatbush.

B.C.—Cultus Lake; Vedder Crossing; Osoyoos; Penticton; Okanagan Falls; Okanagan Landing; Kaslo; West Kootenay; Vernon; Armstrong; Revelstoke; Jesmond; Quesnel; Vancouver I. (Nanaimo Lakes). *N.W.T.*—Ft. Smith; Great Slave Lake (Ft. Resolution; Jones Pt.); Ft. Simpson.

Field notes. The flight period extends from the third week in June to the second or third week in October. This species has been taken emerging at Belleville, Ontario, on June 20 (Walker, 1941b) and at Kaslo, British Columbia, on June 24 (Whitehouse, 1941), but in Ontario emergence reaches its height in July and adults do not become numerous until well on in that month, the season of *S. obtrusum* being about three weeks behind that of *S. internum* (Walker, 1941b). In Quebec the flight period reaches its peak near the end of July and extends into September (Robert, 1963).

In the vicinity of Frank's Bay, Lake Nipissing, Ontario, this species was common in a black-spruce swamp, and mature males were taken over a marshy slough at a bend in a forest stream. This stream is about 30 to 40 feet wide, and very quiet and shallow with a soft mud bottom and a clear channel; there is abundant aquatic vegetation in many places and patches of emergent vegetation along the banks or in mid-stream. In Utah a reared specimen was taken from an ephemeral lake (Musser, 1962). In middle and southern Ontario we have observed that *S. obtrusum* is commonly the only species in waters of high acidity, and this is probably also true in the north (Walker, 1940a).

This is probably the commonest anisopteran in the settled parts of eastern Canada in late summer; and Kennedy (1915b) found it to be the most abundant species of *Sympetrum* in Baker Valley, Oregon, during July.

Whitehouse (1941) once took a mature red male *S. obtrusum* in tandem with a black *S. danae* female.

Sympetrum pallipes (Hagen). (Pl. 30: 14, p. 212; pl. 31: 13, p. 218; pl. 33: 5, p. 228; pl. 43: 2, 9, p. 270)

Diplax pallipes Hagen, Rep. Neur. Colorado, p. 589, 1874.

Sympetrum pallipes Kirby, Cat. Neur. Od., p. 16, 1890.

Structurally very like *S. obtrusum* but usually larger, more robust, and paler in general coloration, with distinct whitish or yellow lateral thoracic stripes.

Male. General colour pale yellowish brown. Face, including labrum and labium, light yellow with a thin growth of dark brown hair. Frons dorsally light brownish yellow; vertex light brown, darker than frons. Occiput metallic green. Rear of head rather light brown with grayish fringe of hair, a fairly broad marginal area, which has a somewhat metallic lustre. Prothorax with a pale fawn fringe. Pterothorax with a pair of pale stripes, often indistinct or obsolete. Lateral stripes whitish or bright yellow, oblique, straight, parallel, the front stripe wider than the hind stripe and widest about the middle. Legs medium or dark brown, paler at the bases, entire surface of fore femora pale yellowish, tibiae yellow externally. Wings hyaline with dusky brown venation, yellow at base; pt. uniform medium brown; membranule amber yellow. Abdomen light yellowish brown.

Sup. apps. (pl. 30: 14, p. 212) with prominent ventral tooth. Hamuli (pl. 31: 13, p. 218) nearly identical with those of *obtrusum*, the stem being bent a little farther from the bifurcation. Genital lobe bluntly angulate as in *obtrusum*.

Female. Generally with a stouter abdomen than in the male and the wings often more heavily and extensively flavescent. Vulvar lamina (pl. 33: 5, p. 228) variable but apparently broader than in *obtrusum* and more distinctly pentagonal. A satisfactory comparison of this structure in these two species requires well preserved material.

Venation. Anx ♂ ♀ 7–9/5–6; pnx ♂ ♀ 8–9/5–6.

Measurements. Total length ♂♀ 32–37; abd. ♂ 24, ♀ 25; h.w. ♂ 24, ♀ 27 (NW: ♂ ♀ 25–28); h.f. ♂ ♀ 6.0–6.5; pt. ♂ ♀ 2.2.

Nymph. Cannot consistently be distinguished from that of *S. obtrusum*. Nymphs of the two species differ slightly in the relative lengths of the lateral spines on abd. seg. 9, and in the relative sizes of the dorsal hooks on segs. 7 and 8 (see key, p. 205).

Eyes moderately prominent, lateral margins of head very oblique, passing into posterior margins without trace of an angle. Prementum of labium narrow at base, the basal width being somewhat less than one-fourth of the greatest width, which is nearly equal to the length; sides moderately concave, diverging in distal fourth at an angle of about 90°, median lobe depressed at an angle of about 30° with the general surface of the prementum; ventral surface of prementum with or without brown spots; premental setae 12 to 15, usually 13, the 4th or 5th from the outside being the longest; palpal setae 10 or 11, usually 10; distal margin of palpus with very shallow crenations, each bearing a group of 3 (occasionally 4) spiniform setae, the 3rd being much longer than the others. Legs pale, each with two dark femoral annuli and sometimes a paler band; wing-sheaths reaching to base of seg. 6 or slightly beyond. Abdomen (pl. 43: 2, 9, p. 270) ovate with greatest width at about seg. 6; dorsal hooks on segs. 4 to 8, barely indicated on 4, increasing in length to 7, a little shorter on 8, all slender and very slightly curved, the last one nearly horizontal; lateral spines on abd. segs. 8 and 9, those on 8 about one-fifth as long as lateral margin of that segment (including the spine); lateral spine on seg. 9 nearly one-third of lateral length of that segment; cerci about two-thirds as long as epiproct and slightly more than one-third as long as paraprocts; epiproct and paraprocts acuminate, ending in very slender apices. Colour greenish brown with obscure dark markings. Abdomen more or less transversely marked with dark and pale spots.

Length 16–18; abd. 10–12*; w.abd. 5.5–6; h.w. 5–6; h.f. 4.6–5; w.hd. 4.5–4.9.

*These measurements are from Musser (1962) and refer to both *S. obtrusum* and *S. pallipes*.

Habitat and range. Small, stagnant pools and ditches. Southern B.C. and Alta., s. in the United States to Calif. and Tex.

Distribution in Canada. B.C.—Vancouver; Harrison Bay; Kawakawa Lake; Hope; Rosedale; Cultus Lake; Chilliwack; Vedder Crossing; Osoyoos; Mabel Lake; Oliver; Penticton; Vaseux Lake; Okanagan Landing; Christina Lake; Nelson; Kaslo; Crawford Bay; Moyie; Wasa; Douglas Lake; Quilchena; Stump Lake; Vancouver I. (Prospect Lake; Langford Lake; Saanich district; Loon Lake; Wellington; Nanaimo district; Gabriola I.). *Alta.*—Waterton Lakes Park (Crooked Creek); Suffield.

Field notes. Whitehouse (1941) gives the limits of the flight period as June 6 and September 24. On Vancouver Island he witnessed emergence between June 6 (Sooke) and July 4 (Florence Lake), and encountered teneral adults on July 25 (Forbes Landing). The Alberta specimens of *S. pallipes* were caught on July 18 and August 2.

Like its near relative *S. obtrusum*, *S. pallipes* prefers small, semi-permanent ponds and puddles, differing in this from *S. costiferum* which more often inhabits shallow, marshy bays and lagoons. We found nymphs in abundance in a small forest pool near Wellington on Vancouver Island. In Utah, Mrs. Musser (1962) obtained nymphs from a permanent pond surrounded by willow trees, frequented by cattle, and with a muddy bottom.

This is the commonest *Sympetrum* in the Nanaimo district of British Columbia.

Genus **Leucorrhinia** Brittinger

S.B. Acad. Wiss. Wien, 4, Mathem.-nat. Klasse: 333, 1850. Type species *L. albifrons* (Burmeister).

These are small, dark, white-faced dragonflies with a predominantly northern distribution. The circumboreal genus comprises about twelve species, seven of which occur in North America (and Canada). *Leucorrhinia* is closely related to *Sympetrum* but easily recognizable by the white face on the otherwise dark pattern varied by yellow or orange spots on the thorax and abdomen and, in mature individuals of some species, the whitish pruinosity which is generally most marked around the bases of the legs. Males are usually larger than females. The flight period is in spring and early summer.

The labrum is yellowish and the vertex mainly black; the ground colour of the thorax is red or yellowish brown, heavily marked with black and supporting a dense growth of hair that may obscure the markings. The sides of the thorax are blackened heavily along the humeral and second lateral sutures. The legs are black. The wings are normally hyaline with a few dark spots at the bases, the largest of these being on the hind wing; the wings are occasionally tinged with smoky brown (rarely amber) for a varying distance from the base, especially in females.

Distinctive venational features (pl. 27: 2, p. 188) are the enlarged heel and short toe of the anal loop, and the short, thick, dark pterostigma which is shorter in the males than in the females and usually scarcely twice as long as wide. Other characters are the complete antenodal cross-veins; the two cross-veins under the pterostigma in both wings, the first one often being aslant like a brace vein; and the shape of the fore-wing triangle in which the anterior side is more than one-half as long as the proximal side.

The nymphs live among water plants and are smooth and greenish with brown or dark brown markings. They are similar to nymphs of *Sympetrum* but somewhat more robust in build. No incisive characters exist for separating nymphs of the two genera because nymphs are variable in the very characters that are most useful in separating them. Nymphs of *Leucorrhinia* may be characterized as follows. The head is somewhat less than twice as broad as long; the eyes are most prominent behind the middle, the lateral margins being very oblique, and curving into the posterior margin with no indication of an angle. The prementum is about as wide as long; the premental setae are usually 13 to 15 (on each side); the distal margins of the palpi have very low crenations, the spiniform setae are in groups of two or three; one seta is much the longest, and sometimes single; the movable hook is slender and scarcely one-half as long as the distal margin of the palpus; the palpal setae number 10 or 11. The abdomen is broader than the head and broadest at segment 6; lateral spines are present only on segments 8 and 9, those on 9 not extending beyond the apices of the anal appendages; dorsal hooks are present or absent, but never present on segment 9; the epiproct is acuminate, distinctly longer than it is wide, and only slightly shorter than the paraprocts; the cerci are about one-half as long as the paraprocts.

The venter of the abdomen either has three broad longitudinal stripes, one median and two lateral, or many parallel rows of small similar dark spots; the former pattern is associated with few or no dorsal hooks and the latter with many and larger hooks.

Key to the Species of Leucorrhinia
Adult Males

1 Abdominal tergites of segs. 4 to 9 black, without yellow or reddish spots 2

   Abdominal tergites of segs. 4 to 9 black, with yellow or reddish spots on at least seg. 7 5

2 Hamuli with inner branch (see pl. 35: 1, p. 242) abruptly curved at tip (pl. 34: 11, p. 238) *glacialis* (p. 249)

   Hamuli with inner branch more or less smoothly curved from base to tip (e.g. pl. 34: 12, p. 238) 3

3 Inf. app. with lateral margins converging posteriorly, at least twice as wide at base as at tip (pl. 34: 6, p. 238) *frigida* (p. 255)

   Inf. app. with lateral margins nearly parallel, distinctly less than twice as wide at base as at tip (pl. 34: 2, 4, p. 238) 4

4 Total length more than 30 mm.; hamuli with outer branch lacking an anteriorly directed process (pl. 34: 12, p. 238; pl. 35: 6, p. 242); sup. apps. in lateral view, with ventral denticles but lacking an ante-apical ventral tooth (pl. 34: 5, p. 238) *proxima* (p. 251)

   Total length less than 30 mm.; hamuli with outer branch forming a curved, anteriorly directed process (pl. 34: 10, p. 238; pl. 35: 5, p. 242); sup. apps., in lateral view, with ventral denticles and an ante-apical ventral tooth (pl. 34: 3, p. 238) *patricia* (p. 246)

5 Yellow or reddish spot only on the tergite of seg. 7; inf. app. with lateral margins strongly divergent posteriorly (pl. 34: 7, p. 238) *intacta* (p. 257)

   Yellow or reddish spot on the tergite of at least segs. 1 to 7; inf. app. with lateral margins nearly parallel or only slightly divergent posteriorly (e.g. pl. 34: 1, p. 238) 6

PLATE 34

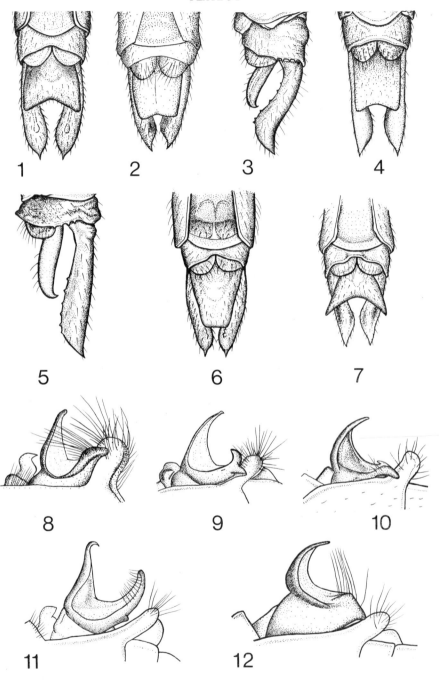

6 Total length less than 34 mm.; hamuli with outer branch forming a
   gently curved, anteriorly directed process (pl. 34: 9, p. 238; pl. 35: 4, p.
   242)   *hudsonica* (p. 244)

   Total length more than 34 mm.; hamuli with outer branch almost
   straight and lacking an anteriorly directed process (pl. 34: 8, p. 238; pl.
   35: 3, p. 242)   *borealis* (p. 240)

Adult Females

1 Yellow or reddish spot on tergite of seg. 7   2

   No yellow or reddish spot on tergite of seg. 7; scales of vulvar lamina
   about one-fourth as long as lateral margin of seg. 9 and separated by a
   distance about equal to their length (pl. 36: 3, p. 248)   *patricia* (p. 246)

2 Pale spot on tergite of seg. 7 narrow and elongate, at least three times as
   long as wide   3

   Pale spot on tergite of seg. 7 broad and short, less than three times as
   long as wide; scales of vulvar lamina separated from their base (pl. 36: 7,
   p. 248)   *intacta* (p. 257)

3 Labium spotted with white at sides; scales of vulvar lamina short, only
   barely visible at the posterior margin of seg. 8 (pl. 36: 5, p. 248)
                                                          proxima (p. 251)
   Labium entirely black   4

4 Scales of vulvar lamina extremely short, little more than a swelling at
   posterior margin of seg. 8 (pl. 36: 4, p. 248)   *glacialis* (p. 249)

   Scales of vulvar lamina prominent, at least one-third as long as lateral
   margin of seg. 9   5

5 Total length more than 33 mm.   *borealis* (p. 240)

   Total length less than 33 mm.   6

6 Length of scales of vulvar lamina at least twice their width at base; scales
   extending posteriorly to about one-half the length of seg. 9 (pl. 36: 6, p.
   248)   *frigida* (p. 255)

   Length of scales of vulvar lamina barely greater than their width at
   base; scales extending posteriorly to less than one-half the length of
   seg. 9 (pl. 36: 2, p. 248)   *hudsonica* (p. 244)

PLATE 34

Leucorrhinia—anal appendages of males, ventral view (except 3, 5, left lateral view): (1) *L.*
hudsonica; (2) *L. patricia*; (3) *L. patricia*; (4) *L. proxima*; (5) *L. proxima*; (6) *L. frigida*; (7) *L. intacta.*
Leucorrhinia—hamulus, right lateral view: (8) *L. borealis*; (9) *L. hudsonica*; (10) *L. patricia*; (11) *L.*
glacialis; (12) *L. proxima.*
(8 after Walker, 1943; 9 modified after Walker, 1940a.)

Nymphs

Nymph unknown *patricia* (p. 247)

1 Dorsal hooks on segs. 3 to 8; venter of abdomen without continuous longitudinal dark bands, although transverse dark bands may be present   2

    Dorsal hooks variable but usually absent on at least segs. 7 and 8; venter of abdomen with three conspicuous (rarely inconspicuous) continuous longitudinal dark bands   4

2 Eyes very prominent; lateral spines on seg. 9 extending back nearly or quite as far as tips of paraprocts (pl. 44: 5, p. 272)   *frigida* (p. 256)

    Eyes less prominent; lateral spines on seg. 9 not extending back as far as tips of paraprocts   3

3 Dorsal hook on seg. 7 as long as mid-dorsal length of seg. 7, that of seg. 8 projecting well over the base of seg. 9; lateral spines on seg. 8 more parallel and directed caudad, their outer margins not continuing the regular curve of the abdominal margin when viewed dorsally (e.g. pl. 44: 6, p. 272); lateral spines of seg. 9 extending beyond tips of cerci *intacta* (p. 257)

    Dorsal hook on seg. 7 shorter than the mid-dorsal length of seg. 7, that on seg. 8 projecting barely over the base of seg. 9; lateral spines on seg. 8 slightly convergent, their outer margins continuing the general curve of the abdominal margin when viewed dorsally (pl. 44: 4, p. 272); lateral spines on seg. 9 not reaching beyond tips of cerci   *proxima* (p. 251)

4 Total length less than 18.5 mm.; lateral spines on segs. 8 and 9 with outer margins slightly divergent (pl. 44: 3, p. 272), dorsal hooks absent, or vestigial on some segments   *hudsonica* (p. 244)

    Total length more than 18.5 mm.; lateral spines on segs. 8 and 9 with outer margins not divergent, but following the curved outline of the abdomen (pl. 44: 2, p. 272); dorsal hooks absent   5

5 Lateral spines on seg. 8 more than one-fourth as long as the lateral margin of seg. 8 (not including the spine)   *glacialis* (p. 250)

    Lateral spines on seg. 8 one-fourth or less as long as the lateral margin of seg. 8   *borealis* (p. 241)

Leucorrhinia borealis Hagen. (Pl. 33: 6, p. 228; pl. 34: 8, p. 238; pl. 35: 3, p. 242; pl. 36: 1, p. 248; pl. 44: 2, p. 272)

Leucorrhinia borealis Hagen, Trans. Amer. Ent. Soc., 17: 232, 1890a.
Leucorrhinia borealis: Muttkowski, Bull. Wis. Nat. Hist. Soc., 6: 113, 1908.
Leucorrhinia borealis: Walker, Can. Ent., 75: 87–89, pl. VII, figs. 1–7, 1943.

This is the largest species of *Leucorrhinia* in North America, being of about the same size as the European *L. rubicunda* (Linnaeus), which it closely resembles, although it is probably more closely related to *L. hudsonica*.

Male. Labrum yellowish white, other parts of face, including frons, greenish. Labium and rear of head black. Thorax (pl. 33: 6, p. 228) dark brown to reddish, variegated with yellowish markings which are often inconspicuous in dried specimens. Thoracic hair very dense. Coxae and trochanters brown, remaining parts of legs black. Wings hyaline with the usual dark spots at bases of both pairs. Abdomen (pl. 33: 6, p. 228) relatively large; segs. 1 to 3 almost entirely red in mature individuals (usually yellowish when dried); spots on remaining segments narrower, each narrowing caudad to a blunt triangle; segs. 9 and 10 wholly black except on the intersegmental areas.

Sup. apps. convergent, very slightly arched, with apices slightly tilted, a ventral prominence at the distal third. Inf. app. about three-fifths of the length of the sup. apps., squarish, with the distal border emarginate. Inner branch of each hamulus (pl. 34: 8, p. 238; pl. 35: 3, p. 242) nearly upright with apices slightly bent back at the tip; outer branch bent mesad, its apex lying close to the genital lobe.

Female. Similar to male in colour pattern and form. Thorax more yellowish. Abdomen somewhat stouter proximally. Terminal segments, especially 9, relatively larger than in the male. Vulvar lamina (pl. 36: 1, p. 248) consisting of two separate lamellae, not or scarcely longer than wide and at least one-third as long as lateral margin of seg. 9.

Venation. Anx ♂ 7–8/6, ♀ 7–9/5–7; pnx ♂ 7–9/8–9, ♀ 7–8/7–11; rows of cells subtended by Rspl 1; ankle cells in anal loop 4 to 8; three rows of post-trigonal cells in fore wing.

Measurements. Total length ♂ 37.2–39.0, ♀ 35.0–38.0; abd. ♂ 25.0–25.3, ♀ 23.2–26.1 (♂♀ 25–27*); h.w. ♂ 29.9–32.9, ♀ 28.6–31.0.
*Combined range as in first author's manuscript.

Nymph. Among the known North American *Leucorrhinia* only this species, *L. glacialis* and *L. hudsonica* have 3 longitudinal dark bands on the venter of the abdomen and lack dorsal hooks on abd. segs. 7 and 8; *L. borealis* resembles *L. glacialis* in its larger body size and in the shape of the lateral spines on abd. segs. 8 and 9; *L. borealis* differs from *L. glacialis* in the relative lengths of these lateral spines.

A large smooth nymph, olivaceous to distinctly yellow. Eyes less prominent than in other North American species of *Leucorrhinia*, the width of the head across the eyes being somewhat less than twice the length (not including the labium); posterolateral margins of head broadly rounded with numerous coarse setae. Prementum of labium subtriangular, the median lobe bluntly obtusangulate; premental setae 13, occasionally 14 or 15, the 3rd to 6th from the outside the longest; palpal setae 11, occasionally 10; distal margin of palpus with broad, shallow crenations, each bearing 1 spiniform seta, or sometimes a second which may or may not be very small. Abdomen somewhat elongate ovate, broadest at seg. 6 or at posterior margin of 5, tapering almost equally proximad and distad; dorsal hooks completely absent; lateral spines (pl. 44: 2, p. 272) on segs. 8 and 9, their outer margins following the curved outline of the abdomen, their inner margins nearly parallel; the spines on seg. 8 about one-seventh to one-fourth the lateral length of that segment (not including the spine); those on seg. 9 about two-fifths to one-half the length of that segment; lateral spines on seg. 9 not extending posteriorly to tips of cerci; epiproct triangular, keeled and somewhat spinulose on distal half above; the relative lengths of cerci, epiproct and paraproct are 1:1.8:2.0.

Dull brownish, more or less distinctly marked with paler yellowish. Head dark, sometimes streaked with pale yellowish behind and beneath the eyes; thorax obscurely mottled; wing-sheaths with more or less distinct pale costal streaks; legs pale with darker annuli, the ante-apical femoral rings being the most distinct and constant. Abdomen dark above with a median line, a pair of dorsolateral spots and the posterolateral angles on most of the segments pale; venter pale with 3 longitudinal dark bands, which may be somewhat obscure but are generally hairy and conspicuous; they are usually narrower but sometimes broader than the intervening pale areas.

Length 19.0–23; abd. 14–15; w. abd. 6.7–7.3; h.w. 6.0–6.5; h.f. 5.0–6; w. hd. 5.25–5.6.

PLATE 35

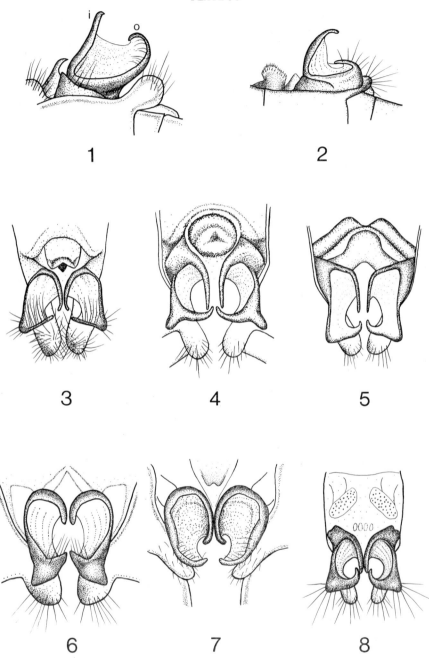

Habitat and range. Prairie ponds, lakelets and bog-holes. Mainly in the western part of the Canadian zone; probably enters the Hudsonian to a slight extent, and is abundant in the northern parts of the prairie Provinces. S. Alaska and Yukon T. e. to n. Ont. Reported from Wyo. (Needham and Westfall, 1955).

Distribution in Canada and Alaska. Alaska—Anchorage; Gulkana; Big Delta. *Yukon T.*—Ross River; Swim Lakes; Mayo; Dawson. *N.W.T.*—Ft. Smith; Slave River (72 miles below Ft. Smith); Great Slave Lake (Ft. Resolution; Gros Cap; Resdelta Channel; Caribou Is.); Norman Wells; Ft. McPherson. *B.C.*—Mayo; Squilax; Canim Lake; 100-mile House; Chilcotin (near Riske Creek); Burns Lake; Ft. Nelson. *Alta.*—Banff; Red Deer; Athabasca River, near Athabasca Lake; 12 miles s. of Ft. Smith. *Sask.*—Indian Head; Pike Lake; Hamton; Lac Vert; Cut Knife (Atten Lake); Prince Albert; Waskesiu Lake. *Man.*—Ninette; Pelican Lake; near Stockton; Aweme; Birds Hill; Dauphin Lake; The Pas. *Ont.*—Kenora district (Patricia Portion: Favourable Lake).

Field notes. Compared with other species of *Leucorrhinia*, *L. borealis* has a flight period that is early and brief. It can extend from the fourth week in May (Red Deer district, Alberta) to the third week in July (Banff, Alberta); most adults are encountered in June, and few before that month. At The Pas in 1931, the "season was apparently nearly over" in the second week of July (Walker, 1933b). Early records (during the first week of June) include captures from Indian Head, Saskatchewan; Aweme, Manitoba; and near Great Slave Lake, Northwest Territories.

At The Pas this dragonfly frequents shallow, mud-bottomed lakes or sloughs with a marginal belt of reeds (*Phragmites*). In Utah, Mrs. Musser (1962) found full-grown nymphs clinging to undercut surfaces at the edge of bog ponds and also in adjacent shallow standing water. At this habitat emergence took place in association with *Somatochlora semicircularis*, *Leucorrhinia hudsonica* and *Aeshna sitchensis*.

Nymphs from ponds near Edmonton, Alberta, examined by Pritchard (1964a), had been feeding on larvae of chironomids and ceratopogonids and on ostracods.

PLATE 35

Leucorrhinia—hamulus, ventral view (except 1, 2, right lateral view): (1) *L. frigida*; (2) *L. intacta*; (3) *L. borealis*; (4) *L. hudsonica*; (5) *L. patricia*; (6) *L. proxima*; (7) *L. frigida*; (8) *L. intacta*. (3 after Walker, 1943; 4, 5, after Walker, 1940a.) (Abbreviations: i-inner branch; o-outer branch.)

Leucorrhinia hudsonica (Selys). (Pl. 34: 1, 9, p. 238; pl. 35: 4, p. 242; pl. 36: 2, p. 248; pl. 44: 3, p. 272)

Libellula hudsonica Selys, Rev. Odon., p. 53, 1850.
Diplax hudsonica: Hagen, Syn. Neur. N. Amer., p. 180, 1861.
Leucorrhinia hudsonica: Hagen, Trans. Amer. Ent. Soc., 17: 223, 1890a.
Leucorrhinia hudsonica: Walker, Can. Ent., 46: 375, 1914 (nymph).

A small species throughout most of its very wide range. It has a variegated pattern of black with irregular spots of red (in males) or yellow (in females).

Male. Labrum creamy yellow; face above greenish white; vertex dull yellowish; rear of head black. Thorax dorsally red beneath a dense coat of blackish hair; sides with a variegated pattern of black and red angular spots. Venter and legs black. Wings hyaline with veins dusky brown; costa reddish. Abdomen with segs. 1 and 2 slightly enlarged and remaining segments nearly equal in size or scarcely more slender; the red spots becoming narrower caudad and mostly longer than wide, broadest in front and tapering caudad to a point on each segment.

Sup. apps. in lateral view arched, the apices obliquely truncate, not produced into a filiform process; ventral denticles distinct; convergent when viewed dorsally. Inf. app. (pl. 34: 1, p. 238) notched at tip in a wide "V." Hamuli (pl. 34: 9, p. 238; pl. 35: 4, p. 242) with inner branches appearing subtriangular in lateral view, leaning and curving caudad, terminating in two small hooks; outer branches gently curved, anteriorly directed and appearing as a low rounded prominence in lateral view, its outer angle produced into a distinct process.

Female. Yellow where male is red. Vulvar lamina (pl. 36: 2, p. 248) divided into two flat scales that meet in a straight middle line, their free lateral margins being convexly curved; length of scales barely greater than their widths at base; scales extending posteriorly to less than one-half the length of seg. 9.

Venation. Anx ♂ 7/5–6, ♀ 7–8/5–6 (NW: ♂♀ hind wing 5–7); pnx ♂ 6–8/7–9, ♀ 7–9/7–9; rows of cells subtended by Rspl 1; ankle cells in anal loop 1 to 4; two or three rows of post-trigonal cells in fore wing.

Measurements. Total length ♂ 27.4–30.2, ♀ 26.7–29.0 (♂♀ 23–32*); abd. ♂ 17.2–19.4, ♀ 17.9–19.5 (♂♀ 18–20*); h.w. ♂ 21.5–24.7, ♀ 21.9–24 (NW ♂♀ 21–27).
*Combined range as in first author's manuscript.

Nymph. Among the known North American *Leucorrhinia* only this species, *L. borealis* and *L. glacialis* have 3 longitudinal dark bands on the venter of the abdomen and lack dorsal hooks on abd. segs. 7 and 8; *L. hudsonica* differs from the other two species in being smaller and in the shape of the lateral spines on abd. segs. 8 and 9 (pl. 44: 3, p. 272).

Similar in form and size to *L. frigida*. Head similar but eyes less prominent and more as in *L. intacta*; posterior margin of head with many long setae. Prementum of labium with sides nearly straight on proximal two-thirds, thence bending outwards so as to be rectangularly divergent; median lobe depressed at an angle of about 30°; premental setae 12 to 14, the 3rd to 5th from the outside being the longest; palpal setae 9 to 11; distal margin of palpus with very shallow crenations each bearing 1 spiniform seta, occasionally 2; rarely a third very small seta on the opposite side of the large one. Abdomen shaped as in *L. borealis*; dorsal hooks remarkably variable and in many nymphs absent altogether; when well developed they occur on 3 to 6 or 4 to 6 with the longest (though all are vestigial) on 5, this reaching the posterior margin of that segment in some specimens and not nearly doing so in others; lateral spines on segs. 8 and 9, their outer margins slightly divergent, their inner margins nearly parallel; the spines on seg. 8 about one-third to two-fifths the lateral length of that segment (not including the spine); those on seg. 9 about two-fifths to three-quarters that segment's length; lateral spines on seg. 9 extending posteriorly to, or slightly beyond, the tips of the cerci; epiproct slender, acuminate, somewhat decurved; the relative lengths of cerci, epiproct and paraprocts are 1:1.8:2.0.

Colour dull brownish above with faint indications of mottling, but no distinct pattern. Venter of abdomen with 3 conspicuous, dark brown, longitudinal bands.
Length 16–18; abd. 11; w. abd. 5.5–6.5; h.w. 5.3–5.5; h.f. 4.0–4.4; w. hd. 5–5.2.

Habitat and range. Cold marshy waters and bog ponds. A boreal species, widely distributed through Alaska, Yukon T. and N.W.T. to Labr., incl. Nfld.; also from B.C. across the Prairies to Mich., Ont., Que., the Maritime Provinces and the New England States; to s. recorded from Calif., Ore., Nebr. and Utah. The only libellulid that is really common in the Hudsonian.

Distribution in Canada and Alaska. Nfld.—Cinq Cerf River; Grand Bruit; Burgeo; Port aux Basques; Spruce Brook; Gander; Bay of Islands; St. Anthony. *Labr.*—Great Caribou Is. (Battle Harbour); St. Lewis Inlet; Goose Bay; North West River (Hamilton Inlet); Cartwright; Hopedale. *N.S.*—Annapolis, Halifax and Pictou counties; Cape Breton I. (Victoria county). *N.B.*—Charlotte and York counties. *P.E.I.*—Souris (New Harmony Road). *Que.*—Hull; Danford Lake (near Kazabazua); Montreal; Lanoraie; Berthierville; Kipawa Lake; Lake Témiscouata; Saguenay River (Cap Jaseux); Godbout; Moisie; Thunder River; Anticosti I.; Magdalen Is.; Little Mécatina I.; Abitibi county; Shefferville; Poste de la Baleine; Ft. Chimo. *Ont.*—Peel, Bruce, Victoria, Carleton (Mer Bleue, Carlsbad Springs), Lanark and Renfrew counties; Muskoka, Nipissing (incl. Algonquin Park), Algoma, Sudbury, Thunder Bay, Cochrane and Kenora (incl. Patricia Portion: n. to Cape Henrietta Maria and Ft. Severn) districts. *Man.*—Sandilands; Winnipeg; Portage la Prairie; Winnipeg Beach; Victoria Beach; Dauphin Lake; The Pas; Norway House; Wabowden; mile 200, 332 and 412, Hudson Bay Rlwy.; Gillam; Herchmer; Churchill. *Sask.*—Saskatoon; Lac Vert; Wallwort (near Melfort); Prince Albert; Waskesiu Lake. *Alta.*—Banff; Laggan (51° 26' N., 116° 11' W.); Red Deer; Nordegg; near Flatbush. *B.C.*—Vancouver; Grouse Mt. (4,500'); Cultus Lake; Hope district; Aspen Grove; Mt. Robson; Ainsworth; Kaslo; Vernon; Alta Lake; Chase; Falkland district; Salmon Arm; Jesmond; Bridge Lake; Canim Lake; Field; Quesnel; Prince George; Vanderhoof; Chilcotin (near Riske Creek); Prince Rupert; Smithers; Atlin; Vancouver I. (Elk Lake; Royal Oak; Alberni; Forbidden Plateau (3,200')). *N.W.T.*—Ft. Smith; Great Slave Lake (Ft. Resolution; Pearson Pt.; Tochatiwi Bay; Gros Cap); Ft. Simpson; Ft. Norman; Norman Wells; Great Bear Lake (Cameron Bay; Port Radium); Ft. McPherson; Reindeer Depot. *Yukon T.*—Swim Lakes; Otter Lake; Dawson. *Alaska*—Ketchikan; Admiralty I.; Juneau; Kukak Bay; Virgin Bay (Prince William Sound); Chitina; Gulkana.

Field notes. The flight period is unusually long, from the first week in May in the west (beginning about one month later in the east) to the fourth week in August. Adults are most numerous from about mid-June to mid-July. Representative early and late dates are June 9 at Fort Smith, Northwest Territories, and August 30 at Hopedale, Labrador.

Whitehouse (1917) recorded tenerals of *L. hudsonica* appearing with *L. borealis* at Red Deer, Alberta, on May 26, and noted that their flight periods were almost identical. At that time *L. borealis* had not yet been taken in Ontario; but when it was discovered in the Kenora district no such similarity between the two species was observed. *L. borealis* appeared much earlier and its flight period was over long before that of *L. hudsonica*. Thus there seems to be some difference in the life history of the same species in the two localities—if these two reports are correct.

Although widely distributed in British Columbia, *L. hudsonica* is said by Whitehouse (1941) not to appear 'at home' in the more southerly localities. Occasional individuals are noted but one "misses the swarming hordes of the insect in its truer habitat."

Near Bend, Oregon, Kennedy (1915b) found this species to be common in sloughs where the grass was more open so that the surface of the water was exposed. "They spent most of their time seated on the tops of aquatic plants. They copulated on the wing, the male picking up the female as she sat on some plant. The flight was short, after which the male dropped the female but hovered near, while she oviposited by tapping the tip of her abdomen repeatedly on the surface of the water."

Pritchard (1964b) records *L. hudsonica* adults feeding heavily on mosquitoes near Flatbush, Alberta, and a teneral itself falling prey to *Cordulia shurtleffi*.

The nymphs fed mainly on chironomid larvae, and predominantly so in earlier instars (Pritchard, 1964a).

Variations. Individuals from the more southern localities such as the vicinity of Vancouver and south of our limits in Oregon and Washington are distinctly larger than are those from most of British Columbia and the more eastern Provinces.

Leucorrhinia patricia Walker. (Pl. 34: 2, 3, 10, p. 238; pl. 35: 5, p. 242; pl. 36: 3, p. 248)

Leucorrhinia patricia Walker, Can. Ent., 72: 12, 1940a.

Similar to *L. hudsonica* but slightly smaller and with a much more slender abdomen, which lacks the dorsal yellow spots posterior to segment 3.

Male. Occipital triangle and vertex black; clypeus and labrum creamy yellow. Postclypeus with a black lateral margin, labrum with free margin very narrowly black; labium and rear of head black, the latter with a small yellow spot above the middle near the outer margin. Prothorax black, anterior lobe with front margin yellow and middle lobe with a pair of median yellow spots; posterior lobe with a fringe of long brownish hairs. Pterothorax black, variegated with yellow spots as in *L. hudsonica*. Legs black. Wings hyaline, the dark brown basal spots as in *hudsonica*, i.e., a small spot in the subcostal space of each wing and a triangular spot in the cubital space extending to the anal area and being much larger on the hind wing than on the front wing. Venation black with the following parts whitish: antenodal cross-veins of both series, nodus, edge of costal vein distal to nodus and a patch of veins distal to pt., including costal vein and several of the parallel longitudinal veins behind it; pt. dark fuscous.

Venation as in *hudsonica;* sectors of arculus in fore wing united for a longer distance than is usual in *Leucorrhinia;* triangle of fore wing crossed, the costal side scarcely three-fourths as long as the proximal side; triangle of hind wing free; no additional cubito-anal cross-veins. Abdomen of same length as in *hudsonica* but more slender, black with yellow markings confined to segs. 1 to 3 and very narrow; seg. 1 with small lateral spot; seg. 2 marked with yellow dorsally and laterally except in front of transverse suture and a median dorsal extension of this area behind the suture and tapering to an acute apex.

Sup. apps. (pl. 34: 3, p. 238) about as long as seg. 9, slightly incurved, subequal, with ventral denticles and an ante-apical ventral tooth; apices viewed from above convergent, in lateral view arched, with short apices slightly tilted. Inf. app. (pl. 34: 2, p. 238) two-thirds the length of sup. apps., distally emarginate. Hamuli (pl. 34: 10, p. 238; pl. 35: 5, p. 242) like those of *hudsonica* but with inner branch a little shorter and more abruptly recurved; outer branch smaller, sharply recurved mesad and cephalad, its outer angle prominent but not produced into a definite process as in *hudsonica*. Genital lobes a little shorter than in *hudsonica*.

Female. Face creamy yellow, labrum shining piceous, labium black, dorsum of frons, vertex and most of occipit black, the latter brownish behind with a tuft of long tawny hairs; rear of head also black with pale brownish hairs. Prothorax black with a double median yellow spot and the median part of the hind margin light yellow; hind lobe with long tawny hairs. Pterothorax black with yellow markings as in male, much more clearly defined than in *hudsonica*; antehumeral stripes nearly divided in the upper fourth; inter-alar tubercles and spots at bases of wings clear yellow. Wings hyaline, costal veins brownish proximally, becoming more yellowish beyond the nodus, clear yellow just beyond the stigma; vein R1 at this level is also touched for a short distance with light yellow. Pt. fuscous, the hind margins slightly convex. Dark brown basal spots much as in *hudsonica;* fore wing with two dark dashes, the second larger, occupying the cubito-anal space as far as the cross-vein, and also the first anal cell; hind wing with an anterior spot occupying basal cell as far as first cross-vein, and a large triangular posterior spot, whose anterior margin extends halfway or more to the base of the triangle. Abdomen black with yellow markings as follows: seg. 1 with lateral spot and a transverse dorsal marginal streak; seg. 2 with a thick L-shaped lateral spot and a large dorsal spot, acute behind; seg. 3 in front of transverse carina and dorsally, behind the carina, in the form of a rounded spot, pointed behind; segs. 4, 5 and 6 each with a median narrow dorsal spot or streak, narrowing caudad, sometimes vestigial on 6; no yellow or reddish spot on tergite of 7. Abdomen a little shorter and stouter than in the male. Scales of vulvar lamina (pl. 36: 3, p. 248) shorter than in *hudsonica*, about one-fourth as long as the lateral margin of seg. 9, and separated by a distance about equal to their length.

Venation. Anx ♂ ♀ 7/5–6; pnx ♂ 7/6–8, ♀ 6–7/7; rows of cells subtended by Rspl 1; ankle cells in anal loop 1 to 4; two rows of post-trigonal cells in fore wing.

Measurements. Total length ♂ 25.0–28.5, ♀ 24.5–27.5; abd. ♂ 17–21, ♀ 18–19; h.w. ♂ 18.5–22.0, ♀ 19–25.

Nymph. Unknown (1972).

Habitat and range. Peat-bog ponds. The known range is restricted to Canada n. of the United States from Lake Mistassini and Mont Tremblant, Que., Fort Albany and Fort Severn, Ont., to Great Slave Lake, N.W.T., and Otter Lake, Yukon T. It will probably be found in Alaska.

Distribution in Canada. Que.—Mont Tremblant; near Lake Mistassini. *Ont.*—Renfrew county (Furtado, 1973); Thunder Bay (Orient Bay, Lake Nipigon), Cochrane (Smoky Falls; Ft. Albany) and Kenora (Patricia Portion: Borthwick Lake (type locality); Cape Henrietta Maria; Ft. Severn) districts. *Man.*—Gillam; Churchill. *N.W.T.*—Great Slave Lake (Outpost I.; Gros Cap;

PLATE 36

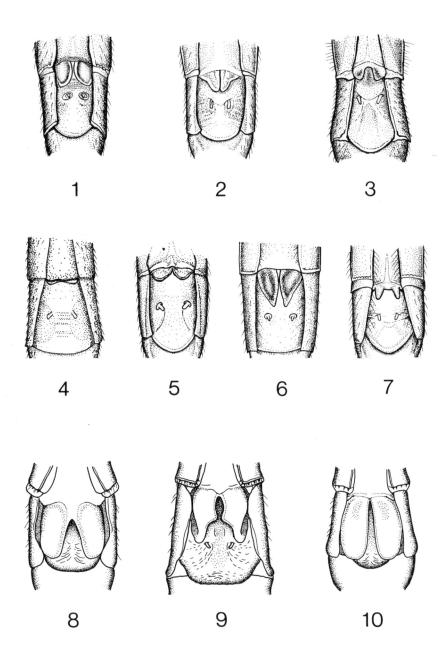

1 2 3

4 5 6 7

8 9 10

Pearson Pt.); Norman Wells; Ft. McPherson. *Yukon T.*—Laforce Lake; Otter Lake.

Field notes. This species flies from mid-June to early August, most records being for July. Tenerals were taken at Norman Wells, Northwest Territories, on June 16 (this also happens to be the northernmost record for *L. patricia*) and at Borthwick Lake, Kenora district, Ontario on June 24. A late recorded date is August 7 at Pearson Point, Great Slave Lake, Northwest Territories. Near Lake Mistassini, Quebec, where males had been mature by July 12, no specimens were seen after mid-August (Robert, 1963).

In Kenora district this species was found at a typical bog or muskeg lake with water of rather high acidity, a floating margin and an innermost zone of *Typha* and *Equisetum.*

Robert (1963) regards *L. patricia* as the "most boreal" of our species of *Leucorrhinia.* The males "wander over moist ground, perch at the tip of some branch, watch the approaching naturalist and often dart away before the net comes near."

Leucorrhinia glacialis Hagen. (Pl. 34: 11, p. 238; pl. 36: 4, p. 248)

Leucorrhinia glacialis Hagen, Trans. Amer. Ent. Soc., 17: 234, 1890a.
Leucorrhinia glacialis: Needham, Bull. N.Y. State Mus., 47: 518, 1901.

A dainty little *Leucorrhinia,* with a ruby-red thorax and jet-black abdomen when alive and mature. The red colour fades rapidly to brown after death.

Male. Labium and vertex black; face pale greenish white; occiput brown. Thorax reddish brown marked with black on sides; black stripe on front narrowed above to a broad triangle; a long red stripe separating antehumeral and humeral black bands in their lower half. Legs black. Wings hyaline with a small brown triangular spot at the base of the hind wing and a trace of the same colour at the base of the fore wing; pt. reddish brown. Abdomen reddish brown on segs. 1 to 3, otherwise shining black. Anal apps. black.

Sup. apps. in lateral view with acuminate tips and about 6 small, equally spaced teeth on ventral surface from near base to two-thirds the length of each sup. app. Hamuli with inner branch abruptly curved at tip (pl. 34: 11, p. 238).

Female. Broadly yellow where the male is red. Wings more flavescent in basal half. Pale spot on tergite of seg. 7 narrow, elongate, at least three times as long as wide. Scales of vulvar lamina (pl. 36: 4, p. 248) extremely short, being little more than swellings on the posterior margin of seg. 8.

Venation. Anx ♂ 7–8/6, ♀ 7–8/6–7; pnx ♂ 7 –10/7–9, ♀ 6–8/7–8; rows of cells subtended by Rspl 2; ankle cells in anal loop 1 to 3; three rows of post-trigonal cells in fore wing.

PLATE 36

Leucorrhinia (1–7) and *Tramea* (8–10)—vulvar lamina, ventral view: (1) *L. borealis*; (2) *L. hudsonica*; (3) *L. particia*; (4) *L. glacialis*; (5) *L. proxima*; (6) *L. frigida*; (7) *L. intacta*; (8) *T. carolina*; (9) *T. lacerata*; (10) *T. onusta*.
(1 after Walker, 1943; 2 after Walker, 1942a; 8–10 after Gloyd, 1958.)

Measurements. Total length ♂ 34.8–37.0, ♀ 34.0–36.1; abd. ♂ 23.8–25.1, ♀ 22.1–24.4; h.w. ♂ 27.1–28.2, ♀ 26.1–28.0 (NW ♂♀ 26–29).

Nymph. We first obtained the nymph of this species in bog-ponds at Prince Rupert, British Columbia in 1926. At that time we remarked (Walker, 1927) that it differed markedly from Needham's (1901) original description and figures. The nymph is described here for the first time, from an exuvia associated with a male adult that died during emergence (Prince Rupert, June 25, 1926, E. M. Walker). The discovery that this exuvia is different from the specimen Needham assigned to *L. glacialis* means that the key we published earlier (Walker, 1916b) requires revision.

Among the known North American *Leucorrhinia* only this species, *L. borealis* and *L. hudsonica* have 3 longitudinal dark bands on the venter of the abdomen and lack dorsal hooks on abd. segs. 7 and 8. *L. glacialis* resembles *L. borealis* in its larger body size and in the shape of the lateral spines on abd. segs. 8 and 9 (pl. 44: 2, p. 272); *L. glacialis* differs from *L. borealis* in the relative lengths of these lateral spines, in this respect being intermediate in position between *L. borealis* and *L. hudsonica*.

In general appearance of body, shape of head and setae on posterolateral margins of head, closely similar to *L. borealis*. Prementum of labium subtriangular, the median lobe bluntly obtusangulate; premental setae 14, the 6th from the outside the longest; palpal setae 10; distal margin of palpus with broad, shallow crenations each with 1 spiniform seta. Abdomen elongate ovate, broadest at segs. 6 or 7; dorsal hooks completely absent; lateral spines on segs. 8 and 9, their outer margins following the curved outline of the abdomen, their inner margins nearly parallel; the spines on seg. 8 about two-fifths or slightly less the lateral length of that segment (not including the spine); those on seg. 9 about one-half that segment's length; lateral spines on 9 almost, but not quite, extending posteriorly to tips of cerci; epiproct triangular, keeled and somewhat spinulose on distal half above; the relative lengths of cerci, epiproct and paraproct are 1:1.8:2.0. Apart from the 3 ventral longitudinal dark bands on the abdomen, no cuticular pattern remains evident on the exuvia.

Length 20.8; abd. 12.6; w.abd. 7.1; h.w. 5.8; h.f. 5.5; w.hd. 5.2.

Habitat and range. Bog lakes and marshes. Occurs in the Canadian and Transitional zones, ranging from Nfld., Me. and Mass., w. through Que., Ont., Mich., Wis., Minn. and the Prairies to B.C.; in w. its s. limits are Calif., Nev., Wash. and Wyo. No records exist for Alaska but it probably occurs there.

Distribution in Canada. Nfld.—Doyles (Codroy Valley). *N.S.*—Annapolis, Halifax and Inverness counties; Cape Breton I.; St. Paul I. *N.B.*—Charlotte and York counties. *Que.*—Covey Hill; Wakefield; Kazubazua; Mont Tremblant Park (Mallard Lake); Saguenay River (Cap Jaseux); Mt. Lyall (near headwaters of Cascapédia River); Gaspé; Godbout. *Ont.*—Middlesex, Norfolk, Peel and Carleton counties; Muskoka, Nipissing (incl. Alonquin Park and Lake Timagami), Algoma, Sudbury, Thunder Bay, Cochrane (Smoky Falls; Lake Abitibi) and Kenora (incl. Patricia Portion: Favourable Lake; Borthwick Lake) districts. *Man.*—Aweme. *Sask.*—Prince Albert National Park (Crean Lake and Kingsmere Lake); Tazin River. *Alta.*—Near Red Deer. *B.C.*—Salmon Arm district; Chase; Horsefly; Quesnel; Prince George (Summit Lake); Prince Rupert; Terrace; Vancouver I. (Loon Lake; Nanaimo district; Forbes Landing).

Field notes. The flight period may extend from the last week in May to the third week in August. Full-grown nymphs were collected in the Kenora district, Ontario, on June 7. In the Laurentides this is one of the first dragonflies to emerge in June (Robert, 1963). Early and late dates are May 31 (Prince Albert National Park, Saskatchewan) and August 19 (Sudbury, Ontario) and 22 (Forbes Landing, British Columbia).

Nymphs in the Kenora district were in a typical bog or muskeg lake with water of rather high acidity, a floating margin, and the usual bog vegetation with an innermost zone of *Typha* and *Equisetum*. A male at Godbout, Quebec, where the species is rare, was captured in an open wood near the pine barrens, about a mile from the nearest lake.

Needham (1901) observes that "during the first week or two of adult life, before age and pruinosity have obscured its remarkably fine coloration, it is a singularly beautiful insect."

Leucorrhinia proxima Calvert. (Pl. 34: 4, 5, 12, p. 238; pl. 35: 6, p. 242; pl. 36: 5, p. 248; pl. 44: 4, p. 272)

Leucorrhinia proxima Calvert, Trans. Amer. Ent. Soc., 17: 38, 1890.

A small black and brown species having the lighter coloration somewhat more extended, and the thoracic hair somewhat less dense than in our other species of *Leucorrhinia*.

Male. Labium black with pale spots; labrum, clypeus and frons white; anteclypeus sometimes brown; vertex and occiput black, hairy. Thorax dark brown, conspicuously fringed with long, white hair; pale stripes on front converge at upper ends to form a truncate squarish mid-dorsal black stripe; at lower ends they are confluent posteriorly with the pale stripe between antehumeral and humeral black stripes; first and third lateral stripes complete and joined by an irregular oblique black band giving the form of a letter "N" on the side. Legs black. Wings hyaline; pt. tawny; the hinder of the two basal brown streaks on hind wing well developed behind along membranule. Abdomen with basal segments swollen, redder dorsally than in our other species of *Leucorrhinia*, thinly pruinose on venter in old individuals; narrow mid-dorsal spots of yellow or red reduced to cover little more than part of carina on segs. 4 to 7; segs. 4 to 10 otherwise black. Anal apps. black.

Sup. apps. (pl. 34: 5, p. 238) blunt and slightly divergent at tips; straight, somewhat club-shaped; scarcely arched in their basal half; with 9 to 11 minute denticles on ventral surface. Inf. app. (pl. 34: 4, p. 238) with lateral margins more or less parallel and posterior margin with a wide, smooth, U-shaped notch. Hamuli (pl. 34: 12, p. 238; pl. 35: 6, p. 242) with outer branch lacking an anteriorly directed process.

Female. Generally similar to male. Labrum black with yellow lateral spots; labium black, spotted with white at sides. Vulvar lamina (pl. 36: 5, p. 248) with scales extremely short, only barely visible at the posterior margin of seg. 8, and appearing as two short, approximated tubercles.

Venation. Anx ♂ 7–8/5–6, ♀ 7–8/6; pnx ♂ 8–10/7–9, ♀ 8–10/8–10; rows of cells subtended by Rspl 1; ankle cells in anal loop 1 to 4; three rows of post-trigonal cells in fore wing.

Measurements. Total length ♂ 33.7–36.5, ♀ 31.3–34.0; abd. ♂ 23.0–24.3, ♀ 21.1–23.7; h.w. ♂ 26.1–28.9, ♀ 25.5–29.0 (NW: ♂ ♀ 24–27).

Nymph. Similar to *L. intacta* but differs in being slightly larger, and in having a shorter dorsal

PLATE 37

1

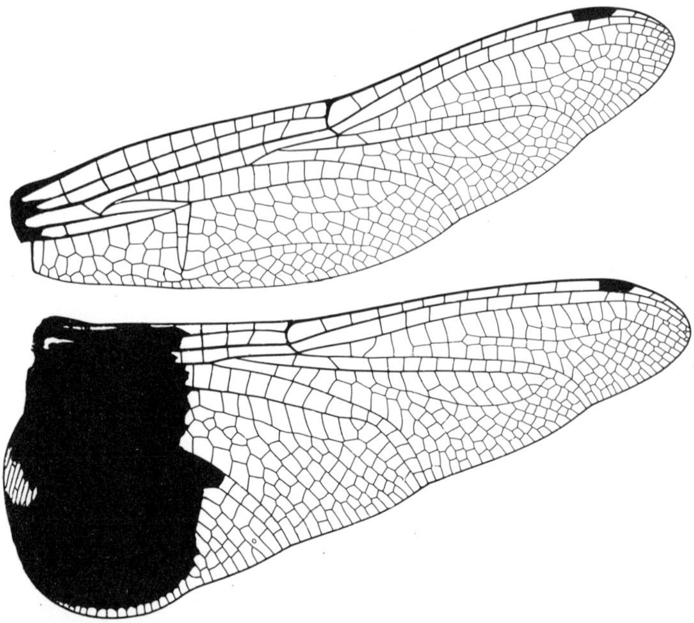

2

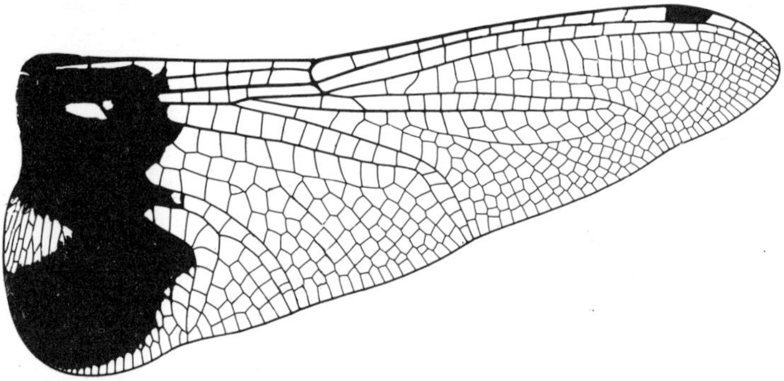

3

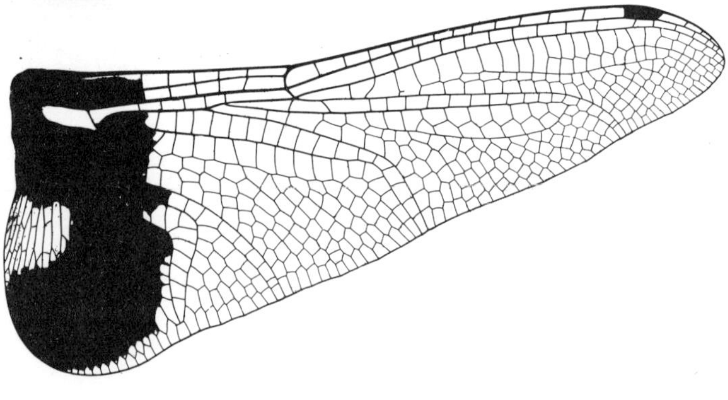

hook on abd. seg. 8, a smaller and less robust dorsal hook on seg. 2, and shorter and less divergent lateral spines on segs. 8 and 9 (pl. 44: 4, p. 272). Needham's (1901) description of *L. glacialis* clearly refers to this species.

Width of head across the eyes about twice the length (not including the labium); posterolateral margins broadly rounded and bearing numerous stiff setae. Median lobe of prementum bluntly obtusangulate; premental setae 11 to 15, the outer 4th to 6th the longest; palpal setae 10 or 11; distal margin of palpus with broad, shallow crenations each bearing 2 (occasionally 1) spiniform setae, one shorter than the other. Lateral spines on abd. segs. 8 and 9; those on seg. 8 about two-fifths to three-sevenths the lateral length of that segment (not including the spine); their outer edges continuing the regular curve of the abdominal margin, their general direction slightly convergent; those of seg. 9 about three-fifths to five-sevenths that segment's length, extending posteriorly to tips of cerci, their outer margins straight, and about in line with those of this segment, their general direction somewhat convergent; epiproct keeled above in distal half, basal width about three-quarters of the length; cerci about one-half the length of epiproct; paraprocts slightly longer than epiproct, with 4 or 5 small spiniform setae on their outer margins.

Abd. segs. with a row of pale spots on each side, nearly midway between the midline and lateral margins; these spots are surrounded by darker cloudings and behind each is a dark spot; they are obsolescent in front of seg. 6; lateral of these are irregular dark annuli with pale centres, and at the lateral margins of most of the segments the darker colour occupies the anterior half of the segment; lateral spines pale, tipped with brown.

Length 19–21; abd. 13; w. abd. 6.5–7.0; h.w. 6; h.f. 5–6.0; w. hd. 5.

Habitat and range. Still marshy or bog waters. Generally distributed through the Canadian zone, ranging from N.S., Me., N.H., Mass. and Pa. w. through Wis. and the Prairies to B.C., extending n. to N.W.T., Yukon T. and Alaska, and s. to Wyo. and Wash.

Distribution in Canada and Alaska. N.S.—Antigonish and Pictou counties; Cape Breton I. (Victoria county). N.B.—Charlotte and York counties. *Que.*—Covey Hill; Hemmingford; Wakefield; Alcove; Mont Tremblant (Lac des Flames); Lake Témiscouata; Saguenay River (Cap Jaseux); St. Anne; Godbout; Gaspé peninsula; Abitibi county. *Ont.*—Welland, Lincoln, Waterloo, York, Dufferin, Simcoe, Victoria, Hastings, Prescott and Carleton (Mer Bleue) counties; Muskoka, Parry Sound, Nipissing (incl. Algonquin Park), Algoma, Sudbury, Thunder Bay, Cochrane (n. to Fort Albany) and Kenora (incl. Patricia Portion: n. to Cape Henrietta Maria) districts. *Man.*—Aweme; Onah; Winnipeg Beach; The Pas; Gillam. *Sask.*—Prince Albert National Park (Waskesiu Lake and Crean Lake). *Alta.*—Banff; Red Deer; Nordegg; Jasper; near Flatbush. *B.C.*—Vancouver (Stanley Park); Princeton; Aspen Grove; Osoyoos; Vernon; Crawford Bay; Kaslo; Revelstoke; Red Pass; Tatla Lake; Smithers; Atlin; Lower Post; Vancouver I. (Victoria; Loon Lake). *N.W.T.*—Fort Smith; Great Slave Lake (Caribou Is.; Christie Bay; Gros Cap; Outpost I.; Pearson Pt.); Ft. McPherson. *Yukon T.*—Watson Lake; Whitehorse; Ross River; Swim Lakes. *Alaska*—Anchorage.

PLATE 37

Wings of *Tramea*: (1) *T. carolina*; (2) *T. lacerata*; (3) *T. onusta*.

PLATE 38

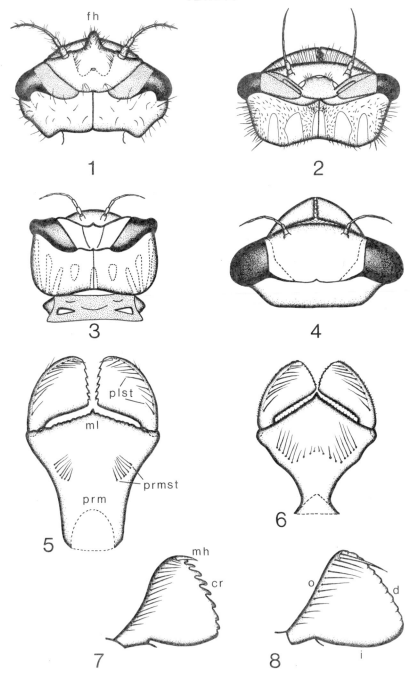

Field notes. Adults are most numerous in July. In the west the flight period extends from the second week in May to the third week in August (Whitehouse, 1948). In Quebec it begins at the end of May in the south, but seldom before mid-June in the Laurentides, and lasts until mid-August (Robert, 1963). Teneral adults were encountered at Borthwick Lake, Kenora district, Ontario on June 18 and 24; and on July 18 at the Lake of the Woods, Kenora district, the old appearance of adults showed that the flight period was nearly over. Late recorded dates are August 10 (Pearson Point, Great Slave Lake, Northwest Territories) and 22 (Forbes Landing, British Columbia).

At Godbout, Quebec, in July 1918 this was the most abundant species of *Leucorrhinia*. It was common over a large bog at the head of a small lake behind the pine barrens. Few individuals of *L. proxima*, or of *L. hudsonica* with which it was flying, were visible as long as the sun was obscured. When the weather cleared, however, they seemed to appear in an instant.

In the Lake Abitibi region of northern Ontario we have taken a female *L. proxima* in copula with a male *L. hudsonica*.

Leucorrhinia frigida Hagen. (Pl. 34: 6, p. 238; pl. 35: 1, 7, p. 242; pl. 36: 6, p. 248; pl. 44: 5, p. 272)

Leucorrhinia frigida Hagen, Trans. Amer. Ent. Soc., 17: 231, 1890a.
Leucorrhinia frigida: Muttkowski, Bull. Wis. Nat. Hist. Soc., 6: 112, 1908.

A small, delicate, blackish species with little or no discernible colour pattern even in teneral specimens. A grayish pruinosity at the base of the abdomen is a characteristic of mature males.

Male. Labium, vertex and occiput black. Thorax brown, becoming black with age, invested with a thin covering of whitish hair. Basal dark spot on hind wing extending far enough to cover first antenodal cross-vein but not reaching triangle; costal margin yellowish; pt. brown. Abdominal segs. 2 to 4 swollen, grayish pruinose dorsally; mid-dorsal pale spots lacking or reduced to short, narrow lines on segs. 5 to 7 or 6 to 7. Anal apps. black.

Inf. app. (pl. 34: 6, p. 238) slender at tip, slightly bent, lateral margins converging posteriorly; at least twice as wide at base as at tip.

Female. Similar to male in colour pattern and form. Wings tinged with brown out to triangles.

PLATE 38

Libelluloidea nymphs—head, dorsal view: (1) *Macromia*; (2) *Libellula quadrimaculata*; (3) *L. lydia*; (4) *Pachydiplax longipennis*.
Libellulidae nymphs—prementum and palpi of labium, dorsal view when flattened: (5) *Libellula* (representative of *L. julia* and *L. lydia*); (6) *Libellula* (representative of remaining Canadian species).
Libellulidae nymphs—left labial palpus, dorsal view when flattened: (7) *Pantala*; (8) *Tramea*.
(1, 4–8 after Wright and Peterson, 1944; 2 after Gardner, 1954; 3 after Levine, 1957.)
(Abbreviations: cr—crenations; d—distal margin; fh—frontal horn; i—inner margin; mh- —movable hook; ml—median lobe; o—outer margin; plst—palpal setae; prm—prementum; prmst—premental setae.)

Abdomen with a yellow streak on dorsal carina of segs. 6 and 7. Vulvar lamina (pl. 36:6, p. 248) comprising two prominent obovate-triangular scales, at least twice as long as their basal width, and extending posteriorly to about one-half the length of seg. 9.

Venation. Anx ♂ 7–9/6–7, ♀ 7–8(10)/6 (NW: ♂♀ hind wing 5–6); pnx ♂ 6–8/7–10, ♀ 7–9/8–10; rows of cells subtended by Rspl 1; ankle cells in anal loop 1 to 4; two rows of post-trigonal cells in fore wing.

Measurements. Total length ♂ 29.9–31.1, ♀ 27.4–29.4; abd. ♂ 18.9–20.6, ♀ 18.0–19.6; h.w. ♂ 23.5–25.1, ♀ 22.9–24.5 (NW: ♂♀ 21–24).

Nymph. Very similar to *L. intacta* but smaller and with more slender legs, more prominent eyes, and lateral spines on abd. seg. 9 extending further posteriorly.

Premental setae 10 to 13, the 4th or 5th from the outside the longest, and the inner 4 smaller than the others; palpal setae 9 or 10; distal margin of palpus with very shallow crenations, each with 1 spiniform seta. Abdomen (pl. 44: 5, p. 272) broadest at seg. 6, scarcely narrowing on 7, slightly on 8 and abruptly on 9; dorsal hooks on segs. 3 to 8, those on 3 and 4 larger, less erect, and more curved than in *L. intacta*, and very slender; those on 5 to 7 with the apices sharp and extending straight back, reaching about the middle of the following segment; on 8 similar to those of the preceding segments but less elevated, directed straight back; lateral spines on segs. 8 and 9; those on seg. 8 about one-half to three-fifths the lateral length of that segment (not including the spine); those on seg. 9 about nine-tenths or slightly more the length of that segment.

On well marked specimens the coloration is so closely similar to that of *L. intacta* that description is unnecessary; usually, however, it is somewhat obscure, although the legs are always distinctly banded.

Length 15–16; abd. 9–10.6; w.abd. 6.0–6.8; h.w. 4.6–4.75; h.f. 4; w.hd. 4.7–4.8.

Habitat and range. Bog lakes and ponds, especially those with floating sphagnum. Evidently a glacial relict. Occurs from N.S. s. through New England States to Va.; extends w. through Que., Ont., Mich., Ind., Wis., Minn., Man., N. Dak. to B.C.

Distribution in Canada. N.S.—Annapolis, Halifax and Queen's counties. *Que.*—Wakefield; Nominingue; Mt. Lyall (near headwaters of Cascapédia River). *Ont.*—Essex (Pt. Pelee), Kent, Waterloo, Peel, York, Simcoe, Victoria, Lennox and Addington, and Renfrew counties; Muskoka, Parry Sound, Nipissing (incl. Algonquin Park), Algoma, Sudbury and Thunder Bay district. *Man.*—Teulon. *B.C.*—Prince Rupert.

Field notes. Adults of *L. frigida* have been encountered from the end of May until the third week in August. The peak of the flight period falls in July. In Ontario tenerals were seen at Snelgrove, Peel county, from May 30 to June 29, and at Go Home Bay, Muskoka district, from June 1 to 24. The latest date recorded in Ontario (Go Home Bay) is August 19; but in Quebec adults disappear abruptly at about the end of July (Robert, 1963).

At Go Home Bay this is one of the common species that are partial to the edges of sphagnum bogs bordering small lakes and ponds (Walker, 1915). Borror (1940) remarks that it "flies low, in a very erratic fashion, through marsh grasses and rushes."

Leucorrhinia intacta Hagen. (Pl. 27: 2, p. 188; pl. 34: 7, p. 238; pl. 35: 2, 8, p. 242; pl. 36: 7, p. 248; pl. 44: 6, p. 272)

Diplax intacta Hagen, Syn. Neur. N. Amer., p. 179, 1861.

Leucorrhinia intacta: Hagen, Trans. Amer. Ent. Soc., 17: 235, 1890a.

A small blackish species, common in farm ponds, and recognizable by the conspicuous yellow or reddish spot on abdominal segment 7.

Male. Labrum yellow; face white with short, sparse hairs; labium black with a large squarish pale spot on each palpus; vertex and occiput black. Thorax thickly clothed with hairs, tawny in front and pale-tipped at sides; obscure pattern on thorax of wide, irregular, broadly confluent, black stripes; thorax becoming wholly black with age. Legs black, coxae lighter. Wings hyaline except for a very small spot at base of fore wing and one occupying 4 to 6 cells at base of hind wing. Abdomen black with pale basal segments and an isolated pale twin-spot on the tergite of seg. 7. An. apps. black.

Inf. app. (pl. 34: 7, p. 238) with lateral margins strongly divergent posteriorly and thus with a broadly forked posterior margin. Hamuli (pl. 35: 2, 8, p. 242) with outer branches curving around and meeting mesally above the inner branches; a pair of minute rounded spinulose lobes on rim of anterior lamina.

Female. Coloration similar to that of male except that the pale twin-spot on seg. 7 is preceded by smaller spots on segs. 4, 5 and 6. Vulvar lamina (pl. 36: 7, p. 248) not platelike but represented by two long, slender scales that lie separated from each other from their base and by a distance almost equal to their own length.

Venation. Anx ♂ 7–8/5–6, ♀ 6–8/5–6; pnx ♂ 7–10/7–10, ♀ 6–9/6–9; rows of cells subtended by Rspl 1; ankle cells in anal loop 1 to 4; two or three rows of post-trigonal cells in fore wing.

Measurements. Total length ♂ 30.0–36.2, ♀ 31.6–33.2; abd. ♂ 19.8–23.1, ♀ 20.6–21.5; h.w. ♂ 24.2–28.8, ♀ 25.2–27.0 (NW: ♂ ♀ 23–25).

Nymph. Differs from *L. frigida* in its greater size, less prominent eyes, and relatively shorter lateral spines on abd. seg. 9; and from *L. proxima* in its longer dorsal hook on seg. 8, larger and more robust dorsal hook on seg. 2, and longer and more divergent lateral spines on 8 and 9 (pl. 44: 6, p. 272).

Premental setae 12 to 14; the 6th from the outside the longest; palpal setae 10 to 12, sometimes with a minute basal setella; crenations on distal margin of palpus very shallow, each with 1, occasionally 2, spiniform setae. Dorsal hooks on abd. segs. 3 to 8, needle-like, those on 3 and 4 slightly arched, the rest more curved, and that on 8 lying almost parallel to the dorsum of abdomen; lateral spines on segs. 8 and 9; those on seg. 8 about one-fourth to two-fifths the lateral length of that segment (not including the spine); those on seg. 9 about two-thirds or slightly less that segment's length; lateral spines on 9 extend posteriorly beyond tips of cerci, but not as far as tips of epiproct or paraproct.

Venter of abdomen usually marked with transverse dark stripes or with rows of spots; dorsum also marked with dark spots and sometimes lateral dark stripes.

Length 14–19; abd. 10–13; w.abd. 6–7.0; h.w. 5; h.f. 4–5; w.hd. 5.*

*Measurements incorporate those of Needham (1901), Garman (1927), Needham and West-fall (1955) and Musser (1962) except those for length which include also records obtained by Kenneth J. Deacon from a sample of 86 living larvae collected near Guelph, Ontario. Deacon's measurements range from 14 to 17 with a well defined mode at 16 mm. Accordingly it is possible that nymphs from the north and east of the species' geographical range may be smaller than those from Utah, where Musser found certain dimensions to be greater (length 18–19 *vs.* 14–17.5; abd. 12–13 *vs.* 10).

Habitat and range. Marshy bays, ponds and slow streams; often in farm

PLATE 39

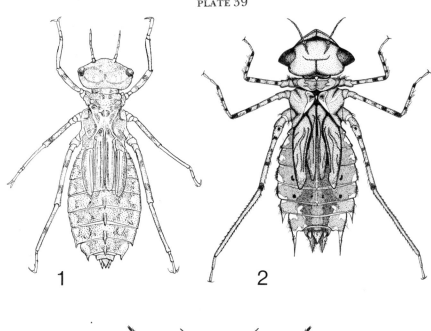

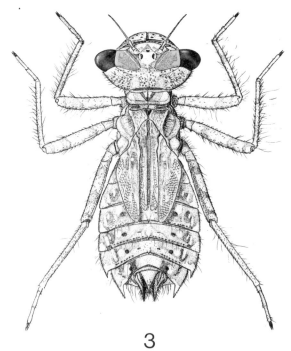

ponds. N.S. s. to Me., w. through Que., Ont., N.Y., Mich., Minn. and across the Prairies to B.C., the s. limit lying in Pa., Ky., Mo., Nebr., Utah and Calif.

Distribution in Canada. N.S.—Annapolis, Digby, Halifax, Pictou and Queen's counties; Cape Breton I. (Victoria county). *N.B.*—Charlotte, Sunbury and York counties. *P.E.I.*—Souris (New Harmony Road). *Que.*—Covey Hill; Hemmingford; near Ste. Chrysostome; St. Paul; Knowlton; Beauharnois; Jacques Cartier; Montreal; Hull; Wakefield; Alcove; Masham; Mt. Albert; Seven Islands *Ont.*—Essex (Pt. Pelee), Kent, Lambton, Elgin, Middlesex, Norfolk, Oxford, Brant, Welland, Lincoln, Wentworth, Waterloo, Halton, Wellington, Peel, York, Dufferin, Ontario, Durham, Prince Edward, Simcoe, Victoria, Peterborough, Hastings, Frontenac, Leeds, Grenville, Glengarry, Prescott, Carleton (incl. Mer Bleue) and Renfrew counties; Muskoka, Parry Sound, Nipissing, Algoma, Rainy River and Kenora (Sioux Lookout; Minaki) districts. *Man.*—Ninette; Stockton; Aweme; Winnipeg Beach; Victoria Beach. *Sask.*—Prince Albert. *Alta.*—Red Deer. *B.C.*—Vancouver; Chilliwack; Vedder Canal; Agassiz; Osoyoos; Oliver; Westbank; Creston; Kaslo; Crawford Bay; Okanagan district; Vernon; Enderby; Salmon Arm; Chase; Vancouver I. (Victoria; Elk Lake; Royal Oak; Prospect Lake; Alberni; Saanich district; Thetis Lake district; Florence Lake; Goldstream Creek; Wellington Lakes).

Field notes. Whitehouse (1948) gives the limits of the flight period as the first week in May and the third week in August. In eastern Canada adults appear nearer to the last week in May (an early record from Toronto, Ontario, is May 20). It is likely that emergence is protracted, since in British Columbia it has been recorded on June 24—seven weeks after the first adults were recorded. The main flight period is in June and July, and "usually only a few are left by the end of August" (Walker, 1941b). A teneral female captured at Lake Simcoe, Ontario, on August 26, 1907 may have been unrepresentative: it had abnormally dark flavescent wings.

At Lake Nipissing, Ontario, *L. intacta* was well represented at a small bog lake with no permanent outlet. The lake was surrounded by an open bog supporting a fairly dense growth of leatherleaf (*Chamae daphne calyculata*), wild rosemary (*Andromeda polifolia*), cranberries, pitcher-plants, and other typical bog plants (Walker, 1932).

Whitehouse (1941) observes that the male "will follow the ovipositing female, hovering some twelve inches above the water and slightly behind his mate, while she busies herself distributing eggs where the water plants touch

---

PLATE 39

Libellulidae nymphs: (1) *Perithemis tenera*; (2) *Celithemis monomelaena*; (3) *Sympetrum occidentale fasciatum.*
(2 after Leonard, 1934; 3 from Musser, 1962.)

PLATE 40

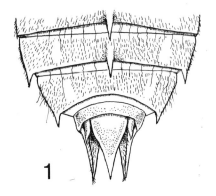

1

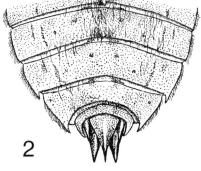

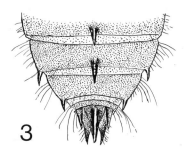

2

3

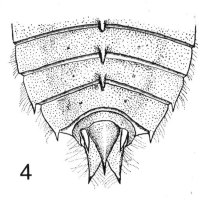

4

the surface. That duty finished for the time being . . . she will shoot perpendicularly twenty or thirty feet into the air." The male "seemed to miss this manoeuvre," remaining on the water and apparently searching for her.

Genus **Tramea** Hagen

Syn. Neur. N. Amer., p. 143, 1861. Type species *Libellula carolina* Linné.

These are large, handsome dragonflies with a worldwide distribution in tropical and warm temperate latitudes. They are exceptionally powerful fliers and several species are known to migrate, an activity for which the greatly enlarged basal area of the hind wing equips them well and makes possible the effortless gliding flight that, on a warm day, is so characteristic of adults of this genus and of *Pantala*.

The body is generally reddish or yellowish brown and there is a large, conspicuous blackish or reddish brown area of varying extent across the base of the hind wing. The latter feature has presumably provided the basis for their vernacular name: "saddle-bags."

The head is large and the eye-seam long; the legs are slender and bear large spines. The fore-wing triangle (pl. 37: 1, p. 252) is usually three-celled and is much further distad than is the hind-wing triangle; the proximal side of the fore-wing triangle is more than twice as long as is the anterior side. The fore-wing subtriangle is large and indistinct, comprising about four to six cells. The radial supplement subtends two rows of cells and turns forward to join IR3, causing the latter to become zigzagged at the junction. In the long anal loop of the hind wing the gaff joins the sole in a rounded curve at the heel. The hind wing has a single cubito-anal cross-vein. The pterostigma of the fore wing is distinctly longer than that of the hind wing. Vein R3 is smoothly curved. The anal area of the hind wing is traversed by many radiating veins that fork repeatedly before reaching the hind margin.

The nymphs are pale greenish with delicate brown markings; like those of many other migratory dragonflies they are active, usually dwell amongst vegetation near the surface, and are capable of very rapid growth. The head is large and the eyes prominent; the labium is large and the movable hook on the palpus is unusually long and straight (for a libellulid). The crenations on the distal margin of the palpus are shallow, and do not resemble those of Corduliidae. The abdomen lacks dorsal hooks and bears large lateral spines on segments 8 and 9, those on 8 being only slightly shorter than those on 9, and those on 9 extending posteriorly beyond the tips of the cerci; the lateral spines are flattened, usually curve mesad, and bear many large spiniform

---

PLATE 40

Libellula nymphs—terminal abdominal segments, dorsal view: (1) *L. luctuosa*; (2) *L. pulchella*; (3) *L. semifasciata*; (4) *L. vibrans*.

setae on their outer margins. The epiproct has a long acuminate tip but is distinctly shorter than the paraprocts.

Three of the six North American species have been recorded from Canada, and only from southwestern Ontario.

This genus is known to some authors as *Trapezostigma* but Gloyd (1972) has stressed the desirability of retaining the older and more widely used name *Tramea*.

Key to the Species of Tramea

Adults

1 Hamuli of male shorter than genital lobes of abdominal seg. 2; inf. app. less than one-half as long as sup. apps. Vulvar lamina of female short (about one-half as long as seg. 9) with a distinct marginal lobe or protuberance at mid-length on each side of the median cleft (pl. 36: 9, p. 248)   *lacerata* (p. 265)

Hamuli of male as long as or longer than genital lobes of abdominal seg. 2; inf. app. more than one-half as long as sup. apps. Vulvar lamina of female more than one-half as long as seg. 9 and without a lobe or protuberance on margin of median cleft (pl. 36: 8, 10, p. 248)   2

2 Hamuli of male only as long as or very slightly longer than the genital lobes of abdominal seg. 2. Vulvar lamina of female two-thirds as long as seg. 9 (pl. 36: 8, p. 248)   *carolina* (p. 262)

Hamuli of male about one-third longer than genital lobes of abdominal seg. 2. Vulvar lamina of female subequal in length to seg. 9 (pl. 36: 10, p. 248)   *onusta* (p. 268)

Nymphs

1 Paraprocts longer than epiproct; lateral spines on seg. 8 directed approximately straight posteriorly   *carolina* (p. 263)

Paraprocts shorter than epiproct; lateral spines on seg. 8 curved inwards (mesad)   2

2 Cerci about six- to seven-tenths as long as epiproct; antennal seg. 4 about one-half as long as antennal seg. 3   *lacerata* (p. 267)

Cerci about eight- to nine-tenths as long as epiproct; antennal seg. 4 about two-thirds as long as antennal seg. 3   *onusta* (p. 269)

Tramea carolina (Linné). (Pl. 36: 8, p. 248; pl. 37: 1, p. 252)

Libellula carolina Linné, Amoenit. Acad., 6: 411, 1763.
Tramea carolina: Hagen, Syn. Neur. N. Amer., p. 143, 1861.

A reddish brown *Tramea* with a black-tipped abdomen.

Male. Face warm buff below to shining dark brown or olive green above, darkening with age; labium with black and brown prementum; labial palpi reddish-brown; labrum and clypeus brown; frons above metallic blue to violet, in front mostly brown or black; vertex brown, broad, apex not concave, with a slight superior spot of metallic blue; occiput brown, hairy; postgenae brown.

General colour of body reddish brown; thorax uniformly reddish brown, without stripes and thickly clothed with tawny hair; caudal margin of the pronotum entire with few or no hairs; sides, including subalar carina a little paler reddish brown; brown (or metallic blue?) in pits of first and third lateral sutures; legs, except basally, black, the coxae, trochanters and basal half of femora brown; tarsal claws with teeth much beyond the middle; fore wing entirely hyaline or slightly yellowish or brownish at extreme bases; wings hyaline or uniformly light amber with reddish veins; hind wing (pl. 37: 1, p. 252) with a large, irregular, reddish brown area basally, usually covering the entire width of the wing from costa to posterior margin and extending for about one-third to one-half the distance between base and nodus, i.e., from base of wing to third or fourth antenodal cross-vein, proximal end of Mspl, to CuP beyond anal loop at ankle, and posterior to ankle to A_1 usually; two clear areas, one posterior to apex of gray membranule, extending from hind-wing margin to about A_3, the other in median space from base to a little beyond the arculus; anal angle of the hind wing much enlarged and with numerous small, narrow cells; in addition to generic venational characters: between veins CuP and 1A are two rows of cells for a distance along a notable concavity in gaff; anal loop with more extra cells than other species, these variously disposed betweeen its two regular marginal rows; stigma reddish or tawny.

Abdomen generally tawny to dark brown or even black; reddish brown dorsally, olivaceous on sides of enlarged basal segments, black on lateral margins and beneath, and deeper black across dorsum of segs. 8 and 9; 10 paler.

Anal apps. paler, brown, with black tips; sup. apps. 3.75 mm. long, about as long as 8 + 9; inf. app. about one-half as long; hamuli without lateral lobes and consisting entirely of hooks which are only as long as or slightly longer than the genital lobes.

Female. Differs from male in having front of frons and tip of vertex orange, with a broad, greenish black brow between the two; rear of vertex and occiput olivaceous; brown area of the hind wing enclosing a somewhat larger pale spot on the mesal margin; vulvar lamina (pl. 36: 8, p. 248) about as long as seg. 9 and bifid for about two-thirds its length.

Venation. Anx ♂ 11–14/7–8, ♀ 11–12/7; pnx ♂ 9–11/10–13, ♀ 8–11/9–12.

Measurements. Total length ♂ 46.5–50.3, ♀ 46.0–50.0; abd. ♂ 28.5–31.5, ♀ 30.2–32.8; h.w. ♂ 40.0–45.0, ♀ 42.5–45.0.

Nymph. Generally resembles *T. lacerata* but is larger and somewhat darker (Needham, 1901). Labrum with distinct brown median spot at upper margin. Antennal seg. 4 three-quarters to three-fourths as long as seg. 3. Premental setae 14 or 15, occasionally 13; palpal setae 10, occasionally 9. Abdomen without dorsal hooks; lateral spines on seg. 8 directed approximately straight posteriorly; epiproct shorter than paraprocts; cerci six- to seven-tenths as long as epiproct.

Length 25; abd. 15; w. abd. 19; h.w. 7–8; h.f. 7.5; w. hd. 7.5.

Habitat and range. Ponds and small lakes, especially those with rooted, submerged vegetation. A southern and eastern species, ranging from the Gulf of Mexico n. to s. Ont., N.Y. and Mass.; its recorded western limits are Wis., Iowa, Kans. and Tex. Adults disperse widely; thus some locality records, especially northern ones, probably do not indicate permanently resident populations.

Distribution in Canada. Ont.—Essex (Pelee I.; Pt. Pelee) and York (Toronto, High Park) counties.

Found only in the extreme southern part of Ontario (the Carolinian zone).

PLATE 41

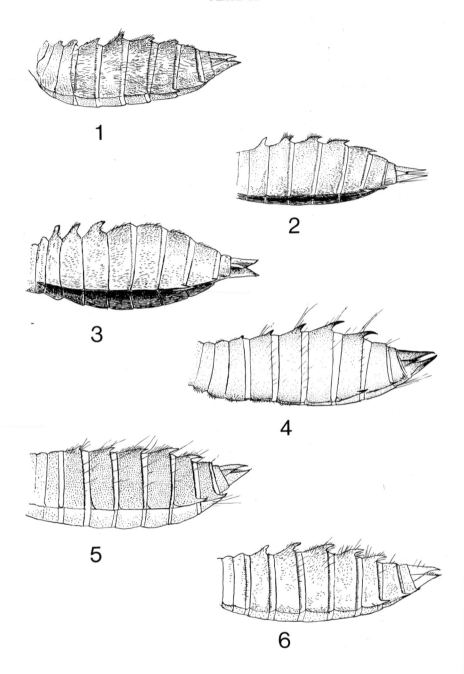

1

2

3

4

5

6

Field notes. Adults of *T. carolina* have seldom been seen in Canada. A fresh male was caught in Toronto on May 24 and specimens have been taken at Point Pelee on June 11 and 16, and in mid- and late August (Root, 1912; Corbet and Eda, 1969). It is inferred that in Ontario there are two periods of emergence (Walker, 1941b). In Ohio this species is 'comparatively common' in summer (Kellicott, 1899) and in Indiana the main flight period lies between late April and mid-June, with occasional adults being encountered up to late August (Montgomery, 1945).

This is principally a southern species in North America and little is known of its biology in Canada or the northern United States. In Florida, where it is common, Byers (1930) records that it is "fond of congregating on small trees and bushes growing along fences between cultivated fields. It will fly up and down the furrows of newly plowed fields, or remain poised for some little time at the tip of a small branch." *T. carolina* is one of the few Anisoptera to have been observed swarm-feeding at times other than at sunset or sunrise (Corbet, 1962). On June 12, 1942 several thousand individuals of this species were darting about in a limited area near New Smyrna Beach, Florida, and catching small insects in flight (Wright, 1944).

Bick (1950) saw many adults in tandem and ovipositing in a large pineland pond near Kiln, coastal Mississippi; and he obtained nymphs from standing water in mud-bottom ponds. In Staten Island, New York, a male *T. carolina* has been recorded as repeatedly releasing his tandem hold on a female and then promptly grasping her again (Davis, 1898).

Tramea lacerata Hagen. (Pl. 36: 9, p. 248; pl. 37: 2, p. 252)

Tramea lacerata Hagen, Syn. Neur. N. Amer., p. 145, 1861.

A large, broad-winged blackish dragonfly. The extensive dark-brown patches at the bases of the hind wings give it a flickering appearance in flight.

Male. Face pale yellow; labrum black; labium pale; frons above dark metallic violet; vertex and occiput black, the latter beset with thin, brown hairs.

Thorax walnut-brown with broad violet stripes on mid-dorsum, mesopleural and metapleural sutures; the whole with a metallic lustre and covered with long, fine, gray hairs; legs mostly black; wings with apical two-thirds at least clear, remainder coloured as follows: fore wing with (in least pigmented examples) a mere trace of brown at base to, at most, a basal blackish-brown streak in costal and subcostal spaces as far as second antenodal cross-vein, and another similar basal spot extending in cubital space to arculus, and posteriorly to apex of gray membranule; median space clear; hind wing (pl. 37: 2, p. 252) with large irregular blackish-brown spot covering entire basal third, reaching from costa to wing margin, or to last row of cells just anterior to posterior margin, and from wing base and hind margin distally to fourth antenodal cross-vein, tip of triangle, and bisector of anal loop posterior to ankle; two clear

PLATE 41

Libellula nymphs—abdomen, left lateral view: (1) *L. quadrimaculata*; (2) *L. julia*; (3) *L. lydia*; (4) *L. semifasciata*; (5) *L. incesta*; (6) *L. vibrans*.

PLATE 42

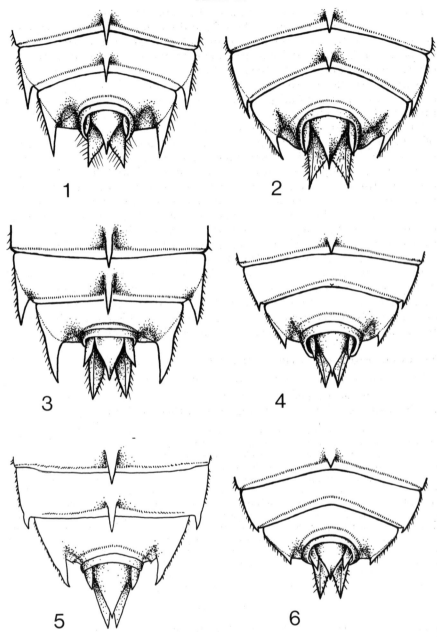

spaces, one at apex of membranule extending from hind wing margin distally for 5 to 7 rows of cells; the other, much smaller, and sometimes infuscated, in median space proximad and slightly distad of arculus; stigma black to tawny.

Abdomen black at maturity, with broad dorsal yellowish spots, somewhat obscure on middle segments, becoming broad cross-bands of brighter yellow on segs. 7 and 8; 9 black; 10 paler. Anal apps. blackish; sup. apps. at least 5 mm. long, as long as segs. 8 + 9, bent just before the middle; inf. app. not quite one-half the length of the sup. apps.

Female. Vertex buff; vulvar lamina (pl. 36: 9, p. 248) with mesal margins sinuate (in dried specimens), the median notch between the two plates extending to the base of the segment.

Venation. Anx ♂ 10–13/6–7, ♀ 9–13/6–7; pnx ♂ 9–10(12)/9–14, ♀ 8–10/10–11.

Measurements. Total length ♂ 45.0–49.2, ♀ 46.8–51.7; abd. ♂ 29.0–31.7, ♀ 30.3–34.8; h.w. ♂ 41.1–44.2, ♀ 42.6–48.5.

Nymph. Eyes placed laterally so that head is widest at posterior margin of eyes. Labrum without distinct brown median spot. Antennal seg. 4 one-half as long as seg. 3. Premental setae 14 or 15; palpal setae 10 or 11. Legs long and slender, pale except for 2 dark rings on each femur. Abdomen without dorsal hooks; lateral spines on seg. 8 curved mesad; epiproct longer than paraprocts; cerci six- to seven-tenths as long as epiproct.

Length 24–25; abd. 14–16; w. abd. 9; h.w. 9; h.f. 7.5–8; w. hd. 7.5.

Habitat and range. Open marshy lagoons and bays. Transcontinental in the southern United States and occurs in Mexico. Recorded n. limits are Wash., Nev., Utah, Nebr., Iowa, Wis., Mich., s. Ont and N.H. Probably does not overwinter in the nymphal stage in the northernmost localities where it has been recorded.

Distribution in Canada. Ont.—Essex (incl. Pelee I. and Pt. Pelee), Kent, Norfolk, Oxford, Brant, Welland, Lincoln, Waterloo, Halton, Peel, York and Simcoe counties.

A summer visitor to the Carolinian zone in southern Ontario.

Field notes. The numerous dates of capture of *T. lacerata* in Ontario fall clearly into two aggregates—one extending from the third week in May to perhaps the third week in July, and centred on the third week in June; and the other extending from perhaps the first week in August to the second week in October and centred on the second week in September. Whereas there is no reason to doubt that these aggregates reflect two periods of emergence (Walker, 1941b), it is likely that most, and perhaps all, of the adults encountered in Ontario in May and June flew into Ontario as sexually mature adults, having emerged further south. (Specimens taken at Rondeau Park on June 20 were fully mature.) The aggregate centred on September,

---

PLATE 42

Sympetrum nymphs—terminal abdominal segments, dorsal view: (1) *S. vicinum*; (2) *S. costiferum*; (3) *S. semicinctum*; (4) *S. danae*; (5) *S. ambiguum*; (6) *S. internum*. (1–4, 6 after Walker, 1917b; 5 after Wright, 1946a.)

by contrast, comprises teneral or fresh adults, some and perhaps all of which have emerged in Ontario. Thus the life-history of *T. lacerata* in Ontario may closely resemble that of the populations of *Anax junius* studied at Ste. Anne de Bellevue, Quebec (Trottier, 1966) and Caledon, Ontario (Trottier, 1971).

This species is known to travel widely. On September 20, 1952 a large flight of dragonflies passed Todd's Point on Long Island Sound near Old Greenwich, Connecticut. "The flight was in progress . . . at noon, and continued for the next two hours. Thousands of dragonflies passed the point during this period, all moving in a southwesterly direction; they passed in groups, and over a hundred were in sight at a time." Of the several species present, about 90 per cent were *Anax junius* and *T. lacerata* (Borror, 1953). At Cape May, New Jersey on September 22, 1945 *T. lacerata* was the principal dragonfly in "migrating hordes"; most of these adults were teneral or young, supporting the view that they migrate soon after emergence (Beatty, 1946).

When large numbers are aggregating there can be competition for roosting sites, such as that described by Corbet and Eda (1969) at Rondeau Park on August 26, 1968. "Immature adults of *T. lacerata* were alighting on insolated parts of trees at heights of 1–17 m. Sparsely leaved branches at 3–5 m. overhanging the beach were preferred. The dragonflies changed position frequently, partly perhaps because the distribution of "hot-spots" on the westfacing foliage was changing rapidly as the sun set, and partly because most adults tried to settle on relatively few branches, causing considerable interaction when an adult would alight on the thorax or abdomen of an earlier arrival. When displaced, adults flew off a distance of about 5 m. and approached the site again upwind and tried to alight again." At Rondeau Park, as on Pelee Island in mid-August 1910 (Root, 1912), *T. lacerata* was the main constituent of the large aggregation of dragonflies.

When flying in tandem the male of *T. lacerata* has been seen to release the female while she darts down to oviposit, and then to seize hold of her again as she rises; in accomplishing this the male sometimes starts by grasping her head with his feet (Montgomery, 1929) but at other times uses his anal appendages alone (Kennedy, 1917c). "This quick release and the almost immediate reclasping of the female was one of the most dexterous performances I had ever observed in dragonflies" (Kennedy, 1917c).

Nymphs of *T. lacerata* were taken in southern Utah from a spring pond containing watercress and other aquatic plants with trees and bushes surrounding the pond; but elsewhere nymphs may occupy ponds without surrounding trees (Musser, 1962).

Tramea onusta Hagen. (Pl. 36: 10, p. 248; pl. 37: 3, p. 252)

Tramae onusta Hagen, Syn. Neur. N. Amer., p. 144, 1861.

Resembling *T. carolina* but smaller and of a lighter tint of red.

Male. Face, including vertex and labrum, pale, becoming red with age; clad with short bristling black hairs; labium warm buff; frons and occiput greenish to rear margin of frons with a tinge of white hairs.

Front of thorax reddish or tawny to dark olive buff, with a crest above it of the same colour; rather thinly hairy; usually dull; sides paler, olivaceous, without black markings (or dark streaks only in the bottom of the lateral sutures; legs pale brown almost to knees and blackish thereafter; wings reddish-veined; stigma tawny; fore wing entirely hyaline; hind wing (pl. 37: 3, p. 252) with basal fourth reddish brown, extending from just above subcosta to last 2 to 3 rows above posterior wing margin, and from base to the second to third antenodal cross-vein, tip of triangle, and bisector of anal loop posterior to ankle; two clear areas, one at apex of membranule, large, extending from median third of hind margin distally to A_2; the other in median space from base of wing connecting distally with clear apical three-fourths by means of a clear space between R_1 and R_2–R_{4+5}; thus separating basal dark area into two regions, a narrow anterior one between roughly Sc and R_1, and a larger posterior one below median space and R_2–R_{4+5}.

Abdomen red-brown to tawny, olivaceus on enlarged basal segments with encircling red carinae; black spot on dorsum of segs. 8 to 10; reddish on dorsum of middle segments.

Anal apps. yellowish red with black tips; sup. apps. 3.5 mm. long, shorter than 8 + 9; inf. app. about one-half as long as superiors.

Female. Differs from male in having the frons and vertex olive buff.

Venation. Anx ♂ (10)11–12(13)/6–7(8), ♀ (10)11–12(13)/(6)7; pnx ♂ (8)9–11/10–12(13), ♀ 8–9/10–11(12).

Measurements. Total length ♂ 43.0–46.0 (NW: ♂♀ 41–49*); abd. ♂ 27.8–29.2 (NW: ♂♀ 29–34*); h.w. ♂ 38.2–39.1 (NW: ♂♀ 38–43).
*These measurements include the abdominal appendages.

Nymph. Generally resembles *T. lacerata* and is less pigmented than *T. carolina*. Labrum without distinct brown median spot. Antennal seg. 4 two-thirds as long as seg. 3. Premental setae 14 to 16; palpal setae 11 or 12 plus a setella. Abdomen without dorsal hooks; lateral spines on seg. 8 curved mesad; epiproct shorter than paraprocts; cerci about eight- to nine-tenths as long as epiproct.

Length 24; abd. 15; w. abd. 9; h.w. 7; h.f. 7.5; w. hd. 8.

Habitat and range. Ponds. Transcontinental in the southern United States and occurs in Mexico, s. to Panama, and in Cuba and Puerto Rico. Recorded n. limits are Nev., Utah, Nebr., Iowa, Ill., s. Ont. and N.C.

Distribution in Canada. Ont.—Essex county (Pt. Pelee) (June 30, 1931, 1 ♂, G. S. Walley). Occasional in the extreme s. of Ontario.

Field notes. T. onusta will probably be encountered only occasionally in Ontario, and presumably well within the limits of its flight period in Ohio —recorded by Borror (1937) as being May 7 and August 26.

Needham and Heywood (1929) have described the oviposition of *T. onusta* in considerable detail. "After copulation the pair, in a wild flight, come dashing downward toward the surface of the pond. When about a foot above the surface the male releases his hold on the head of the female and moves forward a little at that level, marking time, while the female descends and touches the water with the tip of her abdomen. As she rises, without a sign of effort he seizes her again. Her head seems to slip between his claspers with

PLATE 43

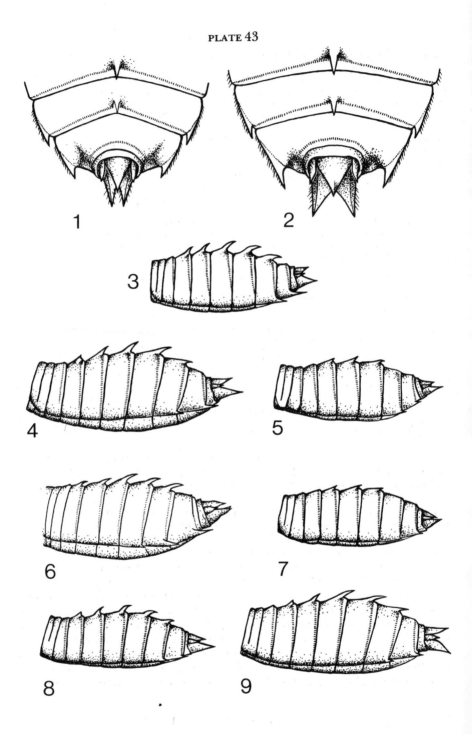

wonderful precision. Without the slightest delay they are coupled together and off on another bound. And this separation and recouplement is repeated at every descent." In Dallas county, Texas, Ferguson (1940) noted that an ovipositing female laid by preference on algal mats at the surface of the water.

Genus **Pantala** Hagen

Syn. Neur. N. Amer., p. 141, 1861. Type species *Libellula flavescens* Fabricius.

Members of this genus are large and robust, with the broad base to the hind wing and the powerful gliding flight that are the hallmarks of the wanderer and migrant. One species, *P. flavescens*, is the best-known migrant among Odonata and is the only dragonfly with a circumtropical distribution.

In this genus the body is brownish orange with a tendency to redden with age. The eye-seam is long and the vertex broad. The fore-wing triangle (pl. 27: 3, p. 188) is two-celled and much further distad than is the hind-wing triangle; the proximal side of the fore-wing triangle is more than twice as long as is the anterior side. The fore-wing triangle is large, indistinct and contains about four to five cells; in the fore wing an extra row of cells lies between the paranal and marginal rows. The anal loop is large and strongly sigmoid, with a low heel and long toe; vein A_2 is angled at the knee beside a very long patella. There are usually two cubito-anal cross-veins in the hind wing. Vein R_3 is strongly undulate; R_{4+5} and MA are bent abruptly at the end of the median supplement.

The nymphs are pale greenish with light brown markings; they are active and able to grow very rapidly. The eyes are prominent. The movable hook is short and robust; crenations on the distal margin of the median lobe are deeper than in most other libellulids and in this respect resemble those of corduliids. The abdomen lacks dorsal hooks and bears large lateral spines on segments 8 and 9, those on 8 being narrower and distinctly shorter than those on 9, and those on 9 not extending posteriorly as far as the tips of the cerci. The epiproct is at least as long as the paraprocts.

There are only two species in North America and both occur regularly in Canada, probably only as summer visitors.

PLATE 43

Sympetrum nymphs—terminal abdominal segments, dorsal view: (1) *S. obtrusum*; (2) *S. pallipes.*
Sympetrum nymphs—abdomen, left lateral view: (3) *S. vicinum*; (4) *S. costiferum*; (5) *S. danae*; (6) *S. ambiguum*; (7) *S. internum*; (8) *S. obtrusum*; (9) *S. pallipes.*
(1–5, 7–9 after Walker, 1917b; 6 after Wright, 1946a.)

PLATE 44

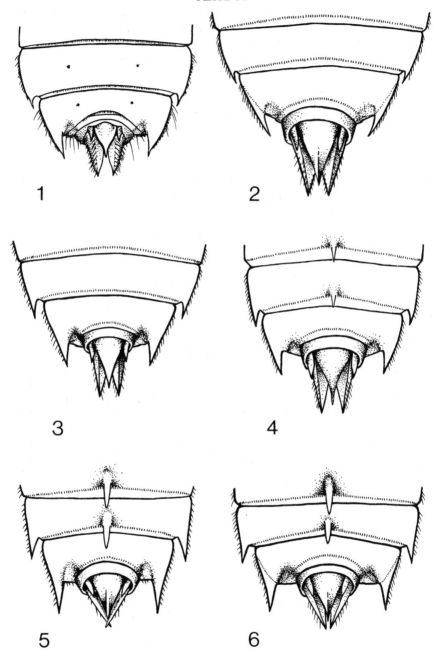

1   2

3   4

5   6

Key to the Species of Pantala
Adults

1 Hind wings with a large brown spot (minimum diameter 3 mm.) near
anal angle   *hymenea* (p. 273)
Hind wings without a brown spot near anal angle   *flavescens* (p. 275)

Nymphs

1 Width of base of lateral spine on seg. 9 more than one-third the length
of that spine; epiproct with rounded dorsal surface when viewed laterally (pl. 45: 5, p. 276)   *hymenea* (p. 274)
Width of base of lateral spine on seg. 9 one-third or less the length of
that spine; epiproct without rounded dorsal surface when viewed laterally (pl. 45: 6, p. 276)   *flavescens* (p. 275)

Pantala hymenea (Say). (Pl. 45: 5, 7, p. 276)

Libellula hymenea Say, J. Acad. Phila., 8: 19, 1839.
Pantala hymenea: Hagen, Syn. Neur. N. Amer., p. 142, 1861.

Closely similar to *P. flavescens* in appearance; distinguished from it by the
dark brown spot at the base of the hind wing.

Male. Similar to *flavescens*, but slightly grayer and browner in general coloration; labium, labrum, anteclypeus and a small area just in front of the ocelli yellow; prominence between the ocelli also yellow; vertex brown; face, including most of frons, pale becoming reddish with age; rear of head warm buff.

Thorax yellowish green or tawny, covered with long dense, tawny or white pubescence; blackish along sutures and above middle coxae; thorax with little pattern but sometimes on sides there are diffuse broad bands of brown on first and third lateral sutures and these may be connected at their lower ends; legs brown or black, with trochanters and at least basal third of femora pale, and tibiae of front and middle legs lined with yellow above; tarsi of all legs black; wings hyaline with tips slightly infuscated; stigma tawny or rufous bordered with black veins; brown spot (usually 5–6 mm. in diameter) on anal angle of hind wing.

Abdomen similar to *flavescens* in most respects; brown, narrowly cross-lined with black on supernumerary carinae of segs. 2 to 5, becoming blackened on sides of 6 and 7 and broadly so on dorsum of 8, 9 and 10. Mid-dorsal carinae on segs. 2 to 6, one transverse carina on tergum 2, three on terga 3 and 4, two on 5 and one each near the caudal margins of terga 6 to 9.

Anal apps. pale brown, the inf. app. about two-thirds as long as the sup. apps.

Female. Similar to the male but lacking the reddish colour of the face; brown spot of the hind wing slightly smaller than that of the male.

Venation. Anx ♂ 13–15/7–8, ♀ 14–15/6–8; pnx ♂ 6–9/8–11, ♀ 7–8/7–10.

PLATE 44

Pachydiplax nymph—terminal abdominal segments, dorsal view: (1) *P. longipennis*.
Leucorrhinia nymphs—terminal abdominal segments, dorsal view: (2) *L. borealis*; (3) *L. hudsonica*;
 (4) *L. proxima*; (5) *L. frigida*; (6) *L. intacta*.
(1 after Wright and Peterson, 1944; 2–6 after Walker, 1916b.)

Measurements. Total length ♂ 46.2–48.9, ♀ 44.1–49.0; abd. ♂ 28.5–30.5, ♀ 28.1–31.2; h.w. ♂ 41.5–43.4, ♀ 39.9–44.1 (NW: ♂♀ 40–45).

Nymph. Premental setae 15 to 18; palpal setae 15 (Musser (1962) records 15 to 18); movable hook twice as long as deepest crenations on distal margin of palpus. Legs long and slender, pale except for 2 broad dark rings on each femur; mid and hind tarsi dark. Minute needle-like dorsal hooks on abd. segs. 3 and 4; prominent, inwardly curved lateral spines on segs. 8 and 9, (pl. 45: 5, p. 276), the width at base of that on 9 being more than one-third its length; epiproct (pl. 45: 7, p. 276) robust, tapering abruptly at tip, and with a rounded dorsal surface, when viewed laterally.

Length 26–28; abd. 17–19; w. abd. 10; h.w. 8.0–8.5; h.f. 7–8; w. hd. 6.5.

Habitat and range. Small, standing waters, incl. temporary and artificial ponds. Transcontinental in southern United States; occurs in Mexico, Cuba and s. to Chile. Recorded n. limits are Nev., Man., Wis., Mich., s. Ont., s. Que. and Me. The Manitoba record is beyond its expected range.

Distribution in Canada. N.B.—Charlotte county (St. Andrews, Sept. 18, 1930, M. W. Smith). *Que.*—Saint Fulgence, Saguenay Region, August 13, 1967 (Fernet and Pilon, 1968b). *Ont.*—Essex county (Pelee I.: Pt. Pelee and York county (W. W. M. Edmonds). *Man.*— Husavick; Victoria Beach.

Occasional in the extreme south of Ontario; rare elsewhere.

Field notes. Adults have been taken in Canada over a wide range of dates—from June 11 to August 27 (Point Pelee). Fullgrown nymphs were collected at St. Andrews, Nova Scotia, on September 18 (Walker, 1933a), which supports the supposition that *P. hymenea* can breed in southwestern Ontario (Walker, 1941b). It remains to be discovered whether or not nymphs can survive the winter in Canada, and it may be that the progenitors of the summer populations comprise adults that have migrated from the south in spring. The adults emerging in Canada in late summer probably migrate southwards soon afterwards: *P. hymenea* was among the Anisoptera encountered at Long Point and Rondeau Provincial Parks and at Point Pelee at the end of August, 1968 (Corbet and Eda, 1969) and was one of the principal species on Pelee Island in mid-August, 1911 (Root, 1912).

This was the most interesting species that we encountered over ponds at Victoria Beach, Lake Winnipeg, Manitoba, at the end of June, 1930 (Walker, 1933b). "A small number . . . were flying back and forth over the water so swiftly that their capture seemed for some time to be practically impossible. In the end, however, three males and two females were netted, including an ovipositing pair. This pair flew erratically, dropping suddenly to the water, which the female would strike only once, the pair then swooping to another spot a few yards away, or off to the next pond. The other female captured was ovipositing unaccompanied. The males chased away other dragonflies from their beat, including the vigorous *L. pulchella* as well as other males of their own species." Many observers have commented on the great difficulty of capturing this powerful and agile dragonfly.

The ponds at Victoria Beach were shallow, with a firm sandy bottom. Nymphs of this species can complete development in less than five weeks in July in Oklahoma (Bick, 1951a) and have been found in pools both with and without vegetation.

Pantala flavescens (Fabricius). (Pl. 27: 3, p. 188; pl. 45: 6, 8, p. 276)

Libellula flavescens Fabricius, Suppl. Ent. Syst., p. 285, 1798.
Pantala flavescens: Hagen, Syn. Neur. N. Amer., p. 142, 1861.

A large reddish-orange dragonfly with hind wings that are unmarked and expanded at the base. Most often seen hovering a few metres above the ground over clearings in hot weather. The only cosmopolitan species of dragonfly.

Male. Labium warm buff; labrum, clypeus and frons yellow to pale buff; vertex yellow to pale buff on sides; occiput shining yellow to olivaceous, bare of hair; face, top of frons and vertex becoming red with age; face with a narrow, black cross stripe through the middle ocellus; rear of head brown and yellow.

Dorsum of thorax tawny yellow, amply covered with fine white or golden pilosity; shading into yellowish green at sides; black streaks in pits of both first and third lateral sutures and a black ring around spiracle; three black crescentic spots above leg bases; legs beyond basal segments blackish, coxae, trochanters, and basal third of femora tawny; tibiae sometimes lined with yellow; wings hyaline except for flavescent tinge on hind wing between A3 and wing margin; tips infuscated; stigma tawny to yellow.

Abdomen rather short and stout; segs. 1 to 4 yellowish green, buff above, swollen, conspicuous blackish transverse carina; segs. 5 to 10 tawny; tergum 2 with two carinae, 3 and 4 with three, 5 with one at middle and another at the caudal margin; dorsum of terga 4 to 8 with a faint brown spot, having lateral arms behind the middle of each spot; tergum 9 with a subcircular spot; dorsum of 10 black; sides of terga 2, 3 and 4 with short, bent, sometimes interrupted dashes or spots below the lateral carinae.

Sup. apps. yellow at bases, darker at tips, with a weak, obtuse, ventral angle a little beyond the middle and with a row of small teeth proximad of the angle.

Female. Colour similar to the male; shoulders of mes- and metepimera black in some specimens; ventral projection of seg. 9 keeled below, bluntly rounded at tip.

Venation. Anx ♂ 13–16/7–8, ♀ 13–15/6–8; pnx ♂ 6–9/8–11, ♀ 7–8/8–9.

Measurements. Total length ♂ 44.6–48.0, ♀ 46.9–48.5; abd. ♂ 28.0–30.1, ♀ 30.3–33.1; h.w. ♂ 39.2–42.9, ♀ 41.9–43.0 (NW: ♂♀ 36–42).

Nymph. Premental setae 15 (Musser (1962) records 17 to 19); palpal setae 12 to 14; movable hook less than twice as long as deepest crenations on distal margin of palpus. Legs long and slender, pale except for 2 broad dark rings on each femur; mid and hind tarsi dark. Abdomen (pl. 45: 6, p. 276) without dorsal hooks; prominent, inwardly curved lateral spines on segs. 8 and 9, the width at base of that on 9 being one-third or less its length; epiproct (pl. 45: 8, p. 276) robust, tapering evenly posteriorly, and without rounded dorsal surface, when viewed laterally. Length 24–26; abd. 15–18; w. abd. 8–9; h.w. 6–7.5; h.f. 7–8; w. hd. 7.

Habitat and range. Small, standing waters, incl. temporary and artificial ponds, sometimes only a few feet in diameter, and often with little or no rooted aquatic vegetation. An inveterate migrant and the only circumtropical dragonfly, occurring n. and s. of the equator to low temperate latitudes

PLATE 45

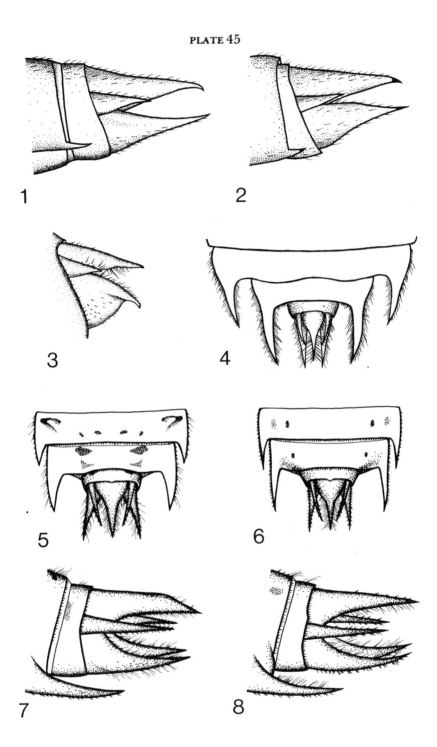

(about 40°) with summer strays extending the range temporarily. Found breeding on all continents except Europe. In North America it occurs n. to Calif. and Nebr. in w. and n. to Mich., s. Ont., s. Que. and Nfld. in e. The most northerly record on the continent is from Husavick, Man. (50°33' N., 97°00' W.), although the one from Cow Head, Nfld. approaches it closely, as does that from the Bouleau River estuary, Que.

Distribution in Canada. Nfld.—Cow Head (Valle, 1955). *Que.*—Ste. Anne de Bellevue (Trottier, 1967); Montreal; Sorel (Robert, 1963); Saguenay River (Cap Jaseux) (Fernet and Pilon, 1969); Bouleau River estuary (near Pigou River) (Alan Hayton). *Ont.*—Essex (Pt. Pelee), Lambton (Pt. Edward), Wentworth (Hamilton), Waterloo (Strasbourg), Durham (Bowmanville), Simcoe (De Grassi Pt., Lake Simcoe) and Carleton (Ottawa) counties. *Man.*—Husavick.

Field notes. In the Cap Jaseux region of the Saguenay River, Quebec, adults have been seen from July 13 to September 10 (Fernet and Pilon, 1969). However most if not all other records of Canadian specimens are for August and September, late captures being from Point Pelee (September 10) and De Grassi Point (September 12).

The observations of Trottier (1967) at Ste. Anne de Bellevue, Quebec make it extremely unlikely that *P. flavescens* can pass the winter at that latitude. Trottier saw emergence occurring from a small pond in late August and in September, and a mature female (that later laid eggs) flying over the pond on September 4. On October 18 the half-grown larvae that he found in the pond were all dead. Thus, whereas this shows that *P. flavescens* can complete a generation as far north as 45°25' in eastern Canada, the observation that larvae died in October, coupled with the knowledge that *P. flavescens* has a high thermal coefficient for larval growth (see Corbet, 1962), suggests that summer populations in Canada are maintained solely by migration from the south each spring.

P. flavescens appears in small numbers among the masses of Anisoptera that assemble at Point Pelee in the fall (Corbet and Eda, 1969). Between August 28 and September 7, 1973 Alan Hayton saw numerous adults "swarming" near the edge of spruce forest on the estuary of the Bouleau River, Quebec.

PLATE 45

Libellula nymphs—anal appendages, left lateral view: (1) *L. incesta*; (2) *L. vibrans*.
Erythemis nymph—anal appendages, left lateral view: (3) *E. simplicicollis*.
Tramea nymph: (4) terminal abdominal segments, dorsal view.
Pantala nymphs—(5, 6) terminal abdominal segments, dorsal view and (7, 8) anal appendages, left lateral view: (5, 7) *P. hymenea*; (6, 8) *P. flavescens*.
(3, 5–8 after Musser, 1962; 4 after Wright and Peterson, 1944.)

ADDENDA AND CORRIGENDA: VOLUMES I AND II

Here we record additions and corrections that have become necessary since Volumes I and II were published. It should be noted that these sections are not the product of exhaustive search or scrutiny.

ADDENDA

SYNONYMY

Aeshna canadensis Walker

As a result of work by Fernet, Pilon and Beique (1968), which should be consulted for details, the following entry must be added:
Anax maritimus Provancher, Nat. Canad., 22: 79, 1895.

DISTRIBUTION

Species new to Canada

Two species should be added to the list of dragonflies recorded from Canada, both on the basis of earlier records that had been overlooked.

Anax longipes Hagen
This magnificent insect was seen at the south end of Pelee Island, Ontario by F. M. Gaige in June, 1918 (Montgomery, 1937b). There is no reason to doubt this record: as Montgomery remarks, the insect's appearance is striking and distinctive; and a male was taken near Sandusky, Ohio (less than 25 miles from Pelee Island) in July, 1913.

Tachopteryx thoreyi (Hagen)
A male was collected by Provancher near Quebec City, Quebec and described by him under the genus *Petalura* (Provancher, 1877b). Fernet and Pilon (1968c), who discovered this, found the specimen to be well preserved and its identity to be unequivocal. This species occurs in the nearby states of New York and New Hampshire, and the record from Quebec slightly extends its known distribution northwards.

Extension of known geographical range

Listed here are records that constitute additions to either or both of the entries headed *Habitat and range* and *Distribution in Canada and Alaska*.

In this section we give prominence to new records from Canada or from contiguous states. In this connection it is appropriate to cite the relevant contributions made to a symposium on the distribution of the Odonata of the North Central Region of the United States held at East Lansing, Michigan in 1967 (see papers cited for 1968 by Corbet, Lawrence, Montgomery and Ries).

Aeshna canadensis Walker

Price (1958) has recorded this species from Ohio. Its known northerly limit in Alberta is now Flatbush (54°42' N.) (Pritchard, 1964b).

Aeshna clepsydra Say

In 1966 a male was captured at Lac Noir, Saint-Jean-de-Martha, Joliette county (Fernet and Pilon, 1968a); this is a new record for Quebec and slightly extends the known northerly limit of the species. *A. clepsydra* has now been recorded from Pennsylvania (Beatty and Beatty, 1968).

Aeshna eremita Scudder

The range now includes Washington (Paulson, 1970).

Aeshna interrupta interna Walker

The known northeastern limit includes Montana and Wyoming (Kormondy 1960).

Aeshna interrupta lineata Walker

The known northerly limit, previously Norman Wells (65°26' N.) is now Sagwon, Alaska (approximately 69°21' N., 148°16' W.) (Gorham, 1972). The southeasterly limit now includes Ohio (Price, 1958).

Aeshna juncea Linné

A new northerly record for Alberta is Flatbush (Pritchard, 1964). It has now been recorded from Washington (Paulson, 1970).

Aeshna palmata Hagen

The known eastern limit in North America includes Montana and Wyoming (Kormondy, 1960).

Aeshna septentrionalis Burmeister

The capture of a male on Herschel Island, Yukon Territory (approximately 69°36' N., 139°0' W.) by D. M. Wood in late July, 1971 (J. E. H. Martin, *in litt.*) has extended the known northerly limits of this species, and may well have revealed the northernmost population of dragonflies in Canada. The specimen was collected from the edge of a pond that fits the description given by Whitehouse (1941) and cited by Walker (1958, p. 97). (A brief account of Herschel Island is given by Harrington (1927).) A new northerly record for Quebec is Poste de la Baleine, Hudson Bay (J.-G. Pilon, *in litt.*).

Aeshna tuberculifera Walker

The record from the Saguenay River (Cap Jaseux) is the first for Quebec (Robert, 1963). The southerly limit now includes Ohio (Restifo, 1972) and Washington (Paulson, 1970).

Aeshna umbrosa Walker

The known northerly limit for Alberta is now Flatbush (Pritchard, 1964b).

New records for Manitoba are from Westbourne, Riding Mountain National Park and Duck Mountain National Park (H. B. White, *in litt.*). Waterloo is a new county record for Ontario (J. Peterson, *in litt.*).

Agrion aequabile (Say)
  The southerly limit now includes Pennsylvania (Beatty and Beatty, 1968), and the easterly limit Newfoundland (Morris, 1969).

Anax junius Drury
  New county records for Ontario are Oxford, Brant and Waterloo (Furtado, 1973).

Anomalagrion hastatum (Say)
  The known northerly range now includes Wisconsin (Ries, 1969).

Argia emma Kennedy
  The easterly limit now includes Nebraska (Pruess, 1968).

Argia translata Hagen
  Alrutz (1961) has recorded this species from Ohio.

Basiaeschna janata (Say)
  The known northerly limit in Quebec is now the Saguenay River (Cap Jaseux) (Fernet and Pilon, 1969).

Boyeria grafiana Williamson
  The known westerly limit is now Michigan (Kormondy, 1958).

Boyeria vinosa (Say)
  The known northwestern limit extends to Wisconsin (Kormondy, 1960).

Chromagrion conditum (Hagen)
  The known northerly limit in Quebec is now the Saguenay River (Cap Jaseux) (Fernet and Pilon, 1969).

Coenagrion interrogatum (Hagen)
  The southerly range now includes Michigan (Kormondy, 1958) and Wisconsin (Ries, 1969).

Coenagrion resolutum (Hagen)
  The range now includes Ohio to the south (Price, 1950), New Hampshire to the southeast (White and Morse, 1973) and Washington (Paulson, 1970). New county records for Ontario are Waterloo and Carleton (Furtado, 1973).

Cordulegaster diastatops (Selys)
  The known northerly limit now includes Michigan (Kormondy, 1958).

Cordulegaster maculatus Selys
  The range now extends to Wisconsin to the northwest (Ries, 1969).

Enallagma aspersum (Hagen)

The known northwesterly limit has been extended to include Michigan (Kormondy, 1958) and Wisconsin (Ries, 1969).

Enallagma boreale Selys

Two males were captured by D. M. Wood on Herschel Island, Yukon Territory in late July, 1971 (J. E. H. Martin, *in litt.*) (see remarks for *Aeshna septentrionalis*). A new northerly record for Quebec is Poste de la Baleine, Hudson Bay (J.-G. Pilon, *in litt.*), and a new county record for Ontario is Renfrew (Furtado, 1973).

Enallagma clausum Morse

Frontenac is a new county record for Ontario (Furtado, 1973).

Enallagma cyathigerum (Charpentier)

The known northerly limit in Quebec is now Poste de la Baleine, Hudson Bay (J.-G. Pilon, *in litt.*). The southerly limit now includes Wisconsin (Ries, 1969).

Enallagma vernale Gloyd

The southeasterly limit now includes New Hampshire (White and Morse, 1973). New county records for Ontario are Oxford, Dufferin and Frontenac (Furtado, 1973).

Gomphus borealis Needham

The known northerly range has been extended to include the Saguenay River (Cap Jaseux), Quebec (Fernet and Pilon, 1969). The species has now been recorded from Pennsylvania (Beatty and Beatty, 1968).

Gomphus descriptus Banks

The range to the southeast now includes New Hampshire (White and Morse, 1973).

Gomphus externus Hagen

Price (1950) records this species from Ohio, and Kormondy (1960) from North Dakota.

Gomphus furcifer Hagen

White and Morse (1973) record this species from New Hampshire. A new county record for Ontario is Renfrew (Furtado, 1973).

Gomphus notatus Rambur

The known range now includes Pennsylvania (Beatty and Beatty, 1968).

Gomphus scudderi Selys

Records now exist for New Hampshire (White and Morse, 1973), Pennsylvania (Beatty and Beatty, 1968) and Wisconsin (Ries, 1969).

Gomphus spicatus Hagen
  The known northerly limit in Quebec is now the Saguenay River (Cap Jaseux) (Fernet and Pilon, 1969). A new county record for Ontario is Oxford (Furtado, 1973).

Gomphus spiniceps (Walsh)
  Along its easterly limit this species occurs north to New Hampshire (White and Morse, 1973).

Gomphus villosipes Selys
  The known range is extended to the north and east to include New Hampshire (White and Morse, 1973).

Gomphus viridifrons Hine
  Michigan now constitutes the known northwesterly limit (Kormondy, 1958).

Hagenius brevistylus Selys
  The known northerly limit in Quebec is now the Saguenay River (Cap Jaseux) (Fernet and Pilon, 1969).

Hetaerina americana (Fabricius)
  Robert W. Adams has found this species in Waterloo county, Ontario. It is now known from New Hampshire (White and Morse, 1973).

Lanthus parvulus (Selys)
  The record for the Saguenay River (Cap Jaseux), Quebec slightly extends the known northerly limit of the range (Fernet and Pilon, 1969).

Lestes congener Hagen
  A record now exists for Wisconsin (Ries, 1969).

Lestes disjunctus australis Walker
  The known northerly limit is now Michigan (Kormondy, 1958).

Lestes unguiculatus Hagen
  The known northerly limit in Quebec is now the Saguenay River (Cap Jaseux) (Fernet and Pilon, 1969).

Nasiaeschna pentacantha (Rambur)
  The range now includes Michigan to the northwest (Kormondy, 1958).

Nehalennia gracilis Morse
  The range now includes Michigan (Kormondy, 1958).

Nehalennia irene (Hagen)
  The range now includes Washington (Paulson, 1970).

Ophiogomphus anomalus Harvey
The known northerly limit in Quebec is now Mont Tremblant Park (Rivière du Diable) (Robert, 1963).

Ophiogomphus aspersus Morse
The known northeasterly limit now includes Wisconsin (Ries, 1969).

Ophiogomphus carolus Needham
The known range now includes Pennsylvania (Beatty and Beatty, 1968) and New Hampshire (White and Morse, 1973) to the south, and Michigan (Kormondy, 1958) to the west.

Ophiogomphus mainensis Packard
The record for the Saguenay River (Cap Jaseux), Quebec slightly extends the known northerly limit of the range (Fernet and Pilon, 1969).

ECOLOGY

Aeshna spp.
The prey of adult *Aeshna eremita*, *A. interrupta lineata* and *A. umbrosa*, studied in nature near Flatbush, Alberta, was found to consist mainly of small Diptera (Pritchard, 1964b). Most of the prey was very small but the aeshnids investigated occasionally took larger insects, such as caddisflies and other dragonflies (e.g. *Sympetrum internum*).

Anax junius Drury
Field studies near Montreal, Quebec and near Toronto, Ontario have provided information on the sex ratio (about 48 per cent males) and temporal pattern of emergence (Trottier, 1966, 1971); the latter study also revealed that populations of *A. junius* in Canada are maintained each year by residents as well as by migrants. Near Toronto a resident population over-wintered as half-grown larvae and developed in approximately 11 months—from mid-July of one calendar year to late June of the next. In the same pond a non-resident population (arising from eggs laid by spring immigrants) developed in approximately 3 months—from June to September. An exceptionally early arrival of adults in southern Ontario was witnessed on April 4, 1974, an event that was probably causally correlated with the arrival in the same locality of a low-pressure system that had originated in Kansas, U.S.A. (Butler, Peterson and Corbet, 1975).

Argia vivida Hagen
As anticipated by Walker (1953), nymphs have been found in warm springs issuing from the baths at Banff, Alberta (Pritchard, 1971). They occur there amongst vegetation in the thermal pools, in streams connecting the pools, and amongst mosses and vascular plants in wet areas at the edges of the pools and streams. The water temperature there in 1970 and 1971 ranged between 26 and 27°C in spite of air-temperature variations from −20°C in February to 32°C in June.

General

At their study site on the Saguenay River, Fernet and Pilon (1970c) have been able to characterize the relationship between the onset of emergence of several species of Odonata and phenological indicators such as leaf-growth on trees.

CORRIGENDA

VOLUME I

Page Line

71	21–22	The reference cited is Williamson (1906b).
84	16	The first alternative in couplet 2 leads to couplet 3.
85	5	Couplet 7. The length of abd. segs. 8–10 (with paraprocts) should be 5 (not 2) mm. in *disjunctus* and 5.75 (not 2.75) mm. in *unguiculatus*.
245	26	Heading is "*Enallagma vesperum.*"
254	last but one	For "fourth" read "fifth."
261	36	"*I. verticalis . . .*"
281	19	"HEYMONS, R. 1904. Die Hinterleibesanhänge . . ."
284	48	For "1904b" read "1940b."

VOLUME II

ix	28	"*Aeshna interrupta lineata* Walker"
xi	33	"*Cordulegaster diastatops* (Selys)"
6	17	For "CuI" read "Cu₁."
54	34	Couplet 5. The second alternative leads to couplet 15.
	38	Couplet 6. The second alternative leads to couplet 9.
71		Missing abbreviations are: o—oblique vein Rspl—radial supplement Sc—subcosta
94	9	"finding"
158	13	The date of Selys' publication is 1854.
201		Replace "*G.*" by "*O.*" throughout the plate caption.
218	9	"part"
247	4	*Gomphus fraternus manitobanus* subsp. n.

In reviewing Volume 2, Walley (1959) pointed out that no holotype had been designated for this subspecies. This I now do.

I designate as the LECTOTYPE a male adult (so labelled) in

the Royal Ontario Museum, Toronto collected by J. B. Wallis at Winnipeg, Manitoba on June 25, 1910.

I designate as PARALECTOTYPES those adults (so labelled) in the Royal Ontario Museum (ROM) and the Canadian National Collection of Insects, Ottawa (CNC), namely:

ROM: Aweme, Manitoba, collected by N. Criddle: one male on June 22, 1909; two males on July 19, 1910. Winnipeg, Manitoba, collected by J. B. Wallis: four males on June 25, 1910.

CNC: Aweme, Manitoba, collected by N. Criddle: four males on (respectively) June 14, 1915; July 3, 1917; June 25, 1920; and June 7, 1923.

BIBLIOGRAPHY

AHRENS, CARSTEN
1938.   A list of dragonflies (Odonata) taken in southern Alaska. Ent. News, **49**: 225–227.
ALRUTZ, ROBERT W.
1961.   Notes and records of Ohio dragonflies and damselflies (Odonata). Ohio J. Sci., **61**: 13–24.
ASAHINA, SYOZIRO, and SHIGEO EDA
1960.   The Japanese libellulas (Odonata, Libellulidae). Kontyû, Tokyo, **28**: 63–64.
BARTENEV, A. N.
1915.   [On the classification and constitution of the genus *Sympetrum* Newm.] Univ. Izviestija Varsava, **46**, 19 pp.
BEATTY, GEORGE H.
1946.   Dragonflies (Odonata) collected in Pennsylvania and New Jersey in 1945. Ent. News, **57**: 1–10; 50–56; 76–81; 104–111.
BEATTY, GEORGE H., and ALICE FERGUSON BEATTY
1968.   Check list and bibliography of Pennsylvania Odonata. Proc. Pennsylvania Acad. Sci., **42**: 120–128.
1969.   United States records of *Williamsonia fletcheri* (Odonata: Corduliidae). Mich. Ent., **2**: 13.
BELYSHEV, B. F.
1973.   The dragonflies of Siberia. 2 vols. "Nauka" Siberian Branch, Novosibirsk. [In Russian.] 1–330 and 331–620 pp.
BEUTENMULLER, W.
1890.   Preliminary catalogue of the Odonata found in the State of New York. *In* Lamborn, R. H., Dragonflies *vs.* mosquitoes, pp. 163–164. Appleton, New York.
BICK, GEORGE H.
1941.   Life-history of the dragonfly, *Erythemis simplicicollis* (Say). Ann. Ent. Soc. Amer., **34**: 215–230.
1950.   The dragonflies of Mississippi (Odonata: Anisoptera). Amer. Midl. Nat., **43**: 66–78.
1951a.   Notes on Oklahoma dragonflies. J. Tenn. Acad. Sci., **26**: 178–180.
1951b.   The early nymphal stages of *Tramea lacerata* Hagen (Odonata: Libellulidae). Ent. News, **62**: 293–303.
BICK, GEORGE H., and JUANDA C. BICK
1961.   An adult population of *Lestes disjunctus australis* Walker (Odonata: Lestidae). Southwestern Nat., **6**: 111–137.
1963.   Behavior and population structure of the damselfly, *Enallagma civile* (Hagen) (Odonata: Coenagriidae). *Ibid.*, **8**: 57–84.
BORROR, DONALD J.
1937.   An annotated list of the dragonflies (Odonata) of Ohio. Ohio J. Sci., **37**: 185–196.
1940.   A list of the dragonflies (Odonata) taken in the region of Muscongus Bay, Maine. Ent. News, **51**: 45–79.
1942.   A revision of the libelluline genus *Erythrodiplax* (Odonata). 286 pp. Ohio State Univ., Columbus.
1945.   A key to the New World genera of Libellulidae (Odonata). Ann. Ent. Soc. Amer., **38**: 168–194.
1953.   A migratory flight of dragonflies. Ent. News, **64**: 204.

BRAUER, FRIEDERICH
  1868. Verzeichniss der bis jetzt bekannten Neuropteren im Sinne Linne's. Verh. Zool.-Bot. Ges. Wien, **18**: 359–416; 711–742.

BRIMLEY, C. S.
  1903. List of dragonflies (Odonata) from North Carolina, especially from the vicinity of Raleigh. Ent. News, **14**: 150–157.

BRITTINGER, C. C.
  1850. Die Libelluliden des Kaiserreichs Österreich. S. B. Acad. Wiss. Wien 4, Mathem.-nat. Klasse: 328–336.

BUCKELL, E. R.
  1937. Some locality records of British Columbia dragonflies, pp. 55–63. Dominion Ent. Lab., Vernon, B.C.

BURMEISTER, H.
  1839. Handbuch der Entomologie. Vol. 2. Enslin, Berlin.

BUTLER, THERESE, JANE E. PETERSON, and PHILIP S. CORBET
  1975. An exceptionally early and informative arrival of adult *Anax junius* in Ontario (Odonata: Aeshnidae). Can. Ent., **107**: in press.

BYERS, C. FRANCIS
  1927a. An annotated list of the Odonata of Michigan. Occas. Pap. Mus. Zool., Univ. Mich., 183, 16 pp.
  1927b. Notes on some American dragonfly nymphs (Odonata, Anisoptera). J. N.Y. Ent. Soc., **35**: 65–74.
  1927c. The nymph of *Libellula incesta* and a key for the separation of the known nymphs of the genus *Libellula* (Odonata). Ent. News, **38**: 113–115.
  1930. A contribution to the knowledge of Florida Odonata. Univ. Florida Publ. Biol. Sci. 1, 327 pp.
  1941. Notes on the emergence and life history of the dragonfly *Pantala flavescens*. Proc. Florida Acad. Sci., **6**: 14–25.

CALVERT, PHILIP P.
  1890. Notes on some North American Odonata, with descriptions of three new species. Trans. Amer. Ent. Soc., **17**: 33–40.
  1893. Catalogue of the Odonata of the vicinity of Philadelphia with an introduction to the study of this group of insects. *Ibid.*, **20**: 152a–272.
  1898. Footnote (*S. elongata* var. *minor*, n.sp.). Ent. News, **9**: 87.
  1905. Fauna of New England, 6. List of the Odonata. Occas. Pap. Bost. Soc. Nat. Hist., 7, 43 pp.
  1926. Relations of a late autumnal dragonfly (Odonata) to temperature. Ecology, **7**: 185–190.
  1929. Different rates of growth among animals with special reference to the Odonata. Proc. Amer. Phil. Soc., **68**: 227–274.

CALVERT, PHILIP P., and WM. SHERATON
  1894. Notes on Nova Scotian dragonflies. Can. Ent., **26**: 317–320.

CHARPENTIER, TOUSSAINT DE
  1840. Libellulinae europaeae, descriptae ac depictae. 180 pp. Voss, Lipsiae.

COOK, CARL
  1950. Notes on a collection of dragonflies (Odonata) from Nova Scotia. Can. Ent., **82**: 190–194.

CORBET, PHILIP S.
  1951. The development of the labium of *Sympetrum striolatum* (Charp.) (Odon., Libellulidae). Ent. Mon. Mag., **87**: 289–296.
  1953. A terminology for the labium of larval Odonata. Entomologist, **86**: 191–196.
  1962. A biology of dragonflies. 247 pp. Witherby, London.

1966. The study of Odonata. Pp. 70–78 *in* Centennial of entomology in Canada 1863–1963. A tribute to Edmund M. Walker (G. B. Wiggins ed.). Univ. of Toronto Press.

1967. A dragonfly new to Canada. Can. Field-Nat., **81**: 230.

1968. The Odonata of Ontario. Proc. North Central Branch Ent. Soc. Amer., (1967) **22**: 116–117.

1969. Dr. Edmund Murton Walker (1877–1969) in memoriam. Tombo, Tokyo, **12**: 2–3.

CORBET, PHILIP S., and SHIGEO EDA

1969. Odonata in southern Ontario, Canada, in August 1968. Tombo, Tokyo, **12**: 4–11.

CORBET, PHILIP S., CYNTHIA LONGFIELD, and NORMAN W. MOORE

1960. Dragonflies. 260 pp. Collins, London.

CORBET, PHILIP S., and G. S. WALLEY

1969. The dragonflies (Odonata) of the Mer Bleue. Can. Field-Nat., **83**: 14–16.

COWLEY, J.

1944. Preliminary report on the Odonata collected by Drs. S. Feliksiak and T. Jaczewski in Canada and Newfoundland. Entomologist, **77**: 123–124.

CRUDEN, R. W., and N. L. CURRIE

1961. Additions to the state and county records of Ohio dragonflies. Ohio J. Sci., **61**: 189–191.

CURRIE, ROLLA P.

1901. Papers from the Harriman Alaska Expedition, XXII. Entomological results (14): The Odonata. Proc. Wash. Acad. Sci., **3**: 217–223.

1905. Dragonflies from the Kootenay Dist. of British Columbia. Proc. Ent. Soc. Wash., **7**: 16–20.

DAVIS, W. T.

1898. Preliminary list of the dragonflies of Staten Island, with notes and dates of capture. J.N.Y. Ent. Soc., **6**: 195–198.

1913. *Williamsonia*, a new genus of dragonflies from North America. Bull. Brooklyn Ent. Soc., **8**: 93–96.

DE GEER, CARL F.

1773. Mémoire pour servir à l'histoire des insectes. Vol. 3, 696 pp. Grefing, Stockholm.

DONNELLY, THOMAS W.

1961. The Odonata of Washington, D.C., and vicinity. Proc. Ent. Soc. Wash., **63**: 1–13.

DONOVAN, EDWARD

1811. The natural history of British insects. Vol. 15. Rivington, London.

DRURY, DREW

1770. Illustrations of natural history. Vol. 1. 130 pp. White, London.

1773. Illustrations of natural history. Vol. 2. 90 pp. White, London.

EVANS, W. F.

1845. British Libellulinae; or, dragonflies. 28 pp. Bridgewater, London.

FABRICIUS, JOHANN CHRISTIAN

1793. Entomologia systematica emendata et aucta, etc. Vol. 2. pp. 373–383. Proft. (Schubothe), Hafniae.

1798. Supplementum entomologiae systematicae. pp. 283–285. Proft. (Schubothe), Hafniae.

FERGUSON, ALICE

1940. A preliminary list of the Odonata of Dallas County Texas. Field and Lab., Dallas, **8**: 1–10.

FERNET, L., and JEAN-GUY PILON

1968a. *Aeshna clepsydra* (Say) (Odonata: Aeshnidae) au Québec. Nat. Canad., **95**: 741–745.

1968b. *Pantala hymenea* (Say) (Odonata: Libellulidae) au Québec. *Ibid.*, **95**: 767–771.

1968c. Une première mention oubliée pour le Québec et pour le Canada. *Ibid.*, **95**: 1033.

1969. Inventaire des Odonates de la région du Cap Jaseux, Saguenay. Ann. Ent. Soc. Québ., **14**, 85–102.
1970a. Inventaire préliminaire des odonates de la Gaspésie. Phytoprotection, **51**: 52–62.
1970b. Les odonates, indicateurs de la nature du milieu. Nat. Canad., **97**: 401–420.
1970c. Relation entre le début de l'émergence des Odonates, la croissance des feuilles des arbres et la température de l'eau, au Saguenay. Ann. Ent. Soc. Québ., **15**: 164–168.

FERNET, LUC, JEAN-GUY PILON, and RENÉ BEIQUE
1968. Note sur *Anax maritimus* Provancher (Odonata: Aeshnidae). Nat. Canad., **95**: 1081–1084.

FRASER, F. C.
1957. A reclassification of the Order Odonata. Roy. Zool. Soc. N. S. W. Handbook, 12, 133 pp.

FURTADO, JOSÉ I.
1973. Annotated records of some Canadian dragonflies (Odonata) especially from the vicinity of Waterloo, Ontario. Can. Field-Nat., **87**: 463–466.

GARDNER, A. E.
1951. The life history of *Sympetrum danae* (Sulzer) = *S. scoticum* (Donovan) (Odonata). Ent. Gaz., **2**: 109–127.
1954. A key to the larvae of the British Odonata Part II. Suborder Anisoptera. *Ibid.*, **5**: 193–213.

GARMAN, PHILIP
1927. Guide to the insects of Connecticut. Part V, the Odonata or dragonflies of Connecticut. Conn. Geol. and Nat. Hist. Surv. Bull., 39, 331 pp.

GIBBS, ROBERT H., and SARAH PREBLE GIBBS
1954. The Odonata of Cape Cod, Massachusetts. J.N.Y. Ent. Soc., **62**: 167–184.

GLOYD, LEONORA K.
1932. Four new dragonfly records for the United States. (Odonata.) Ent. News, **43**: 189–190.
1938. Notes on some dragonflies (Odonata) from Admiralty Island, Alaska. *Ibid.*, **49**: 198–200.
1939. A synopsis of the Odonata of Alaska. *Ibid.*, **50**: 11–16.
1958. The dragonfly fauna of the Big Bend Region of Trans-Pecos Texas. Occ. Pap. Mus. Zool., Univ. Michigan, 593, 23 pp.
1959. Elevation of the *Macromia* group to family status (Odonata). Ent. News, **70**: 197–205.
1972. *Tramea, Trapezostigma*, and time (Anisoptera: Libellulidae). A nomenclatural problem. Odonatologica, **1**: 131–136.

GLOYD, LEONORA K., and MIKE WRIGHT
1959. Odonata. Chapter 34 (pp. 917–940) *in* H. B. Ward and G. C. Whipple, Freshwater Biology (W. T. Edmondson ed.) second edition. Wiley, New York.

GORHAM, J. R.
1972. Studies of the biology and control of arthropods of health significance in Alaska. 4. Ecological studies of biting flies on the north slope of Alaska: 1970. Arctic Health Research Center, Dept. of Health, Education and Welfare, Fairbanks, Alaska, 62 pp. Mimeographed.

HAGEN, H. A.
1861. Synopsis of the Neuroptera of North America, with a list of the South American species. Smithson. Misc. Coll., 4 (1862), 347 pp.
1867. Revision der von Herrn Uhler beschriebenen Odonaten. Stett. Ent. Zeitung, **28**: 87–95.
1873. Report on Mr. S. H. Scudder's Odonata from the White Mountains, after an examination of the types. Proc. Bost. Soc. Nat. Hist., **15**: 376–377.

1874. Report on the Pseudo-neuroptera and Neuroptera collected by Lieut. W. L. Carpenter in 1873 in Colorado. Ann. Rep. U.S. Geol. and Geog. Surv. Terr., 1873: 571–606.

1875. Synopsis of the Odonata of North America. Proc. Bost. Soc. Nat. Hist., **18**: 20–96.

1890a. A synopsis of the odonat genus *Leucorhinia* Britt. Trans. Amer. Ent. Soc., **17**: 229–236.

1890b. Descriptions of some North American Cordulina. Psyche, **5**: 367–373.

1890c. Notes and descriptions of some North American Libellulina. Synopsis of Neuroptera, second edition. *Ibid.*, **5**: 383–387.

HÄMÄLÄINEN, MATTI
1967. *Somatochlora sahlbergi* Trybom (Odon., Corduliidae) Utsjoki (In L). Luonnon Tutkija, **71**: 25.

HAMRUM, CHARLES L., ROBERT EVANS CARLSON, and ARTHUR W. GLASS
1965. Identification and distribution of Minnesota *Leucorrhinia* species (Odonata, Libellulidae). J. Minn. Acad. Sci., **33**: 23–26.

HARRINGTON, RICHARD
1972. Herschel Island. Can. Geographical J., **85**: 172–177.

HARVEY, FRANCES LEROY
1892. A contribution to the Odonata of Maine, II. Ent. News, **3**: 116–117.

1901. A contribution to the Odonata of Maine, IV. *Ibid.*, **12**: 269–277.

1902. A catalogue and bibliography of the Odonata of Maine, with an annotated list of their collectors. Univ. Maine Studies, 4, 16 pp.

JACOBS, MERLE E.
1955. Studies on territorialism and sexual selection in dragonflies. Ecology, **36**: 566–586.

JOHNSON, CLIFFORD
1962a. A study of territoriality and breeding behavior in *Pachydiplax longipennis* Burmeister (Odonata: Libellulidae). Southwestern Nat., **7**: 191–197.

1962b. Breeding behavior and oviposition in *Calopteryx maculatum* (Beauvais) (Odonata: Calopterygidae). Amer. Midl. Nat., **68**: 242–247.

KARSCH, F.
1889. Beiträge zur Kenntnis der Arten und Gattungen der Libellulinen. Berl. Ent. Zeits., **33**: 347–392.

KELLICOTT, D. S.
1899. The Odonata of Ohio. Ohio Acad. Sci. Spec. Paper, 2, 116 pp.

KENNEDY, CLARENCE HAMILTON
1913. Notes on the Odonata or dragonflies of Bumping Lake, Washington. Proc. U.S. Nat. Mus., **46**: 111–126.

1915a. Interesting western Odonata. Ann. Ent. Soc. Amer., **8**: 297–303.

1915b. Notes on the life history and ecology of the dragonflies (Odonata) of Washington and Oregon. Proc. U.S. Nat. Mus., **49**: 259–345.

1917a. The Odonata of Kansas with reference to their distribution. Univ. Kansas Dep. Ent. Bull., **11**: 129–145.

1917b. A new species of *Somatochlora* (Odonata) with notes on the *cingulata* group. Can. Ent., **49**: 229–236.

1917c. Notes on the life history and ecology of the dragonflies (Odonata) of central California and Nevada. Proc. U.S. Nat. Mus., **52**: 483–635.

1922. The morphology of the penis in genus *Libellula*. Ent. News, **33**: 33–40.

1923. The naiad of *Pantala hymenea* (Odonata). Can. Ent., **55**: 36–38.

1924. Notes and descriptions of naiads belonging to the dragonfly genus *Helocordulia*. Proc. U.S. Nat. Mus., **64**: 1–4.

KIRBY, W. F.
1889. A revision of the subfamily Libellulinae with descriptions of new genera and species. Trans. Zool. Soc. London, **12**: 249–348.

1890.   A synonymic catalogue of the Neuroptera Odonata or dragonflies with an appendix on the fossil species. 202 pp. Gurney and Jackson, London.

KLUGH, A. B., and E. G. McDOUGALL
1924.   The faunal areas of Canada. Handbook of Canada, British Association Meeting, Toronto.

KORMONDY, EDWARD J.
1958.   Catalogue of the Odonata of Michigan. Misc. Publ. Mus. Zool. Univ. Michigan, 104, 43 pp.
1959.   The systematics of *Tetragoneuria*, based on ecological, life history, and morphological evidence (Odonata: Corduliidae). *Ibid.*, 107, 79 pp.
1960.   New North American records of anisopterous Odonata. Ent. News, **71**: 121–130.

LAMB, LAURA
1925.   A tabular account of the differences between the earlier instars of *Pantala flavescens*. Trans. Amer. Ent. Soc., **50**: 289–312.

LAWRENCE, VINNEDGE M.
1968.   The distribution of Odonata in Indiana and Ohio. Proc. North Central Branch Ent. Soc. Amer., (1967) **22**: 117–120.

LEACH, W. E.
1815.   Entomology. *In* Brewster's Edinburgh Encyclopaedia **9**: 52–172. Edinburgh.

LEONARD, JUSTIN W.
1934.   The naiad of *Celithemis monomelaena* Williamson (Odonata: Libellulidae). Occ. Pap. Mus. Zool., Univ. Mich., 297, 5 pp.

LEVINE, HARRY R.
1957.   Anatomy and taxonomy of the mature naiads of the dragonfly genus *Plathemis* (family Libellulidae). Smithson. Misc. Coll., 134 (11), 28 pp.

LIEFTINCK, M. A.
1926.   Odonata Neerlandica. Tweede gedeelte: Anisoptera. Tijdschr. Ent. **69**: 85–226.

LINNÉ, CAROL
1758.   Systema naturae. Editio decima reformata. Holmiae.
1763.   Amoenitates academicae seu dissert. variae. etc. Vol. sextum. Holmiae 1763. CXXI. Centuria Insectorum, etc. 32 pp. Johansson, Upsaliae.

LONGFIELD, CYNTHIA.
1960.   *In* P. S. Corbet, C. Longfield, and N. W. Moore (1960). Dragonflies. 260 pp. Collins, London.

LUTZ, PAUL E.
1968.   Life history studies on *Lestes eurinus* Say (Odonata). Ecology, **49**: 576–579.

LUTZ, PAUL E., and CHARLES E. JENNER
1964.   Life-history and photoperiodic responses of nymphs of *Tetragoneuria cynosura* (Say). Biol. Bull., **127**: 304–316.

MACLACHLAN, R.
1886.   Two new species of Corduliina. Ent. Mon. Mag., **23**: 104–105.

MARTIN, RENÉ
1906.   Cordulines in: Collections zoologiques du Baron Edm. de Selys Longchamps. 98 pp. Hayez, Impr. des Académies, Brussels.

McMAHAN, ELIZABETH A., and I. E. GRAY
1957.   Variations in a local population of the dragonfly *Helocordulia*. Ann. Ent. Soc. Amer., **50**: 62–66.

MILLER, LEE A., CHARLES L. HAMRUM, and MYRON A. ANDERSON
1964.   Identification and distribution of *Sympetrum* in Minnesota (Libellulidae, Odonata). Proc. Minn. Acad. Sci., **31**: 116–120.

MONTGOMERY, B. ELWOOD

1929. Records of Indiana dragonflies, III. 1927–1928. Proc. Ind. Acad. Sci., **38**: 335–343.

1937a. Oviposition of *Perithemis* (Odonata, Libellulidae). Ent. News, **48**: 61–63.

1937b. Records of Indiana dragonflies, IX, 1935–1936. Proc. Ind. Acad. Sci., **46**: 203–210.

1941. Records of Indiana dragonflies, X. 1937–40. *Ibid.*, **50**: 229–241.

1943a. *Sympetrum internum*, new name for *Sympetrum decisum* auct., nec Hagen (Odonata, Libellulidae). Can. Ent., **75**: 57–58.

1943b. *Williamsonia fletcheri* Williamson (Odonata: Corduliidae) from New England. Ent. News, **54**: 1–4.

1945. The distribution and relative seasonal abundance of the Indiana species of Corduliidae and Libellulidae (Odonata). Proc. Ind. Acad. Sci., **54**: 217–224.

1950. Notes and records of Indiana Odonata, 1941–1950. *Ibid.*, **60**: 205–210.

1968. Geographical distribution of the Odonata of the North Central States. Proc. North Central Branch Ent. Soc. Amer., (1967) **22**: 121–129.

MOORE, NORMAN W.

1960. *In* P. S. Corbet, C. Longfield, and N. W. Moore (1960). Dragonflies. 260 pp. Collins, London.

MORRIS, RAY F.

1969. Occurrence of the damselfly, *Agrion aequabile*, in Newfoundland. Can. Ent., **101**: 163.

MÜNCHBERG, PAUL

1932. Beiträge zur Kenntnis der Biologie der Libellenunterfamilie der Cordulinae Selys. Internat. Revue der gesamten Hydrobiol. und Hydrographie, **27**: 265–302.

MUSSER, R. JEAN

1962. Dragonfly nymphs of Utah (Odonata: Anisoptera). Univ. Utah Biol. Series, 12 (6), 66 pp.

MUTTKOWSKI, RICHARD A.

1908. Review of the dragonflies of Wisconsin. Bull. Wis. Nat. Soc., **6**: 57–127.

1910a. Catalogue of the Odonata of North America. 207 pp. Bull. Mus. Milwaukee.

1910b. New Records of Wisconsin dragonflies. Bull. Wis. Nat. Hist. Soc., **8**: 53–59.

1911a. New records of Wisconsin dragonflies, II. *Ibid.*, **9**: 28–41.

1911b. Studies in *Tetragoneuria* (Odonata). *Ibid.*, **9**: 91–134.

NEEDHAM, JAMES G.

1897. *Libellula deplanta* of Rambur. Can. Ent., **29**: 144–146.

1901. Aquatic insects in the Adirondacks. Order Odonata. Bull. N.Y. State Mus., **47**: 429–540.

1903. On aquatic insects in New York State. Life-histories of Odonata. *Ibid.*, **68**: 218–279.

1904. New dragon-fly nymphs in the United States National Museum. Proc. U.S. Nat. Mus., **27**: 685–720.

NEEDHAM, JAMES G., and HORTENSE BUTLER HEYWOOD

1929. A handbook of the dragonflies of North America. 378 pp. Chas. C. Thomas, Springfield, Ill., Baltimore, Md.

NEEDHAM, JAMES G., and MINTER J. WESTFALL

1955. A manual of the dragonflies of North America (Anisoptera), including the Greater Antilles and the Provinces of the Mexican Border. 615 pp. Univ. of Calif. Press, Berkeley and Los Angeles.

NEVIN, F. REESE

1929. Larval development of *Sympetrum vicinum*. Trans. Amer. Ent. Soc., **55**: 79–102.

NEWMAN, EDWARD

1833. Entomological notes. Ent. Mon. Mag., **1**: 505–514.

OSBURN, R. C.

1905. The Odonata of British Columbia. Ent. News, **16**: 184–196.

PACKARD, A. S.
1867. The dragonfly. Amer. Nat., **1**: 304–313.

PAULSON, DENNIS R.
1970. A list of the Odonata of Washington with additions to and deletions from the state list. Pan-Pacific Ent., **46**: 194–198.

PEARSE, A. S.
1932. Animals in brackish water ponds and pools at Dry Tortugas. Publ. Carnegie Inst. Wash., **435**: 125–142.

PRICE, H. F.
1950. Notes on the dragonflies (Odonata) of northwest Ohio. Ohio J. Sci., **50**: 71–78.
1958. Additional notes on the dragonflies of northwest Ohio. *Ibid.*, **58**: 50–62.

PRITCHARD, GORDON
1964a. The prey of dragonfly larvae (Odonata; Anisoptera) in ponds in northern Alberta. Can. J. Zool., **42**, 785–800.
1964b. The prey of adult dragonflies in northern Alberta. Can. Ent., **96**: 821–825.
1971. *Argia vivida* Hagen (Odonata: Coenagrionidae) in hot pools at Banff. Can. Field-Nat., **85**: 187–188.

PROVANCHER, L'ABBÉ L.
1875. Une excursion à St. Hyacinthe. Nat. Canad., **7**: 205-219; 232–252.
1877a. Faune entomologique du Canada. II, Fasc. I, Les orthoptères et les névroptères. 157 pp. Quebec.
1877b. Faune canadienne. Nat. Canad., **9**: 38–43, 84–90, 118–123, 173–176, 201–205, 209–217, 241–244, 257–269.

PRUESS, NEVA C.
1968. Checklist of Nebraska Odonata. Proc. North Central Branch Ent. Soc. Amer., (1967) **22**: 112.

RAMBUR, M. P.
1842. Histoire naturelle des insectes neuroptères. 534 pp. Libraire Encyclopédique de Roret, Paris.

RESTIFO, ROBERT A.
1972. *Aeshna tuberculifera* Walker, a new Ohio record. Ohio J. Sci., **72**: 183.

RIES, MARY DAVIS
1968. Present state of knowledge of the distribution of Odonata in Wisconsin. Proc. North central Branch Ent. Soc. Amer., (1967) **22**: 113–115.
1969. Odonata new to the Wisconsin state list. Mich. Ent., **2**, 22–27.

RIS, F.
1909. Collections zoologiques du Baron Edm. de Selys Longchamps. Fasc. IX, Libellulinen 1. pp. 1–120. Imprimeries des Académies, Hayez, Brussels.
1910. Collections zoologiques du Baron Edm. de Selys Longchamps. Fasc. XI, Libellulinen. pp. 245–384. Imprimeries des Académies, Hayez, Brussels.
1911. Collections zoologiques du Baron Edm. de Selys Longchamps. Fasc. XIII, Libellulinen. pp. 529–700. Imprimeries des Académies, Hayez, Brussels.
1930. A revision of the libelluline genus *Perithemis* (Odonata). Misc. Publ. Mus. Zool., Univ. Mich., 21, 50 pp.

ROBERT, ADRIEN
1944. Premier aperçu sur les Odonates du Comté d'Abitibi. Nat. Canad., **71**: 149–171.
1953. Les odonates du Parc du Mont Tremblant. Can. Ent., **85**: 316–339.
1954a. Observations sur les odonates du parc du Mont Tremblant en 1953. Annales de l'Assoc. can.-franç. pour l'Avanc. des Sci., **20**: 113–118.
1954b. Un nouveau *Somatochlora* subarctique (Odonates, Corduliidae). Can. Ent., **86**: 419–422.

1963. Les libellules du Québec. Minist. Tourisme, Chasse et Pêche, Prov. Québec, Serv. Faune, Bull. 1, 223 pp.

ROBERT, P.-A.
1958. Les libellules (odonates). Delachaux et Niestlé, S. A., Neuchâtel.

ROOT, F. M.
1912. Dragon flies collected at Point Pelee and Pelee Island, Ontario, in the summers of 1910 and 1911. Can. Ent., **44**: 208–209.

SAY, T. (Posthumous)
1839. Descriptions of new North American neuropterous insects and observations on some already described by (the late) Th. Say. J. Acad. Nat. Sci. Phila., **8**: 9–46.

SCHIEMENZ, H.
1953. Die Libellen unserer Heimat. Urania, Jena.

SCUDDER, S. H.
1866. Notes on some Odonata from the White Mountains of New Hampshire. Proc. Bost. Soc. Nat. Hist., **10**: 211–222.

SELYS LONGCHAMPS, E. DE
1850. Revue des odonates ou Libellules d'Europe. Mém. Soc. Sci. Liège, 6, 408 pp.
1871. Synopsis des cordulines. Bull. Acad. Roy. Sci. Belg., (2) **31**: 238–316; 519–565.
1874. Additions au synopsis des cordulines. *Ibid.*, (2) **37**: 16–34.
1878. Secondes additions au synopsis des cordulines. *Ibid.*, (2) **45**: 183–222.
1884. Revision des *Diplax* paléarctiques. Ann. Soc. Ent. Belg., **28**: 29–45.

SHIFFER, CLARK N.
1968. Homeochromatic females in the dragonfly *Perithemis tenera*. Proc. Pennsylvania Acad. Sci., **42**: 138–141.
1969. Occurrence and habits of *Somatochlora incurvata*, new for Pennsylvania (Odonata: Corduliinae). Mich. Ent., **2**: 75–76.

SMITH, RAY F., and A. EARL PRITCHARD
1971. Odonata. Chapter 4 (pp. 106–153) *in* Aquatic insects of California (R. L. Usinger ed.). Univ. of Calif. Press, Berkeley and Los Angeles.

SNODGRASS, R. E.
1954. The dragonfly larva. Smithson. Misc. Coll., 123(2), 38 pp.

STÖHR, L. M.
1918. Odonates des environs de Saint-Alexandre, Ironside, P.Q. Nat. Canad., **45**: 81–85.

SULZER, J. H.
1776. Abgekürzte Geschichte der Insekten nach der Linnéischen Form. Steiner, Winterthur.

TROTTIER, ROBERT
1966. The emergence and sex ratio of *Anax junius* Drury (Odonata: Aeshnidae) in Canada. Can. Ent., **98**: 794–798.
1967. Observations on *Pantala flavescens* (Fabricius) (Odonata: Libellulidae) in Canada. Can. Field-Nat., **81**: 231.
1969. A comparative study of the morphology of some *Sympetrum* larvae (Odonata: Libellulidae) from eastern Canada. Can. J. Zool., **47**: 457–460.
1971. Effect of temperature on the life-cycle of *Anax junius* (Odonata: Aeshnidae) in Canada. Can. Ent., **103**: 1671–1683.
1973a. A controlled temperature and humidity cabinet for recording the emergence behaviour of aquatic insects. *Ibid.*, **105**: 971–974.
1973b. Influence of temperature and humidity on the emergence behaviour of *Anax junius* (Odonata: Aeshnidae). *Ibid.*, 105: 975–984.

TRYBOM, FILIP
1889. Trollsländor (Odonater), insamlade under den Svenka Expeditionen till Jenisei 1876. Kongl. Svenska Vetensk.-Akad. Handlingar, **15**, Afd. IV (4): 3–21.

UHLER, PHILIP R.
  1857. Contributions to the Neuropterology of the United States. Proc. Acad. Nat. Sci. Phila., 1857: 87–88.

VALLE, K. J.
  1931. Materialien zur Odonatenfauna Finlands. II. *Somatochlora sahlbergi* Trybom. Notul. Ent., **11**: 41–51.
  1952. Die Verbreitungsverhältnisse der ostfennoskandischen Odonaten. Acta Ent. Fenn., **10**: 1–87.
  1955. Odonata from Newfoundland. Ann. Ent. Fenn., **21**: 57–60.

WADSWORTH, MATTIE
  1898. Fourth addition to the list of dragonflies (Odonata) of Manchester, Kennebec Co., Maine. Ent. News, **9**: 111.

WALKER, E. M.
  1907. A new *Somatochlora*, with a note on the species known from Ontario. Can. Ent., **39**: 69–74.
  1912. The Odonata of the Prairie Provinces of Canada. *Ibid.*, **44**: 253–266.
  1913. New nymphs of Canadian Odonata. *Ibid.*, **45**: 163–170.
  1914. New and little-known nymphs of Canadian Odonata. *Ibid.*, **46**: 349–357; 369–377.
  1915. Notes on the Odonata of the vicinity of Go Home Bay, Georgian Bay, Ont. Contrib. Can. Biol., 1911–1914, Fasc. 2. Suppl. to 47th Ann. Rep., Dept. Marine and Fisheries, Fisheries Branch, Ottawa, pp. 53–94.
  1916a. Popular and practical entomology. A few days in Newfoundland. Can. Ent., **48**: 217–221; 257–261.
  1916b. The nymphs of the North American species of *Leucorrhinia. Ibid.*, **48**: 414–422.
  1917a. Some dragonflies from Prince Edward Island. *Ibid.*, **49**: 117–119.
  1917b. The known nymphs of the North American species of *Sympetrum* (Odonata). *Ibid.*, **49**: 409–418.
  1918. On the American representatives of *Somatochlora arctica* with descriptions of two new species. *Ibid.*, 50: 365–375.
  1923. Notes on the Odonata of Godbout, Quebec. *Ibid.*, **55**: 5–12.
  1924. The Odonata of the Thunder Bay District, Ont. *Ibid.*, **56**: 170–176; 182–189.
  1925. The North American dragonflies of the genus *Somatochlora*. Univ. Toronto Studies, Biol. Ser., 26, 202pp.
  1927. The Odonata of the Canadian Cordillera. Prov. Mus. Nat. Hist., B.C., 1927, 16 pp.
  1928. The Odonata (dragonflies) of the Lake Abitibi region. Univ. Toronto Studies, Biol. Series, **32**: 37–44.
  1932. The Odonata (dragonflies) of Lake Nipissing. *Ibid.*, **36**: 225–246. Publ. Ont. Fisheries Res. Lab., 48.
  1933a. The Odonata of the Maritime Provinces. Proc. Nova Scotian Inst. Sci., **18**: 106–128.
  1933b. The Odonata of Manitoba. Can. Ent., **65**: 57–72.
  1934. A preliminary list of the insects of the Province of Quebec. Part IV, the Odonata. Quebec Soc. Prot. Plants, Suppl. Rept., 26, 12 pp.
  1937. A new *Macromia* from British Columbia. Can. Ent., **69**: 5–13.
  1940a. Odonata from the Patricia portion of the Kenora District of Ontario, with description of a new species of *Leucorrhinia. Ibid.*, **72**: 4–15.
  1940b. A preliminary list of the Odonata of Saskatchewan. *Ibid.*, **72**: 26–35.
  1941a. New records of Odonata from Manitoba. *Ibid.*, **73**: 35–36.
  1941b. List of the Odonata of Ontario with distributional and seasonal data. Trans. Roy. Can. Inst., **23** (2): 201–265.
  1941c. The nymph of *Somatochlora walshii* Scudder. Can. Ent., **73**: 203–205.
  1942a. The female of *Leucorrhinia patricia* Walker, with further notes on the male. *Ibid.*, **74**: 74–75.
  1942b. Additions to the list of Odonata of the Maritime Provinces. Proc. Nova Scotian Inst. Sci., **20**: 159–176.

1943. The Subarctic Odonata of North America. Can. Ent., **75**: 79–90.

1947. Further notes on the Subarctic Odonata of North America. *Ibid.*, **79**: 62–67.

1950. Notes on some Odonata from the Kenora and Rainy River Districts, Ont., *Ibid.*, **82**: 16–21.

1951a. *Sympetrum semicinctum* Say and its nearest allies (Odonata). Ent. News, **62**: 153–163.

1951b. The Odonata of the Northern Insect Survey. Can. Ent., **83**: 269–278.

1952. New or noteworthy records of Canadian Odonata. *Ibid.*, **84**: 125–130.

1953. The Odonata of Canada and Alaska. Volume **1**. Part I: General. Part II: The Zygoptera-damselflies. 292 pp. Univ. of Toronto Press.

1958. The Odonata of Canada and Alaska. Volume II. Part III: The Anisoptera–four families. 318 pp. Univ. of Toronto Press.

1966a. Autobiographic sketch. Pp. 14–34 *in* Centennial of entomology in Canada 1863–1963. A tribute to Edmund M. Walker (G. B. Wiggins ed.). Univ. of Toronto Press.

1966b. On the generic status of *Tetragoneuria* and *Epicordulia* (Odonata: Corduliidae). Can. Ent., **98**: 897–902.

WALKER, E. M., and W. E. RICKER

1938. Notes on Odonata from the vicinity of Cultus Lake, B.C. Can. Ent., **70**: 144–151.

WALLENGREN, H. D. J.

1894. Ofversikt af Scandinaviens Pseudoneuroptera. Ent. Tijdskr., **15**: 235–270.

WALLEY, G. STUART

1959. The Odonata of Canada and Alaska. Volume II. (Book review.) Can. Ent., **91**: 291–292.

WALSH, BENJAMIN D.

1862. List of the Pseudoneuroptera of Illinois contained in the cabinet of the writer, with descriptions of over forty new species, and notes on their structural affinities. Proc. Acad. Nat. Sci. Phila., 1862: 361–402.

WEITH, R., and JAMES G. NEEDHAM

1901. The life-history of *Nannothemis bella*, Uhler. Can. Ent., **33**: 252–255.

WHEDON, A. D.

1914. Preliminary notes on the Odonata of southern Minnesota. Rep. Minn. State Entomologist, pp. 77–103.

WHITE, HAROLD B.

1969. Two species of Odonata previously unreported from New England. Ent. News, **80**: 88.

WHITE, HAROLD B., and WALLACE J. MORSE

1973. Odonata (dragonflies) of New Hampshire: an annotated list. Res. Rep. Univ. N.H. Agric. Exp. Station, Durham, 30, 46 pp.

WHITE, HAROLD B., and RUDOLF A. RAFF

1970. The nymph of *Williamsonia lintneri* (Hagen) (Odonata: Corduliidae). Psyche, **77**: 252–257.

WHITEHOUSE, F. C.

1917. The Odonata of the Red Deer district, Alberta. Can. Ent., **49**: 96–103.

1918a. A week's collecting on Coliseum Mountain, Nordegg, Alta. *Ibid.*, **50**: 1–7.

1918b. The Odonata of the Red Deer district. *Ibid.*, **50**: 95–100.

1941. British Columbia dragonflies (Odonata), with notes on distribution and habits (with descriptions of two new nymphs by E. M. Walker). Amer. Midl. Nat., **26**: 488–557.

1948. Catalogue of the Odonata of Canada, Newfoundland and Alaska. Trans. Roy. Can. Inst., **27** (57): 3–56.

WIGGINS, GLENN B. (ed.)

1966. Centennial of entomology in Canada 1863–1963. A tribute to Edmund M. Walker. 94 pp. Univ. of Toronto Press.

WILLIAMSON, E. B.
  1900.   The dragonflies of Indiana. Rep. Ind. State Geologist, pp. 233–333; 1003–1010.
  1902.   Dragonflies (Odonata) from the Magdalen Islands. Ent. News, **13**: 144–146.
  1905.   Oviposition of *Tetragoneuria* (Odonata). *Ibid.*, **16**: 255–257.
  1906a.   Dragonflies (Odonata) collected by Dr. D. A. Atkinson in Newfoundland, with notes on some species of *Somatochlora*. *Ibid.*, **17**: 133–139.
  1906b.   The copulation of Odonata. *Ibid.*, **17**: 143–148.
  1907a.   Two new North American dragonflies (Odonata). *Ibid.*, **18**: 1–7.
  1907b.   A new *Somatochlora* with a note on the species known from Ontario. Can. Ent., **39**: 69–74.
  1909a.   Some corrections in *Somatochlora* (Odonata—dragonflies). Ent. News, **20**: 77–79.
  1909b.   The North American dragonflies (Odonata) of the genus *Macromia*. Proc. U.S. Nat. Mus., **37**: 369–398.
  1910.   A new species of *Celithemis* (order Odonata). Ohio Nat., **10**: 153–160.
  1917.   An annotated list of the Odonata of Indiana. Publ. Mus., Univ. Mich., 2, 13 pp.
  1922a.   Notes on *Celithemis* with descriptions of two new species (Odonata). Occas. Pap. Mus. Zool. Univ. Mich., 108, 22 pp.
  1922b.   Indiana Somatochloras again. Ent. News, **33**: 200–207.
  1923a.   Odonatological results of an auto trip across Indiana, Kentucky and Tennessee. *Ibid.*, **34**: 37–40.
  1923b.   A new species of *Williamsonia* (Odonata, Corduliinae). Can. Ent., **55**: 96–98.
WILSON, CHARLES BRANCH
  1912.   Dragonflies of the Cumberland Valley in Kentucky and Tennessee. Proc. U.S. Nat. Mus., **43**: 189–200.
WRIGHT, MIKE
  1943.   The effect of certain ecological factors on dragonfly nymphs. J. Tenn. Acad. Sci., **18**: 172–196.
  1944.   Notes on dragonflies in the vicinity of New Smyrna Beach, Florida. Florida Ent., **27**: 35–39.
  1946a.   A description of the nymph of *Sympetrum ambiguum* (Rambur), with habitat notes. J. Tenn. Acad. Sci., **21**: 135–138.
  1946b.   Notes on nymphs of the dragonfly genus *Tarnetrum*. *Ibid.*, **21**: 198–200.
WRIGHT, MIKE, and ALVAH PETERSON
  1944.   A key to the genera of anisopterous dragonfly nymphs of the United States and Canada (Odonata, Suborder Anisoptera). Ohio J. Sci., **44**: 151–166.
WRIGHT, MIKE, and C. S. SHOUP
  1945.   Dragonfly nymphs from the Obey River drainage and adjacent streams in Tennessee. J. Tenn. Acad. Sci., **20**: 266–278.

SUBJECT INDEX

Listed here are terms and topics likely to be of interest to the reader. Where a term is defined or explained in the text (usually in the Introduction) the corresponding page reference is designated thus: **16**. The topics cited here are mentioned in the text mainly in the sections devoted to field notes.

AUTHOR INDEX

Names listed are those of persons whose observations (published or not) have been cited in the text. Exceptions are: authors of references that appear in the synonymy for each genus or species; and persons whose records have been used to amplify parts of the morphological description (e.g. size and venational characters) for each species.

ODONATA INDEX

Species are listed under the generic name only. For each taxon (e.g. family, genus or species) page references have been designated in sequence thus:
(1) definitive description (**16**);
(2) key separating the taxon from related taxa and, for all taxa except species, followed by key subdividing the taxon further (*16*);
(3) other references (16).